Intracranial and Inner Ear Physiology and Pathophysiology

Intracranial and Inner Ear Physiology and Pathophysiology

edited by

ANDREW REID,[1] ROBERT J. MARCHBANKS[2] and ARNE ERNST[3]

[1]Head of Audiology, Royal United Hospital NHS Trust, Bath, UK

[2]Non-invasive Intracranial Pressure Assessment (NIPA) Unit, Southampton University Hospital, Southampton, UK

[3]Direktor der HNO–Klinik, Department of Otolaryngology, Unfall Krankenhaus, Berlin, Germany

W

Whurr Publishers Ltd
London

First published 1998 by
Whurr Publishers Ltd
19b Compton Terrace, London N1 2UN, England

British Library Cataloguing in Publication Data
A catalogue record for this book is available from the British Library.

ISBN: 1 86156 066 4

Printed and bound by CPI Antony Rowe, Eastbourne

Contents

Foreword

The Second International Conference on Intracranial and Inner Ear Fluids, held in June 1997 in Bath, has successfully built on the foundations laid by the first meeting in Seeheim, Germany in 1995 and has demonstrated that there is much fertile ground between clinicians and scientists who have specialised in intracranial pressure and haemodynamics and those exploring the fluid systems of the inner ear. There are fascinating parallels between the mechanisms of formation, volume control and drainage of the two fluid systems. Disturbance of one may be reflected symptomatically in the other.

Knowledge of such interactions is as important at the bedside and in the clinic as it is in the laboratory. Benign intracranial hypertension (pseudo-tumour cerebri), idiopathic normal pressure hydrocephalus, cranial surgery, intracranial hypotension, the Tullio phenomenon, inner ear disturbances of pregnancy and Ménière's disease are all examples of such interplay between fluid pressure disturbance and central nervous system symptomatology.

This book reflects the programme's happy mix of keynote lectures, original papers and posters with meaningful debate and discussion. The first two sessions were devoted to the anatomy, pathophysiology and examples of disordered volume control of both the intracranial and inner ear fluids. Methods of assessment of their pressure and fluid dynamics were followed by novel reports of pressure measurement results utilising the tympanic membrane displacement technique, which have generated such interest. Aspects of hearing loss attributable to disturbance of inner ear pressure were debated, together with some insights into the genetic basis of some disorders. There then followed keynote lectures addressing further aspects of the hydrodynamics of the intracranial and inner ear fluids, mixed with some fascinating presentations of functional brain imaging of inner ear disease.

By the end of the meeting, new collaborations were already becoming evident. Clearly there have been significant advances since

Seeheim. The organisers are to be congratulated both on their foresight in forging this bridge between the intracranial and inner ear communities, each of which has so much to learn from the other, and on their industry in securing such early publication of the Proceedings. We look forward to the Third International Conference on Intracranial and Inner Ear Fluids (Fluids 2000) to be held in Florida, USA.

November 1997

Professor Aage R. Møller
Margaret Fonde Jonsson Professor
University of Texas at Dallas
Callier Center for Communication Disorders
1966 Inwood Road
DALLAS
Texas 75235-7298
USA

Professor John D Pickard
Bayer Professor of Neurosurgery
Neurosurgery Unit
University of Cambridge Clinical School
Addenbrooke's Hospital
Hill's Road
CAMBRIDGE
CB2 2QQ
UK

Preface

This book is a result of a meeting entitled 'The Second International Conference on Intracranial and Inner Ear Fluids' held in June 1997. This conference was held under the auspices of the British Society of Audiology, in the World Heritage city of Bath Spa, UK. Bath, an elegant Georgian city based around a Roman spa, is the only city in Britain with a natural hot spring and its association with fluids goes back to 850 BC. King Bladud was apparently cured of his leprosy by bathing in a patch of warm bubbling mud and so Bath Spa was born. The healing properties of the waters reached a zenith in the eighteenth century, when it was claimed that it could even cure *deafness*.

The meeting was opened by Dr Carlborg from Sweden. He felt that combining the clinical and research aspects of fluid and pressure regulation in the intracranial and intralabyrinthine compartments as well as discussions on the clinical implications of abnormal fluid pressure was an excellent idea. The two topics of the meeting have much in common. Maintaining both the intracranial pressure and the intralabyrinthine pressure within normal limits is important for the normal functions of both the central nervous system and the ear. Communication between the intracranial space and the intralabyrinthine space and their interrelationships have not been fully appreciated. These interrelationships allow the possibility to measure intracranial pressure non-invasively by measuring the intralabyrinthine pressure.

This meeting was valuable in disseminating knowledge about research results and clinical experience. Many of the participants departed with new knowledge and new ideas that will contribute to their future work – as well as a bottle of Bath mineral water each! In addition, we would like to acknowledge the financial support given to the meeting by the Wellcome Trust and the administrative help of Mrs Ann Allen at the British Society of Audiology.

We hope you find this book stimulating and that you come away with at least one or two new ideas that may influence either your future research plans or your clinical thoughts. See you at 'Fluids 2000'!

November 1997

Andrew Reid, Bath, UK.
Robert J Marchbanks, Southampton, UK.
Arne Ernst, Berlin, Germany.

Part 1
Intracranial Physiology

Chapter 1
The anatomy and physiology of intracranial fluids

MALCOLM B SEGAL, Sherrington School of Physiology, UMDS, St Thomas' Hospital, London, UK

ABSTRACT. There are two major compartments of free fluid in the brain: the interstitial fluid (ISF), whose source and composition are largely unknown, and the cerebrospinal fluid (CSF). The capillaries of the brain are different from any other vascular bed in that they are sealed by tight junctions and are largely impermeable to small molecules. This is the physical basis of the blood-brain barrier (BBB). This restricted permeability is advantageous in that it protects the brain from fluctuations in the blood of molecules that would affect neurotransmission. However, this resistance to flow prevents the Starling mechanism being used to form brain ISF, so the cerebral capillaries' vessels must in some way secrete this fluid. Evidence from the fact that different-sized molecules move at the same rate within the brain parenchyma has shown that there is a bulk flow of ISF from the brain into CSF, which Davson termed 'the sink action of CSF'.

CSF is secreted by the leaf-like choroid plexuses into the ventricles of the brain. The CSF then flows through the ventricular system, over the pial surface of the brain, before returning to the vascular system via a number of different routes. The interaction between the CSF and brain ECF will be discussed and the properties of the ocular and auditory fluids will be outlined.

Key words: brain, interstitial fluid, cerebrospinal fluid, blood-brain barrier, choroid plexuses, aqueous humour.

Introduction

The fluids that surround and permeate the brain are part of a group of so-called transcellular fluids, which are separated from free exchange with the plasma by a system of barriers. These fluids which include the brain interstitial fluid (ISF); the cerebrospinal fluid (CSF), the aqueous humour of the eye and the two fluids of the inner ear, the perilymph and endolymph, all of which are formed by the process of active secretion. This energy-requiring mechanism has to be used since passive ultrafiltration, the process by which the bulk of the extracellular fluid (ECF) of the body is created, is obstructed by the low permeability of the barriers that surround these fluids.

The formation of ECF

In the majority of capillary beds the hydrostatic pressure within the lumen drives water and salts across the capillary walls to form the ECF. This flow of fluid is partially balanced by an opposing force generated by the colloid osmotic (or oncotic) pressure derived from the plasma proteins – the Starling equilibrium. The excess fluid and plasma proteins, some of which are filtered to a small extent, are returned to the circulation via the lymphatic system (Levick 1996). The endothelial cells that make up the capillary walls in most tissues are joined together by adhering and gap junctions, which form point contacts between the cells, so leaving a pathway for small molecules and water to pass easily around the cells. However, although proteins can also pass through these gaps, they do not escape into the ECF to any degree since the capillary is enveloped in an outer basement membrane or lamina made up of negatively charged fibrils, which impedes the movement of these macromolecules and acts as the ultrafilter (Michel 1985). This turnover of fluid by the total capillary bed represents some 8 litres per day, which reflects the balance between the forces of the Starling equilibrium and the overall relatively low permeability of the capillary wall, which is small when one considers the vast surface area of these vessels and the enormous flow of 5 litres per minute through the capillary beds.

The blood-brain barrier (BBB)

In the central nervous system the capillaries are very different as the endothelial cells are joined together by occluding bands of tight junctions (Figure 1.1). These structures effectively seal off the gaps between the endothelial cells forming the high resistance barrier, which constitutes the site of the blood-brain barrier (BBB) (Brightman and Tao-Cheng 1993). This means that for polar molecules the only major pathway for entry into the brain is directly across the cell walls of these vessels.

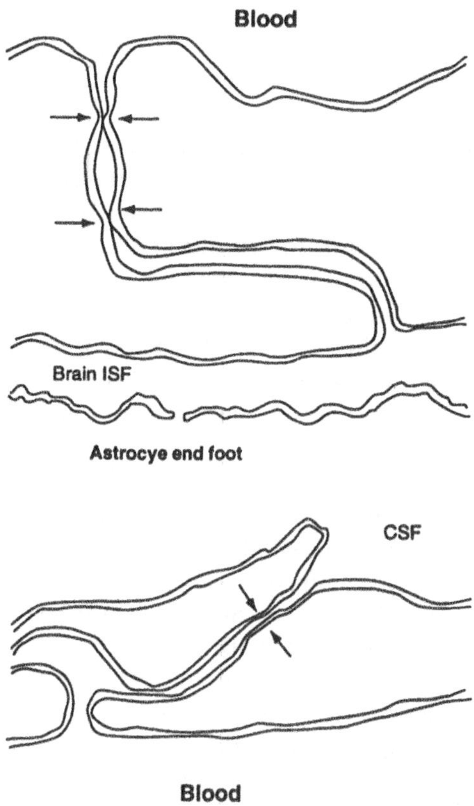

Figure 1.1: the upper figure is a tracing from an electron micrograph showing the interface between two cerebral endothelial cells. Between the two sets of arrows the walls become occluded by the tight junctions that form bands around the cells. The lower figure is a trace of the intercellular junctions of a 'leaky' capillary of a choroid plexus. In this case the walls at the arrow heads come close together but the junctions are not occluded and do not extend around the cell (from Brightman and Tao-Cheng 1993).

The BBB thus impedes the passive movement of fluid, so the Starling mechanism cannot operate in this situation and the brain ISF has to be formed by a process of active transport of ions across the walls of these endothelial cells, which will then draw water across the cells by a process of osmosis (Cserr and Patlak 1991; Davson and Segal 1996; Ennis et al. 1996).

The BBB mainly functions to protect the brain ISF from the normal fluctuations which occur in plasma of ions, amino acids, peptides and neurotransmitters. Ions such as potassium and calcium obviously have a direct effect on the electrical activity of the brain; however several amino acids act not only as the building blocks of protein but also as powerful neurotransmitters such as glycine and GABA. In addition many peptide hormones and neurotransmitters have completely different actions in the CNS compared to their activity in the rest of the body. Examples of these are adrenalin, gastrin, ADH and TRH so the BBB is essential to keep these molecules in the blood separate from the CNS (Segal and Zlokovic 1990).

In order for essential molecules to gain access to the brain, the BBB has a series of transporters that facilitate the entry of sugars, amino acids and nucleosides into the brain. Each transporter is highly specific for its own group of molecules and, for amino acids, there are at least four classes of transporter (Smith et al. 1987). There are also transporter molecules that can remove compounds from the brain ISF to the blood on the abluminal face of the BBB. Until recently it was considered that peptides could not be transported across the BBB, but now a number of transport mechanisms have been identified for these large molecules (Zlokovic 1995).

Although the tight junctions between endothelial cells form the passive aspect of the BBB, there are also enzymes attached to the surface of cerebral endothelia which destroy peptides and hormones during a single pass of the cerebral vasculature (Pardridge 1986). In addition there is a mechanism to deal with unwanted lipid soluble molecules that can 'bypass' the BBB and travel directly through the capillary cell walls. These molecules are stopped by specialised protein transporters such as P-glycoprotein in the cells of the BBB, which have a specificity to a broad range of molecules and transport them back across the BBB into the blood. This explains why drugs like cyclosporin, which are highly lipid soluble, fail to enter the brain (Begley et al. 1996). The BBB therefore has multiple functions both as a barrier to prevent the entry from blood for many molecules, as a system to separate transmitters between the blood and brain and as an active tissue metabolising many compounds, all of which act to maintain the homeostasis of the brain environment.

Brain ISF

The volume of the brain ISF was estimated from the steady-state distribution of ^{14}C sucrose introduced into the blood and the CSF – that is, into both 'sides' of the BBB (Oldendorf and Davson 1967). From these studies the volume of brain ISF has been calculated to be between 200 and 300 ml or some 15 to 20% of the brain weight assuming an average value of some 1500 g. Although it is a relatively large volume of fluid, the nature of brain ISF is largely unknown since it is spread as a thin sheet around neurones and glia and is most difficult to sample. Dialysis probes inserted into the brain do give some idea of the composition. These probes, made with a tip of permeable renal dialysis tubing with properties similar to a systemic capillary wall, will equilibrate with the ISF in the area they are placed. Fluid is pumped through the probe and if the composition of the inflowing fluid is adjusted so it is the same as that coming out of the probe, it is reasonable to argue the brain ISF must have a closely similar composition (Ungerstedt 1984).

Although we can postulate that brain ISF is a secreted fluid, the exact mechanism has not been elucidated. The reason for this resides in the fact that the area of the cerebral capillary interface is vast, yet the actual secretion of fluid is very small. Cserr and her colleagues have estimated that the rate of secretion is only some 150 to 200μl/min (Cserr and Patlak 1991). By studying the movement of molecules with a range of molecular sizes which were microinjected into the brain, it was found that they all moved at the same speed which must reflect bulk flow of brain ISF and not simple diffusion, which would be affected by their molecular weight. The fluid flows slowly from the vessels of the BBB through neuropile of the brain and then across either the permeable ependyma lining ventricle, into the CSF or across the permeable pia-arachnoid into the subarachnoid CSF (Cserr et al. 1981). The small volume of ISF secretion explains why Hugh Davson and I failed to demonstrate any effect of sodium transport inhibitors on the entry of sodium into the brain in the intact rabbit. The active transport of sodium is usually the process by which fluids are secreted (Davson and Segal 1970). Later studies using isolated brain perfusion in the rat confirmed that acetazolamide and bumetanide had no effect on sodium uptake by the brain and that an amiloride analogue that inhibits Na/H exchange only caused a 20% reduction of Na uptake (Ennis et al. 1996). A model of the transport processes identified across the BBB is shown in Figure 1.2.

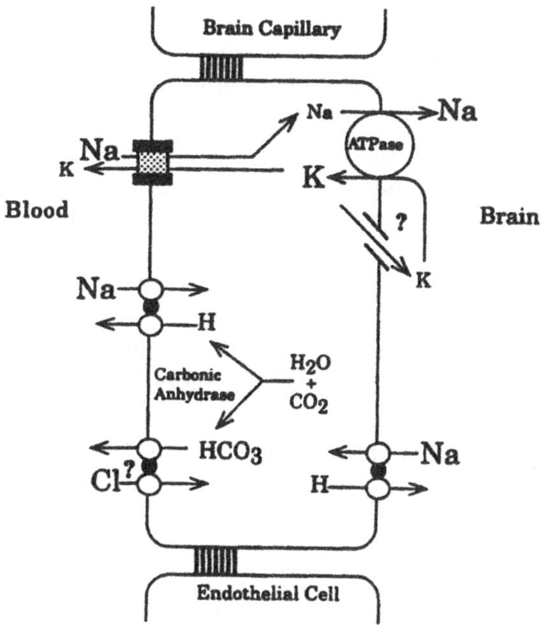

Figure 1.2: the current ionic transport systems that have been identified at the BBB (from Ennis et al. 1996).

CSF

The most accessible of brain fluids is CSF, which can be sampled by lumbar puncture and differs only slightly from ventricular CSF. Most CSF is secreted by the choroid plexuses (CP), which are highly vascular leaflike structures that float in the ventricles. The two largest CPs are found in the lateral ventricles with small extensions into the third ventricle. There is also a plexus that forms the roof of the fourth ventricle beneath the cisterna magna (Figure 1.3). These CPs have permeable fenestrated capillaries so that small molecules can pass easily from the plasma into the extracellular space of the choroid plexuses. However, the CPs are covered by an epithelial-like ependyma whose cells are joined together by occluding bands of tight junctions on the CSF side of the tissue. These junctions form the site of the blood-CSF barrier (BCSFB). Although tight junctions of the BCSFB are not as 'tight' as those of the BBB, they do prevent the rapid movement of small molecules such as mannitol, a polar sugar alcohol, from blood to CSF (Thomas and Segal 1996).

There is also a second BCSFB at the interface between the dura, which has permeable blood vessels and the arachnoid. The arachnoid is a double-layered structure with CSF between the layers. The cells of the outer layer are sealed by tight junctions, so restricting entry into the CSF across this tissue (Figure 1.3).

In man CSF is secreted at some 0.5 ml/minute and has a total volume of about 150 ml. The majority of the fluid being found in the subarachnoid space with some 23 ml (16%) in the ventricles and some 30 ml around the spinal cord.

Cerebrospinal fluid is an amazingly clear and bright fluid, broadly similar in composition to plasma, but with a much lower protein composition of 280 mg/l (plasma = 70 g/l), with 50% less glucose and amino acids. Careful studies on the distribution of ions between plasma and CSF sampled at the same time has revealed small but significant differences that could only be explained if CSF were produced by active transport of ions across the choroid plexuses (Table 1.1) (Davson and Segal 1996).

The size distribution of proteins in CSF is similar to that in plasma but at a much lower concentration (Felgenhauer 1974). There are, however, a number of proteins missing from CSF, or present at very low levels, such as fibrinogen, $ß_1$-lipoprotein and other $ß_2$ globulins. There are also special proteins that are only found in the CSF such as γ_c globulin and $ß_{trace}$. The presence of the prealbumin transthyretin, which is synthesised by the CP and is involved in thyroxine transport (Schreiber et al. 1990), is important. It is of interest that the concentration of protein in CSF increases as it flows along the neuroaxis, and lumbar CSF has a higher concentration than that sampled from the ventricles so the use of lumbar puncture samples may not always reflect what is occurring in the cortex. The protein concentration in the CSF of the neonate is also higher in the first few months of life – see Davson and Segal (1996) for a review on this topic.

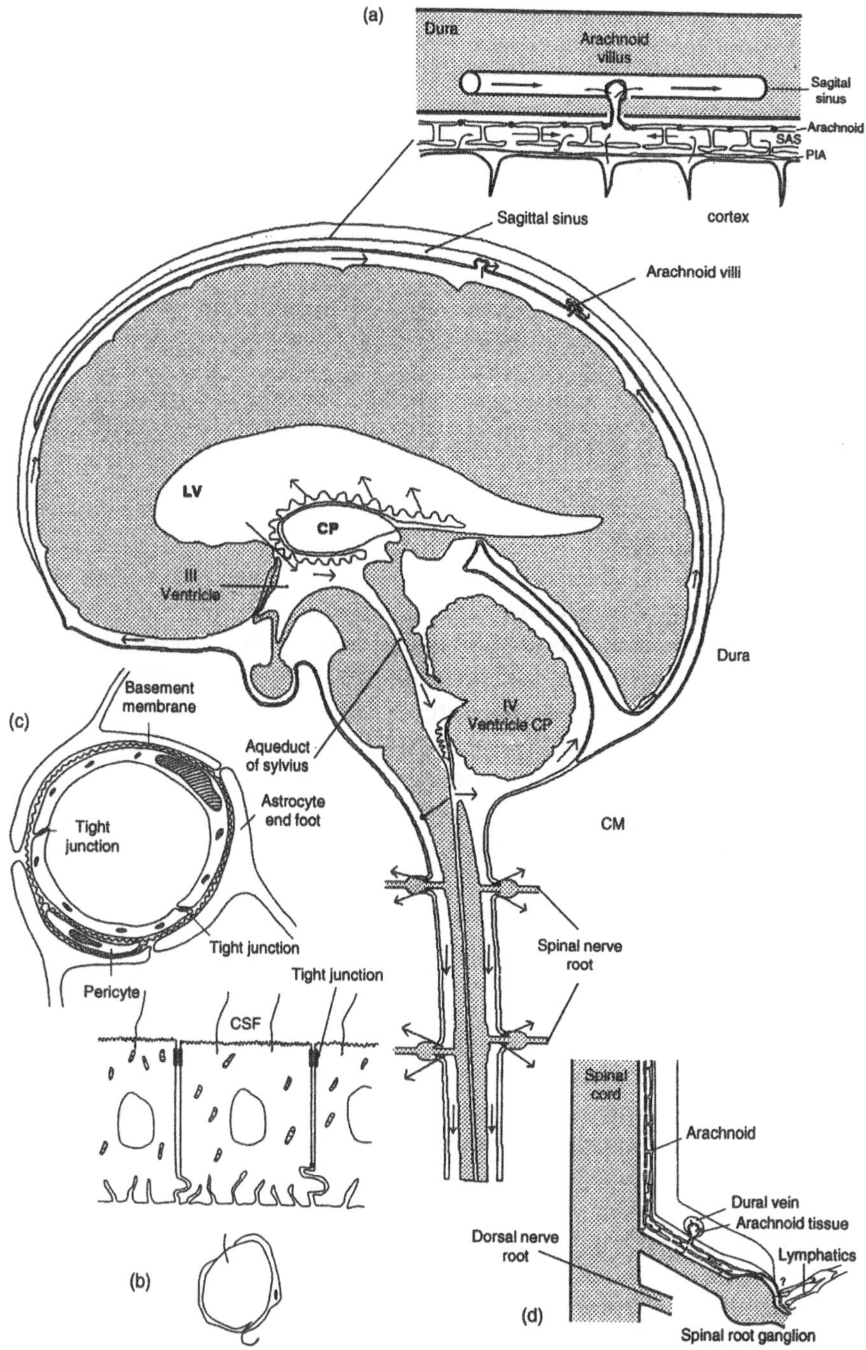

Figure 1.3: cross-section of the brain showing the circulation and drainage of CSF. The insets show the sites of the blood-CSF barrier at (a) the outer arachnoid layer and (b) the choroid plexus. The blood-brain barrier is shown in inset (c) at the level of the cerebral capillary. The drainage routes for CSF are shown in inset (a) at the arachnoid villus and (d) the spinal nerve route. There is also a route via the olfactory bulb, which is not shown. CP – chord plexus, CM – cisterna magna.

Table 1·1: concentrations of various solutes (mEq/kg H_2O) in plasma and lumbar cerebrospinal fluid of human subjects.

Substance	Plasma	CSF	RCSF
Na	150·0	147·0	0·98
K	4·63	2·86	0·62
Mg	1·61	2·23	1·39
Ca	4·70	2·28	0·49
Cl	99·00	113·00	1·14
HCO_3	26·80	23·30	0·87
Br	2·45	0·90	0·37
Inorg.P (mg/100 ml)	4·70	3·40	0·73
Amino acids	2·62	0·72	0·27
Osmolality	289·00	289·00	1·00
pH	7·397	7·307	–
PCO_2 (mmHg)	41·10	50·5	–

(Davson and Segal 1996)

Cerebrospinal fluid secretion can be studied by perfusing the ventriculo-cisternal system with a mock CSF containing the large molecular weight blue dextran. The dilution of this molecule by newly secreted CSF can then be used to determine the rate of CSF secretion (Pappenheimer et al. 1962; Bradbury and Davson 1964). From such studies, using the effect of various inhibitors of transport, it has been shown that an Na/K/ATPase is involved, as is carbonic anhydrase, since both ouabain and acetazolamide cause a 70% inhibition of CSF secretion (Davson and Segal 1970). Other transport inhibitors such as amiloride (Na/H) and frusemide (NaK2Cl) inhibit CSF secretion to some degree, which points to a complex process involving a variety of transport proteins and ion channels (Johanson and Murphy 1990; Johanson et al. 1990). Work with patch-clamp techniques has demonstrated that Cl channels are also central to the process (Garner and Brown 1992) (Figure 1.4). Although most aspects of the mechanism of ion transport are now known there is not a model to explain how ion transport is coupled to the secretion of CSF. However it is hoped that the ideas of Fischarg (1997) may lead to a new hypothesis involving the water channel proteins, the aquaporins.

Functions of CSF

The CSF has several key functions. Firstly, it acts as a fluid cushion reducing the brain's relative weight from 1500 g to 50 g and protects the brain from the effect of physical insults to the skull. Secondly, it acts as a 'sink', removing waste products from the CNS by the steady flow of brain ISF across the brain into CSF, which then passes back into the blood. Thirdly, it seems most likely that the CSF can act a circulation moving various transmitters and trophic factors from one brain region to another. Finally it acts to control the local environment of the brain with the aid of the choroid plexuses which can transport amino acids and peptides from CSF to the blood (Preston and Segal 1990; Banks and Kastin 1995).

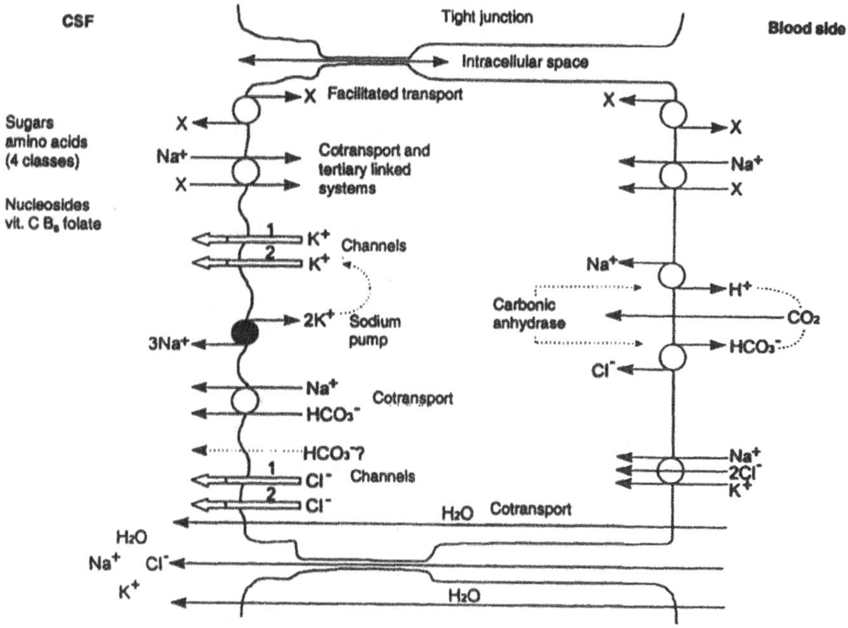

Figure 1.4: diagram of the current ion transport processes identified at the choroid plexus (modified from Johanson and from Garner and Brown).

The distribution of CSF around the brain was initially thought to be a slow process but recent studies by Ghersi-Egar and Fenstermacher have shown that molecules placed in the ventricles rapidly travel to many areas of the brain and a major pathway for penetration into the brain has been identified via the subarachnoid velae (Ghersi-Egea et al. 1996).

CSF pressure and drainage

The hydrostatic pressure of CSF is some 10 cmH_2O; values above 20 cmH_2O are considered abnormal. This pressure is the net force derived from the active secretion of CSF, cerebral blood flow and the resistance of the CSF drainage mechanism. CSF flows from the lateral ventricles into the third ventricle then via the narrow aqueduct of Sylvius into the fourth ventricle; from here CSF passes out of the ventricular system through various foramina and fills the subarachnoid space. CSF then drains back into the circulation by the valve-like arachnoid villi found in the sagittal sinus, by the spinal nerve routes and down the olfactory tract into the lymphatic system (Figure 1.3). The major site of drainage resistance of this complex pathway has not been determined but if there is a congenital deficit or an obstructing mass, the continued secretion of CSF leads to a raised intracranial pressure, severe headaches and eversion of the optic disc. In neonates, as the sutures in the skull are not closed, this pressure leads to expansion of the ventricles and the swollen head of untreated hydrocephalus. In the adult the raised pressure leads to expansion of the ventricles at the expense of neural tissue. However, with the advent of NMR and CAT scans, cases have been identified with enlarged ventricles but without a raised CSF pressure. One suspects that the chronic raised CSF pressure has lead to the opening of other CSF drainage routes with the system adapting to a new steady state (Davson, Welch and Segal 1987).

Acute treatment usually involves a drug to inhibit CSF secretion, for example acetazolamide, but a long-term solution for congenital problems requires the insertion of a drainage tube into the lateral ventricle with a one-way valve to drain the fluid back to a suitable body cavity. Although these tubes often remain patent for many years ingrowth of tissue and infection can occur which leads to complications, so this mechanical system is only a partial solution. Ideally a drug that is specific for CSF secretion with few side effects would be a most useful adjunct to the treatment of this condition.

Aqueous humour

Aqueous humour bears many similarities to CSF. It is secreted by the ciliary body of the eye and fills the anterior and posterior chamber either side of the lens (Figure 1.5). The ciliary body has permeable fenestrated capillaries enclosed within two membranes, an inner pigmented layer joined by a mixture of junctions and attached to an outer layer of cells sealed by tight junctions, the blood-aqueous barrier. The mechanism of secretion also depends on the active transport of ions across the blood-aqueous barrier and many inhibitors that inhibit CSF have a similar effect on the secretion of aqueous humour (Davson 1990). The active nature of the secretion of aqueous humour, the blood flow and the resistance to drainage generates the intraocular pressure (IOP) within the eye of some 15 mmHg. This pressure is essential for the maintenance of the shape of the eye and its role as an optical system. The site of the resistance to drainage of aqueous humour is still a matter of much argument. There is no doubt that fluid drains back into the circulation via the angle of the eye, the canal of Schlemm and the aqueous veins, but which of these structures is the prime site of resistance is a matter of debate. However, destruction of the trabecular meshwork of the canal of Schlemm with a laser has a major effect on the facility for drainage outflow and reduces the IOP. Changes in the angle of drainage in the eye occlude the canal of Schlemm and leads to a raised intraocular pressure (glaucoma) and eventually retinal detachment.

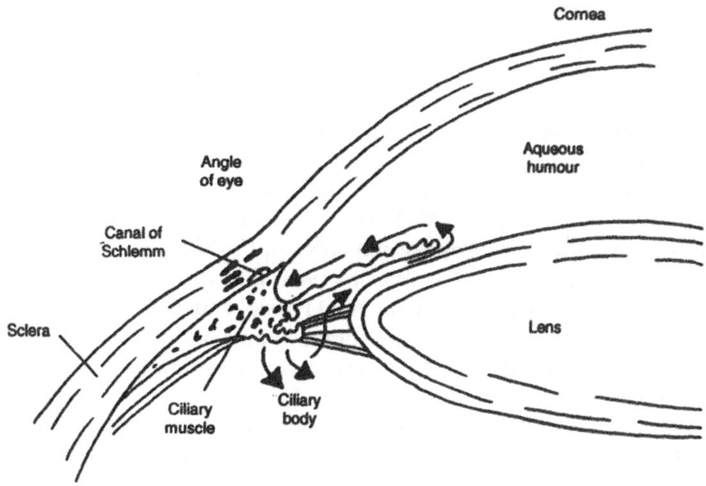

Figure 1.5: cross-section of the eye showing the fluids of the eye and the drainage route.

Aqueous humour is secreted at about 2.5 μl/min and has a similar ionic composition to CSF with a low protein content but there is a high level of ascorbic acid and xanthine. The differences in the ionic distribution between aqueous humour and plasma confirm that this is an actively secreted fluid like CSF (Davson 1990).

Fluids of the inner ear

The fluids of the inner ear, the perilymph and endolymph, although of very low volume, form a critical part of the mechanism of hearing. Perilymph is very similar to CSF whereas endolymph is most unusual and has a high level of potassium – a value that greatly exceeds that of any other mammalian fluid. These fluids will be discussed in detail in later chapters of this book.

Conclusion

The fluids of the brain are enclosed by various barrier systems that restrict their exchange with the plasma. These barriers function to maintain the special composition of these fluids, which in turn plays a critical role in the function and the homeostasis of the brain, eye and ear.

References

Banks WA, Kastin AJ (1995) Peptide transport system–1. In: Greenwood J, Begley DJ, Segal MB (eds) New Concepts of a Blood-Brain Barrier. New York and London: Plenum Press, pp. 111–18.

Begley DJ, Lechardeur D, Chen Z-D, Rollinson C, Bardoul M, Roux F, Scherman D, Abbott NJ (1996) Function expression of P-glycoprotein in an immortalised cell line of rat brain endothelial cells, RBE4. J Neurochem 67: 988–95.

Bradbury WM, Davson H (1964) The transport of urea, creatinine and certain monosaccharides between blood and fluid perfusing the cerebral ventricular system of rabbits. J Physio 170: 195–211.

Brightman MW, Tao-Cheng JH (1993) Tight junctions of brain endothelium and epithelium. In: Pardridge WM (ed.) The Blood-Brain Barrier: Cellular and Molecular Biology. New York: Raven Press, pp. 107–26.

Cserr HF, Cooper DN, Suri PK, Patlak CS (1981) Efflux of radiolabelled polyethylene glycols and albumin from rat brain. Am J Physiol 240: F319–28.

Cserr H, Patlak CS (1991) Regulation of brain volume under isomotic and anisomotic conditions. In: Gilles R (ed.) Advances in Comparative and Environmental Physiology 9. Berlin: Springer Verlag, pp. 61–80.

Davson H (1990) Physiology of the Eye. Basingstoke: Macmillan.

Davson H, Segal MB (1970) The effects of some inhibitors and accelerators of sodium transport on the turnover of ^{22}Na in the cerebrospinal fluid and the brain. J Physiol 209: 131–53.

Davson H, Segal MB (1996) Physiology of the CSF and Blood-Brain Barriers. Boca Raton, Florida: CRC Press.

Davson H, Welch K, Segal MB (1987) The Physiology and Pathophysiology of the Cerebrospinal Fluid. Edinburgh: Churchill Livingstone.

Ennis SR, Ren X-D, Betz AL (1996) Mechanisms of sodium transport at the blood-brain barrier studied with in situ perfusion of rat brain. J Neurochem 66: 756–63.

Felgenhauer K (1974) Protein size and cerebrospinal fluid. Klin Wochenschr 52: 1158–64.

Fischarg J (1997) Mechanism of fluid transport across corneal endothelium and other epithelial layers: a possible explanation based on cyclic volume regulatory changes. Bri J Ophtalmol 81: 85–9.

Garner C, Brown PD (1992) Two types of chloride channel in the apical membrane of the rat choroid plexus epithelial cells. Brain Res 591: 137–45.

Ghersi-Egea J-F, Finnegan W, Chen J-C, Fenstermacher JD (1996) Rapid distribution of intraventricularly administered sucrose into cerebrospinal fluid cisterns via subarachnoid velae in rat. Neurosci 75: 1271–88.

Johanson CE, Murphy VA (1990) Acetazolamide and insulin alter choroid plexus epithelial cell [Na^+] pH and volume. Am J Physiol 258: F1538–46.

Johanson CE, Sweeney SM, Parmalee JT, Epstein MH (1990) Co-transport of sodium and chloride by the adult mammalian choroid plexus. Am J Physiol 258: C211–16.

Levick J (1996) An Introduction to Cardiovascular Physiology. London: Butterworths.

Michel CC (1985) The Malpighi Lecture. Vascular permeability – the consequence of Malpighi's hypothesis. Int J Microcirc Clin Exp 4: 265–84.

Oldendorf WH, Davson H (1967) Brain extracellular space and sink action of cerebrospinal fluid. Arch Neurol 17: 196–205.

Pappenheimer JR, Heisey SR, Jordan EF, Downer J de C (1962) Perfusion of the cerebral ventricular system in unanaesthetized goats. Am J Physiol 203: 763–74.

Pardridge WM (1986) Mechanisms of neuropeptide interaction with the blood-brain barrier. Ann NY Acad Sci 231–48.

Preston JE, Segal MB (1990) The steady-state amino acid fluxes across the perfused choroid plexus of the sheep. Brain Res 525: 275–97.

Schreiber G, Aldred AR, Jaworoswski A, Nilsson C, Achen M, Segal MB (1990) Thyroxine transport from blood to brain via transthyretin synthesis in choroid plexus. Am J Physiol 258: 338–45.

Segal MB, Zlokovic BV (1990) The Transport of Amino Acids and Proteins across the Blood. Lancaster: Kluwer.

Smith QR, Mommo S, Aoyagi M, Rappoport SI (1987) Kinetics of neutral amino acid transport across the blood-brain barrier. J Neurochem 49: 1651–8.

Thomas SA (née Williams), Segal MB (1996) Identification of a saturable uptake system for deoxyribonucleosides at the blood-brain and blood-cerebrospinal fluid barriers. Brain Res 741: 230–9.

Ungerstedt U (1984) Measurement of neurotransmitter release by intracranial microdialysis. In: Marsden CA (ed.) Measurement of Neurotransmitter Release in Vivo. New York: Wiley 81–107.

Zlokovic BV (1995) Cerebrovascular permeability to peptides: manipulations of transport systems at the blood-brain barrier. Pharm Res 12: 1395–406.

Correspondence to Malcolm B Segal, Sherrington School of Physiology, UMDS, St Thomas' Hospital, Lambeth Palace Road, London, SE1 7EH, UK.
Tel. (+44) (0)171 928 9292, ext. 2531. Fax (+44) (0)171 922 8317. E-mail: m.segal@umds.ac.uk.

Chapter 2
Hydromechanics of the intracranial and labyrinthine fluids – experimental study in cats

BJÖRN CARLBORG, Department of Otolaryngology, University Hospital of Malmö, Sweden, and KONRÁD S KONRÁDSSON, Department of Otolaryngology, University Hospital of Lund, Sweden

ABSTRACT. Perilymph pressure is influenced by a multitude of factors. The effect of various pressure changes on labyrinthine fluid pressure depends on the rate of pressure change and the patency of the pressure transfer routes between the cerebrospinal fluid compartment and the labyrinth. The aim of these studies on cats was to investigate the emergence of pressure gradients and the regulation of perilymphatic pressure when the labyrinth is exposed to pressure transfer mediated from the cerebrospinal fluid, vascular influence and ambient pressure changes. The conclusion was that the cochlear aqueduct patency is the single most important factor with respect to the pressure transfer from the cerebrospinal fluid to the perilymph. It has the most profound impact on the duration and amplitude of the perilymphatic response to ambient pressure changes. The wider the cochlear aqueduct the faster the pressure equilibration and shorter the duration of the pressure gradient between the perilymph and the cerebrospinal fluid compartment. It seems that a patent aqueduct mediated pressure equilibration for the negative cerebrospinal fluid pressures more rapidly than for positive ones. This may suggest an easier flow of fluid towards the cochlea than towards the cerebrospinal compartment. The effect of the rate of ambient and middle ear-related pressure changes is limited when the cochlear aqueduct is blocked. However, when the cochlear aqueduct was patent the rates of pressure changes were crucial for the level of the perilymph response. To extrapolate conclusions from experiments on healthy animals directly to human pathophysiology may prove fallible. In this review of our previous studies we have discussed

views relevant to explain the mechanisms of pressure transmission between the cerebrospinal fluid and the labyrinth.

Key words: intracranial, labyrinth, fluid hydromechanics, perilymph, endolymph, cerebrospinal fluid, pressure, cochlear aqueduct.

Introduction

The perilymph pressure is influenced by a multitude of factors. It is affected by external ambient pressure changes as well as changes in the middle ear, cerebrospinal fluid, vascular pressures and blood flow. The effect of these pressure changes on the labyrinthine fluid pressure depends on the rate of pressure change and the patency of the pressure transfer routes between the cerebrospinal fluid compartment and the labyrinth – that is, the cochlear aqueduct, primarily, and the perineural/perivascular spaces and the endolymphatic duct and sac, secondarily, as shown by several authors (Kerth and Allen 1963; Martinez 1968; Beentjes 1970; Carlborg et al. 1981, 1982, 1983, 1990; Kishimoto et al. 1983, Konrádsson et al. 1994).

Increased knowledge is important regarding the emergence of the different pressure gradients from the labyrinth towards its surroundings and how these factors affect the labyrinth. This is especially valuable for the understanding of the pathophysiological mechanisms for perilymphatic fistulae, alternobaric vertigo and endolymphatic hydrops. The aim of the present studies was to investigate the emergence of pressure gradients and the regulation of perilymphatic pressure when the labyrinth is exposed to pressure transfer mediated from the cerebrospinal fluid, vascular influence and ambient pressure changes.

Methods

For these studies healthy cats weighing 2.5 to 3.5 kg were used (Carlborg et al. 1980, 1981, 1982, 1983, 1990; Konrádsson 1994, 1997). The cats were intravenously anaesthetised with ketamine and phenobarbital. Ventilation was maintained by a respirator through a tracheostomy. The chamber (Pch) as well as the perilymph (Ppl), middle ear (Pm), cerebrospinal fluid (Pcsf), central venous (Pcv) and arterial pressures (Pa) were continuously recorded. The Ppl was recorded from the labyrinthine vestibule leaving the ear canal, ear drum and middle ear physiologically intact according to a method developed previously (Carlborg et al. 1980).

In some animals the cochlear aqueduct was surgically transected and blocked with dental amalgam. The approach was through a small opening in the tympanic bulla. This opening was sealed air tight during the experiments. Miller micro-tip transducers were used for pressure recordings.

The results are expressed as mean values ±SD. Student's *t*-test and the Wilcoxon signed rank sum test were used accordingly, depending upon the normality of the data. A probability level of $p < 0.05$ was considered as statistically significant.

Ambient pressure changes

Hypobaric chamber pressure

In these experiments 18 cats were studied in five different experimental conditions, numbered 1 through 5 (Carlborg 1990; Konrádsson 1994, 1997). In condition 1 the tympanic cavity was always open to the atmosphere, but the cochlear aqueduct was either patent or blocked. In conditions 2 through 5 the tympanic cavity was physiologically intact. The animals were anaesthetised and placed in a chamber where the pressure was reduced relative to the atmospheric pressure. The pressure was reduced stepwise to levels of –25 to –55 mmHg for 4–6 minutes. In conditions 2 and 4 the eustachian tube was surgically blocked. During these conditions the manually controlled chamber pressure was reduced by approximately –40 mmHg relative to the prevailing atmospheric pressure. The pressure level was then kept stable for 5 minutes after which it was further reduced to a level of –50 mmHg. After a stable level was maintained for another 5 minutes, the chamber pressure was equilibrated to the atmospheric pressure.

All the hydrostatic pressures reported here are relative to the prevailing atmospheric pressure (mean pressure 745 mmHg) approximately 135 m above sea level (Konrádsson 1997).

The effect of hypobaric pressure on bodily pressures was studied in four cats with the tympanic bulla open (condition 1). This prevented any build-up of a pressure gradient between the middle ear and the chamber. During hypobaric pressure and the observation period of 5 minutes, all the pressures recorded changed parallel to the Pch and no changes in pre-experimental pressure gradients were recorded. This was true regardless of the patency of the cochlear aqueduct.

In all conditions (1 through 5) the reduced chamber pressure induced a fall of the Pcsf. This fall was always simultaneous and parallel to the Pch change.

With the eustachian tube obstructed and the cochlear aqueduct patent (condition 2; 7 cats) the chamber pressure was lowered to –40 (±2) mmHg at the rate of 8 (±2) mmHg (Figure 2.1). The simultaneous change of Pm was restricted and a fall to a level of –6 (±2) mmHg was recorded. During the remainder of the first 5-minute period the Pm was further reduced by 3 mmHg. When the Pch was further reduced to a level of –52 (±1) mmHg the initial effect on the Pm was slight and a slow fall of the Pm occurred during the remaining 5 minutes.

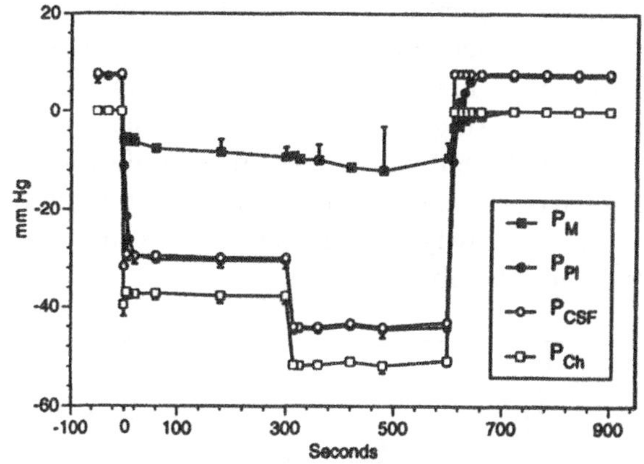

Figure 2.1: changes in chamber (Pch), middle ear (Pm), perilymph (Ppl) and cerebrospinal fluid (Pcsf) pressures relative to atmospheric pressure. Eustachian tube is surgically blocked, cochlear aqueduct physically open. Mean and SD values are based on 32 experiments on seven cats (after Konrádsson et al. 1997).

The reduced chamber pressure also induced a fall of the Ppl. The initial fall of the Ppl was delayed, but within 30 seconds the Ppl reached the Pcsf level. Similar pressure changes were observed during the second reduction of chamber pressure. When the Pch was normalised, a delay of the Pm and Ppl rise to pre-experimental levels was recorded.

In condition 3, when the eustachian tube and the cochlear aqueduct were patent (4 cats) the initial changes of the Pm were similar to those observed with the tube blocked. In all experiments, the physiological opening of the tube occurred spontaneously within 1 minute, resulting in a partial pressure equilibration between the Pm and the Pch. There was individual variation in the timing and the pressure levels at which the tube opened. Consequently, the mean values including pressure gradients were analysed but not statistically evaluated. In one recording from a representative animal the Pch was reduced to –56 mmHg in three steps. Repeated openings of the eustachian tube brought the Pm successively down to the level of –44 mmHg, effectively educing the pressure gradient towards the chamber. The initial changes of the Ppl were similar to those observed in experiments with the eustachian tube blocked and the cochlear aqueduct patent. In a representative animal the first physiologic opening of the tube caused a significant but transient decrease of the Ppl of 4 mmHg below the Pcsf – a difference rapidly equilibrated. This phenomenon was observed in all experiments and also during the second and third lowering of the Pch.

In condition 4 (Figure 2.2) with the eustachian tube and cochlear aqueduct surgically blocked (five cats) the Pm changes were similar to those seen in previous experiments with the tube blocked. During the first minute the Pm rapidly fell to a level of –7 (±4) mmHg equalling 60%

of the total pressure reduction. The Ppl fall was limited compared to the experiments with the cochlear aqueduct intact. Fifty per cent of the total fall occurred within the first minute of the experiment. Both the Pm and Ppl fall continued although the Peh was kept stable. In every experiment, however, the Ppl fall was statistically significantly larger than the corresponding Pm decline. When the chamber pressure was normalised, both the Pm and Ppl slowly approached, but did not reach, pre-experimental levels during the observation period.

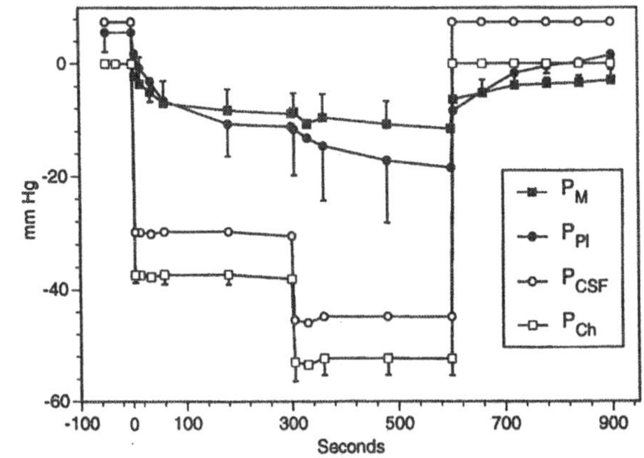

Figure 2.2: changes in chamber (Pch), middle ear (Pm), perilymph (Ppl) and cerebrospinal fluid (Pcsf) pressures relative to atmospheric pressure. Eustachian tube and cochlear aqueduct surgically blocked. Mean and SD values are based on 10 experiments on five cats (after Konrádsson et al. 1997).

With the eustachian tube patent and the cochlear aqueduct blocked (condition 5) the changes were initially similar to the changes observed with both channels blocked. However, in all experiments on the three animals, the tube was forced open within 2 minutes with individual timing and pressure levels. In one representative animal the partial pressure equilibration lowered the Pm by 12 mmHg bringing it closer to the Pch levels and thereby reducing the pressure gradient between the chamber and middle ear. The initial changes of the Ppl were limited and not significantly different from those seen in experiments with the eustachian tube and the cochlear aqueduct obstructed. The partial pressure equilibration of the Pm caused a definite fall of the Ppl (range 6 to 15 mmHg). When the Pch was normalised, the Pm and Ppl slowly approached but did not reach pre-experimental levels during the observation period.

Pressure transfer from the cerebrospinal fluid

I Artificial cerebrospinal fluid infusion

A cannula was located in the lateral ventricle of the brain. Artificial cerebrospinal fluid was infused through this. The Pcsf was recorded in the contralateral ventricle. Two different kinds of infusion experiments were performed:

- 10 cats with the cochlear aqueduct patent and 10 cats with the aqueduct closed were used in each experiment. The volume of the infusion was 0.5 ml and the rate was 0.2 ml/min;
- seven cats with the cochlear aqueduct blocked were used. In this experiment the infusion rate was adjusted to induce a stable elevated level of Pcsf for 8 to 12 minutes.

With the cochlear aqueduct patent a simultaneous increase in Pcsf and Ppl was seen. At no time during the experiments was there a relevant difference between the two pressures (Figure 2. 3). The peak Ppl (16.0 ± 1.8mmHg) was slightly delayed compared to the peak Pcsf (15.9 ± 1.7mmHg). The mean time lag was 1.5 s and did never exceed 8 s.

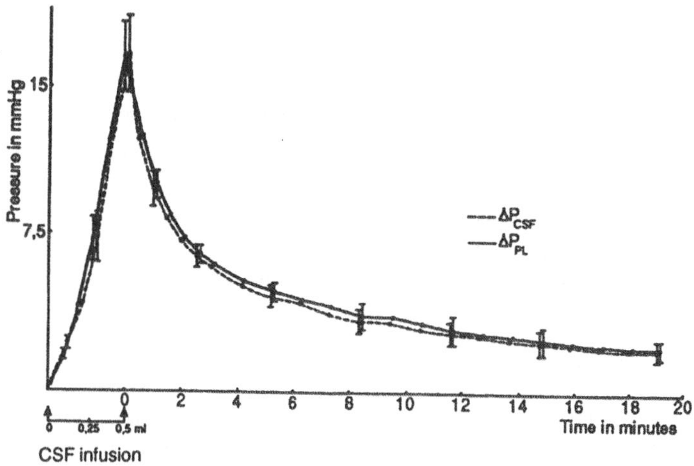

Figure 2.3: changes in cerebrospinal fluid and perilymph pressure during and after subarachnoid infusion (indicated by arrows) in cats with the cochlear aqueduct patent. Mean values (± SD) of the pressure changes are plotted for ten cats (from Carlborg B, Farmer JC (1983) *Am J Otolaryngol* 4: 275. Reproduced with permission).

With the cochlear aqueduct blocked some rise of the Ppl was recorded but it was much slower and smaller than that of the Pcsf (Figure 2.4). At the end of the cerebrospinal fluid infusion the Ppl rise was only 2.2±0.3 mmHg i.e. 14% of the maximal Pcsf increase ($p < 0.001$). Approximately 2 minutes after the Pcsf peak, the Ppl stabilised at a maximum level of 3mmHg. The Pcsf remained however significantly higher ($p < 0.05$) for another 8.5 minutes. Thus a definite increase in the Ppl was recorded during Pcsf pressure decline. The Pa and Pcv were not significantly affected.

In the second type of infusion experiment three different levels of the Pcsf increase were induced: 6.8, 8.7 and 13.4mmHg. These levels were maintained and kept stable for 8 and 12 minutes.

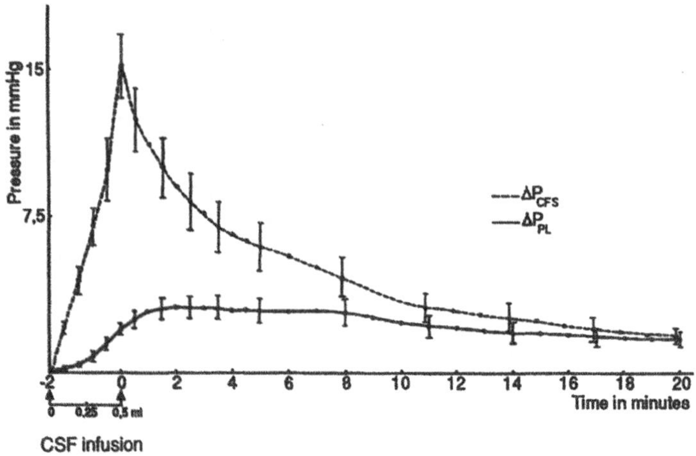

Figure 2.4: changes in cerebrospinal fluid and perilymph pressure during and after subarachnoid infusion (indicated by arrows) in cats with the cochlear aqueduct blocked. Mean values (±SD) of the pressure changes are plotted for 10 cats (from Carlborg and Farmer, Am J Otolaryngol 1983; 4:277. Reproduced with permission).

With the cochlear aqueduct blocked the rise of the Ppl was slow and limited. Compared to the peak Pcsf the peak Ppl was delayed by 4.9 ± 0.3 min and did then stabilise. The peak Ppl seemed to increase with higher Pcsf levels, but this correlation was not clear. The peak Ppl increase was only 39% of the Pcsf increase – significantly smaller ($p < 0.01$). Rapid reduction of the Pcsf induced a slow decline of the Ppl.

II Body posture changes

Seven cats with the cochlear aqueduct patent were used in these experiments. The body posture was changed from the horizontal plane by –15°, +10°, +20°, +30° and +60° respectively. Similar experiments were performed with the cochlear aqueduct blocked (Carlborg et al. 1981).

The results showed that with the cochlear aqueduct patent the Ppl, Pcsf and Pcv increased by approximately 2mmHg for every change of 10° of body position between –15° to +60°. No significant difference in pressure change of the Pcsf and the Ppl was observed ($p > 0.05$). With the cochlear aqueduct blocked, preliminary results indicate that the Ppl increase was much slower and lower than the increase of the Pcsf and Pcv.

Arterial blood gas changes

- Hypercapnia from normal $PaCO_2$ levels (30–39mmHg) was induced by breathing air supplemented with 5% CO_2 via the respirator to the animals. In these experiments, six cats with the cochlear aqueduct patent and six cats with the aqueduct blocked respectively were used.

- Supplying N_2O to the breathing air via the respirator induced *hypoxia* from normal PaO_2 levels (97–109 mmHg). In these experiments, three animals with the cochlear aqueduct patent were used.
- In five cats with the cochlear aqueduct patent *hyperventilation* was induced by increasing the ventilation frequency of the respirator (Carlborg et al. 1981).

The results showed that in animals with the cochlear aqueduct patent any changes in blood gas levels significant enough to induce Pcsf changes resulted in simultaneous and equal changes of the Ppl.

Induction of *hypercapnia* by increasing the $PaCO_2$ to 6.5 kPa (increase by 18 mmHg) and 63 mmHg (increase by 34 mmHg) respectively, resulted in an equal mean Ppl and Pcsf rise of 11.1 to 29.6 mmHg. The PaO_2 was not significantly changed. With the cochlear aqueduct blocked our results indicated a comparable increase in the Ppl as in the Pcsf. *Hypoxia* resulted in an equal and simultaneous increase in Ppl and Pcsf without any significant change of Pcv or Pa. No change in the $PaCO_2$ was recorded. *Hyperventilation* of the animal from normal blood gas and Pcsf levels did not result in any significant reduction of the Ppl nor of the Pcsf levels. However, a significant and equal reduction of the Ppl and Pcsf levels was noted in case of Pcsf hypertension.

Discussion

The inner ear is connected to the middle ear through the round and oval window membranes. The middle ear is in turn directly exposed to ambient pressure changes via the eardrum and the eustachian tube. The perilymph and the cerebrospinal fluid compartments are directly linked by the cochlear aqueduct as well as the perineural/perivascular spaces. An indirect connection exists via the endolymphatic sac and duct. These are the channels through which cerebrospinal fluid pressure changes affect the labyrinth. Through these channels pressure equilibration occurs when the labyrinth is exposed to ambient pressure changes mediated by the middle ear. The characteristics of the pressure transfer determine the quality of the hydrodynamic pressure relation between the perilymph and the cerebrospinal fluid compartments.

We examined the effect of ambient hypobaric pressure on the labyrinth (Carlborg 1990; Konrádsson 1994, 1997). We also examined the effects on the labyrinth of pressure transfer from the cerebrospinal fluid caused by:

- infusion of artificial cerebrospinal fluid (Carlborg 1983a);
- body posture changes (Carlborg 1983b);
- arterial blood gas changes (Carlborg 1981) were also examined.

The characteristics of the pressure transfer to the perilymph are elucidated before and after surgical obstruction of the cochlear aqueduct.

The relative significance of the aqueduct and other communications is also evaluated. These are important factors for understanding the pathophysiology of perilymph fistula, alternobaric vertigo and endolymphatic hydrops. The studies show that the patent cochlear aqueduct is the single most important pressure communication channel to the labyrinth. It always mediates sufficient pressure equilibration to keep the Ppl and the Pcsf equal at resting conditions. This conclusion is not controversial (Martinez 1968; Beentjes 1970; Carlborg et al. 1980, 1982, 1983; Kishimoto et al. 1983; Yoshida et al. 1985).

The effect of hypobaric pressure on the animal was studied in a pressure chamber (Carlborg et al. 1990; Konrádsson et al. 1994, 1997). We found the closed cavities of the middle and inner ear to be the source of the different pressure gradients emerging during hypobaric pressure. With the tympanic bulla open no change in pre-experimental pressure gradient was observed during hypobaric pressure. We found no support for a selective cerebrospinal fluid or vascular influence on the labyrinth at the pressure levels tested (equivalent to an altitude trip from sea level to an altitude of 450 to 650 m). No fundamental difference was observed in the perilymph response to equivalent levels of pressure induced locally in the middle ear compared to hypobaric pressure (Carlborg et al. 1982, 1990; Konrádsson 1994, 1997). The pressure equilibration of the eustachian tube and the cochlear aqueduct was elucidated as well as the effect of blocking these channels. During ascent the physiologic opening of the eustachian tube reduced the pressure gradients across the tympanic membrane. The patent cochlear aqueduct equilibrated the perilymph pressure to the cerebrospinal fluid compartment levels with the result of a considerable and potentially harmful pressure gradient across the oval and round windows. With the aqueduct blocked the pressure decrease within the labyrinth and the tympanic cavities was limited, resulting in large pressure gradients toward the chamber and the cerebrospinal fluid compartments. Under these circumstances the pressure gradient toward the tympanic cavity is limited, but the gradient toward the cerebrospinal fluid compartment is generally large. As the endolymphatic sac in the posterior fossa is directly exposed to the Pcsf the membranous labyrinth is continuously affected by this large pressure gradient during hypobaric pressure.

Regarding the pressure transfer between the cerebrospinal fluid and perilymph we find that the cochlear aqueduct patency is the single most important factor. It has a profound influence on the duration and amplitude of the perilymph response to ambient pressure changes. Consequently, it has been of interest to quantitatively study the patency of the cochlear aqueduct (Carlborg et al. 1982). Such calculations have revealed a definite individual variation of the patency. The wider the cochlear aqueduct the shorter the duration of the pressure difference between the perilymph and the cerebrospinal fluid. Further, the calculations indicated that a patent aqueduct mediated pressure equilibration for negative pressures more rapidly as opposed to positive pressure transfer to the cerebrospinal fluid. This may suggest an easier flow of liquid into the cochlea than out towards the subarachnoid space. Whether these feline findings are directly applicable to humans is unclear.

The effect of the rate of ambient and middle ear-mediated pressure changes is limited when the cochlear aqueduct is blocked. However, when the cochlear aqueduct was patent the rate of the pressure changes was crucial for the level of the peak perilymph pressure response. The cochlear aqueduct mediates an effect similar to that of the low pass filter because of its small lumen, the characteristics of this filter being decided by the patency of the cochlear aqueduct. This is why the peak Ppl was lower when the rate of the pressure stimuli was low.

Examining direct pressure transfer from the cerebrospinal fluid allowed us to qualitatively evaluate the patency of the cochlear aqueduct. These changes were directly induced by intracranial infusion and indirectly by changes in cerebral blood flow caused by hypercapnia and hypoxia. By changing the body posture we achieved a combination of cerebrospinal fluid and vascular influence.

With the cochlear aqueduct patent the transfer of pressure changes to the labyrinth was linear, with a time delay of less than a second. The pressure changes of the perilymph and the cerebrospinal fluid were so closely related that influence of the vast cerebrospinal reservoir obscured the physiologic effects on the labyrinth of other sources than that of the cerebrospinal fluid itself. This close relation has not been found in earlier experiments by others.

With the cochlear aqueduct blocked an elevated Pcsf pressure always resulted in a significant Ppl increase, but this was smaller and much delayed compared to the Pcsf change. Even when a continuous and stable increase of Pcsf was maintained, the Ppl never reached the same level of increase. The Ppl stabilised after approximately 4 to 6 minutes and only reached a maximum change of 20 to 60% of the Pcsf increase. This suggests an indirect pressure transfer from the Pcsf via the endolymphatic duct and sac. Any pressure change in the endolymph will be fully mediated to the perilymph in the case of cochlear aqueduct obstruction.

The effect of the parallel channels – the perineural/perivascular spaces – is unclear. It has been suggested that these channels may mediate a direct communication between the perilymph and the cerebrospinal fluid (Holden and Schuknecht 1968) but the functional patency of these spaces has never been thoroughly investigated. Had the pressure transfer in our study been mediated by direct communication routes the Ppl would expectedly have reached same levels as the Pcsf. Thus direct pressure transfer via the perineural/perivascular spaces does no seem to explain our results.

Changes in labyrinthine blood flow attributable to cerebrospinal fluid hypertension is not a plausible explanation since the results were similar whether the animals were dead or alive (Carlborg et al. 1983).

The tightness of the blocking of the cochlear aqueduct was tested at the end of the experiments without any signs of leakage.

Our results suggest that it is the endolymphatic duct and sac that mediate the slow, limited and non-linear pressure transfer to the labyrinth when the cochlear aqueduct is blocked. This conclusion is derived from acute experiments without surgical artefacts associated with endolymphatic duct obstruction.

Thus, the results on acute as well as chronic obstruction of the cochlear aqueduct have proved the cochlear aqueduct as the only channel by which rapid and substantial pressure changes in the cerebrospinal fluid can be transferred to the labyrinth (Kishimoto et al. 1983; Carlborg et al. 1990; Konrádsson et al. 1994).

A significant vascular influence has been revealed in experiments with the cochlear aqueduct blocked (Carlborg et al. 1982; Kishimoto et al. 1983; Carlborg and Farmer 1983). Cochlear aqueduct obstruction is indeed a prerequisite for detecting pressure effects in the perilymph caused by other factors than the influence of the Pcsf. A direct correlation was found between the changes in systemic Pa and Ppl (Carlborg and Farmer 1983). An increase of Pa induced a progressively greater Ppl. Such an influence was not seen in the cerebrospinal fluid because of the efficient auto regulation of the cerebral blood flow (Carlborg et al. 1982). Thus Ppl regulation with the cochlear aqueduct obstructed seems more dependent on systemic Pa influence than on the influence from the Pcsf or Pcv. Thus the lack of autoregulation of labyrinthine fluid pressure may indicate a certain lack of auto regulation of the labyrinthine blood flow.

Furthermore, despite the blocked cochlear aqueduct, hypercapnia caused an increase in Ppl parallel and similar to that in the Pcsf. This close relation can best be explained by labyrinthine vasodilation inducing a Ppl increase, analogous to the effect of cerebral vasodilation exerted on the Pcsf. Induction of even moderate levels of hypercapnia raised the Ppl definitely more than any change in body posture. This observation may be clinically useful regarding perilymph fistula in man.

With the cochlear aqueduct patent, the Ppl was regulated by cerebrospinal fluid influence. With the cochlear aqueduct blocked, the cerebrospinal fluid influence was mainly via the endolymphatic sac and duct. It was limited and the peak pressure changes were delayed. On the other hand, vascular influence could mediate rapid and substantial pressure changes to the labyrinth.

Conclusions

We conclude that the cochlear aqueduct patency is the single most important factor with respect to the pressure transfer from the cerebrospinal fluid to the perilymph. It has the most profound impact on the duration and amplitude of the perilymphatic response to ambient pressure changes. This is an individual variation of the cochlear aqueduct patency. The wider the cochlear aqueduct the faster the pressure equilibration and shorter the duration of the pressure gradient between the perilymph and the cerebrospinal fluid compartment. Furthermore, calculations indicated that a patent aqueduct mediated pressure equilibration for the negative cerebrospinal fluid pressures more rapidly than for positive ones. This may suggest an easier flow of fluid towards the cochlea than towards the cerebrospinal compartment (Carlborg et al. 1982).

The effect of the rate of ambient and middle ear-related pressure changes is limited when the cochlear aqueduct is blocked. However, when the cochlear aqueduct was patent the rate of pressure changes was crucial for the level of the Ppl response.

A morphological study by Wlodyka (1978) indicated varying patency of the cochlear aqueduct in humans. Little is indeed known about the function and functional patency of the human adult aqueduct but its patency is believed to be poor.

To extrapolate conclusions from experiments on healthy animals directly to human pathophysiology may prove fallible. In this review of our previous studies we have nevertheless allowed ourselves to discuss relevant views to explain the mechanisms of pressure transmission between the cerebrospinal fluid and the labyrinth.

Acknowledgements

This study was supported by grants from the Swedish Society for Medical Research, Medical Faculty of the University of Lund, Sweden, Swedish Silent School Foundation, the Helga Hjerpstedt Foundation, the Swedish Federation of Hard of Hearing People (HRF), Fogarthy International Research Fellowships No 1F05TW0350-01 IARG and National Institutes of Health grant HL 07896.

References

Beentjes BIJ (1970) On the Pressure of the Endolymph and the Perilymph in the Cat. Amsterdam: North-Holland Publishing, pp. 1–68.

Carlborg B (1981) On physiological and experimental variation of the perilymphatic pressure in cat. Acta Otolaryngol 91: 19–28.

Carlborg B, Densert O, Densert B (1982) Functional patency of the cochlear aqueduct. Ann Otol Rhinol Laryngol 91: 209–15.

Carlborg B, Densert O, Stagg J (1980) Perilymphatic pressure in the cat – description of a new method for study of the inner ear hydrodynamics. Acta Otolaryngol 90: 209–17.

Carlborg B, Farmer JC Jr (1983a) Relationship of labyrinthine fluid pressures and blood flow. Laryngoscope 93: 998–1003.

Carlborg B, Farmer JC Jr (1983b) Transmission of cerebrospinal fluid pressure via the cochlear aqueduct and endolymphatic sac. Am J Otolaryngol 4(4): 273–82.

Carlborg B, Farmer JC Jr, Carlborg A (1990) Effects of hypobaric pressure on the labyrinth – cochlear aqueduct patent. Acta Otolaryngol (Stockh) 110: 386–93.

Holden HB, Schuknecht HF (1968) Distribution pattern of blood in the inner ear following spontaneous subarachnoid haemorrhage. J Laryngol Otol 82: 321–9.

Kerth JD, Allen GW (1963) Comparison of the perilymphatic and cerebrospinal fluid pressures. Arch Otolaryngol 77: 581–5.

Kishimoto S, Nagahara K, Fisch U, Dillier N (1983) Inner ear pressure measurements. Effects of obstruction of the cochlear aqueduct and endolymphatic duct on the perilymphatic pressure. Otolaryngol Clin North Am 16(1): 21–35.

Konrádsson KS, Carlborg BI, Farmer JC Jr (1997) Pressure gradients affecting the labyrinth during hypobaric pressure – experimental study. Ann Otol Rhinol Laryngol 106: 495–502.

Konrádsson KS, Carlborg AH, Farmer JC Jr, Carlborg BI (1994) Perilymph pressure during hypobaric conditions – cochlear aqueduct obstructed. Acta Otolaryngol (Stockh) 114(1): 24–9.

Martinez DM (1968) Simultaneous measurements of endolymphatic and perilymphatic fluid pressures before and during anaphylaxis and associated changes in cerebrospinal fluid, venous and arterial pressure. Acta Otolaryngol 238(Suppl): 1–53.

Wlodyka J (1978) Studies on the cochlear aqueduct patency. Ann Otol Rhinol Laryngol 87: 22–8.

Yoshida M, Lowry LD, Liu JJC (1985) Effects of hyperosmotic solutions on endolymph pressure. Am J Otolaryngol 6: 297–301.

Correspondence to Björn Carlborg, Department of Otolaryngology, University Hospital S 205 02 Malmö, Sweden. Tel. (+46) 40332761. Fax (+46) 40336317. E-mail: bjorn.carlborg@oron.mas.lu.se.

Chapter 3
Dynamics of inner ear pressure change caused by cerebrospinal fluid pressure manipulation in guinea pigs

EO THALEN, HP WIT, JM SEGENHOUT, FWJ ALBERS, Department of Otorhinolaryngology, University Hospital Groningen, Netherlands

ABSTRACT. Prior studies have shown that pressure changes in the cerebrospinal fluid are transmitted to the inner ear. The main route for pressure transfer, from the cerebrospinal fluid towards the inner ear, is the cochlear aqueduct. In the present study, the cerebrospinal fluid pressure in guinea pigs was changed in a step-like way using an infusion system filled with Ringer's solution, driven by an electromagnetic system. Simultaneously the hydrostatic pressure was monitored, with a miniature solid-state pressure transducer, in both the peridural space and in the perilymphatic or endolymphatic compartment of the inner ear. The endolymphatic or perilymphatic pressure of the cochlea was directly measured by means of a servo-controlled micropipette system. A bevelled glass micropipette with a tip diameter of 3 to 5 micrometres was inserted into scala tympani by perforating the round window membrane, or into scala media by subsequent perforation of the basilar membrane. While measuring the inner-ear pressure, the DC-potential could be recorded simultaneously with the micropipette to ensure proper placement of the pipette tip. During manipulation, virtually identical pressure changes were recorded in both the endolymphatic and perilymphatic compartments. Pressure changes in the perilymph following manipulated cerebrospinal fluid pressure were similar in pattern and magnitude to those seen in the endolymph. Inner ear pressure changes caused by step-like increase or decrease of cerebrospinal fluid pressure could in first approximation be described by a simple exponential function, rising or falling with a time constant in the range of 2 to 3 seconds. However, upon closer examination, the pressure equalization process turned out to be non-linear; possibly caused by pressure-dependent compliances or flow resistances.

Key words: cerebrospinal fluid pressure, inner ear fluid pressure, cochlear aqueduct, micropipette.

Introduction

A better understanding of the hydromechanical interactions between the intracochlear and intracranial fluids, their interdependence and relationship, is essential to our understanding of both normal and pathological physiology of the cochlea. In the normal labyrinth, the hydrostatic pressure in the endolymphatic and perilymphatic compartment is the same (Böhmer 1993). A pressure change in the cerebrospinal fluid compartment is transmitted to the inner ear (Beentjes 1972; Carlborg 1992). The main route for pressure transfer between the cerebrospinal fluid compartment and the inner ear is the cochlear aqueduct (Beentjes 1972; Kerth 1963). Other routes for pressure transduction are the perivascular and perineural spaces and indirectly the endolymphatic sac and duct (Carlborg 1983, 1992).

The cochlear aqueduct extends from the basal turn of the scala tympani, near the round window membrane, to the subarachnoid space. Within this canal is a loose network of connective tissue, continuous with the arachnoid membrane (Jackler 1993). Until now, details of the dynamic properties of the cochlear aqueduct in equalizing the cerebrospinal fluid and inner ear pressure were not well understood.

We therefore recently started an experiment investigating the dynamics of inner ear pressure change following cerebrospinal fluid pressure manipulation in the guinea pig. The cerebrospinal fluid pressure was changed in a step-like way by means of an infusion system driven with an electromagnetic system. Simultaneously, the hydrostatic pressure was measured in both the peridural space, and in the perilymphatic or endolymphatic compartment of the inner ear by means of a servo-controlled micropipette system.

Materials and methods

In our study we used eight female albino guinea pigs, weighing 650–750 grams and having normal pinnal reflexes. General anaesthesia was induced by intramuscular administration of ketamine/xylazine (60 mg/kg / 3.5 mg/kg), succinylcholine (2.5 mg/kg) was given for muscle relaxation. Ventilation was maintained by a respirator through a tracheotomy. Blood pressure was monitored by means of a catheter, connected to a pressure transducer and inserted into a femoral artery. The animals were placed on a homeothermic blanket, which kept the body temperature at approximately 38°C. Each animal was placed in the supine position, with the head stationary.

In order to manipulate and measure the cerebrospinal fluid pressure, two holes were drilled in the cranium without causing damage to the dura. A miniature solid state pressure transducer was placed in one hole. Through the other hole the cerebrospinal fluid pressure was changed in a step-like way, by means of a subarachnoidal infusion system, filled with Ringer's solution and driven with an electromagnetic system.

The hydrostatic pressure in the inner ear was measured with a servo-controlled micropipette system (WP Instruments, Sarasota, USA, model 900A). This system allows hydrostatic pressure recordings through a micropipette, with a tip diameter of approximately 5 micrometres (Fox 1972). By a retroauricular approach, the left bulla was opened, and the round window membrane was exposed. The bevelled micropipette was inserted through the round window membrane into the scala tympani where the perilymphatic pressure was recorded. By perforating the basilar membrane, the electrode was placed into scala media, for subsequent registering of the endolymphatic pressure.

The DC-potential, recorded simultaneously through the same pipette, monitored localization of the pipette tip and allowed for measurement of the endocochlear potential in the basal cochlear turn. Before and after the measurements were performed, the baseline value for the electric potential was checked and pressure calibration was done at the beginning of each experiment.

Thus, we could increase and decrease the cerebrospinal fluid pressure in a stepwise manner, and simultaneously monitor the hydrostatic pressure in both the peridural space and the perilymphatic or endolymphatic compartment of the inner ear.

Results

The simultaneous measurement of the endocochlear potential, the inner ear pressure and the pressure in the subarachnoidal space, as performed in our experiments, is represented in Figure 3.1. The endocochlear potential shows the localization of the micropipette tip, which is in the endolymph during the first part, and in the perilymph during the second part of this recording.

In all measurements, the pressure changes in the perilymph following manipulated cerebrospinal fluid pressure were similar in pattern and magnitude to those seen in the endolymph. During these pressure changes the endocochlear potential demonstrated no measurable change. The stepwise fast change of the cerebrospinal fluid pressure is obvious. The onset of the pressure equalization process in the inner ear, following cerebrospinal fluid pressure change, is immediate. There is no measurable time lag in the onset of pressure

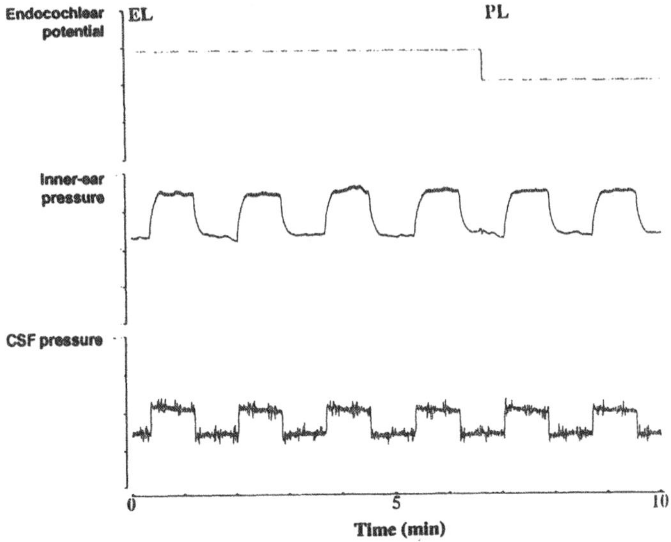

Figure 3.1: results of one measurement: the endocochlear potential and the changes of inner ear pressure, following manipulated cerbrospinal fluid pressure changes. The endocochlear potential shows the localization of the micropipette, which is in the endolymphatic compartment during the first part, and in the perilymphatic compartment during the second part of this recording (EL = endolymph, PL = perilymph).

change of the inner ear, compared with the onset of pressure change of the cerebrospinal fluid. The inner ear pressure stabilizes within a few seconds.

In the inner ear, the pressure equalization process itself seems to be slower for pressure decrease compared to pressure increase. We have tried to quantify this pressure equalization process of the inner ear. In a first approximation, the pressure equalization processes could be described by a simple exponential function rising or falling with a time constant in the range of 2 to 3 seconds. Figure 3.2 shows a rising part of the pressure equalization of the inner ear during increase of the cerebrospinal fluid pressure, fitted with this exponential function. The time constant of this fit is 2.08 seconds.

In our experiments, all time constants found, both for rising and falling of inner ear pressure, are in the range of 2 to 3 seconds. All time constants for the fits of the pressure equalization curve during pressure increase are smaller than the time constants for pressure decrease in the same guinea pig.

Figure 3.2 shows besides the oscillatory pressure changes due to the breathing of the guinea pig, that the lines of the fitted exponential function and the measured pressure are close, but both curves do not coincide. This systematic difference between the measured and fitted curve was a consistent observation, in both rising and falling of the inner ear pressure, indicating that the pressure equalization process of the inner ear is non-linear.

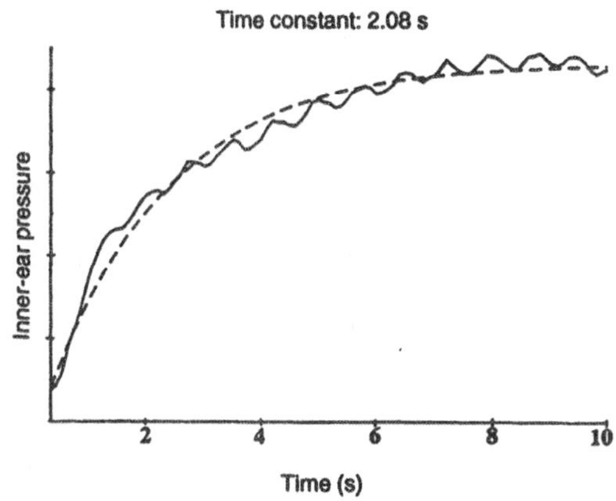

Figure 3.2: inner ear pressure versus time: fit with a simple exponential function (dashed line) of the inner ear pressure equalization process following a step-like increase of the cerebrospinal fluid pressure.

Conclusions

The preliminary results of our experiments support the following conclusions:

- Transmission of cerebrospinal fluid pressure increase to the inner ear compartments appears to be faster than the transmission of cerebrospinal fluid pressure decrease. The cochlear aqueduct might provide a greater resistance to fluid motion in one direction than in the other.
- The pressure equalization of the inner ear turned out to be non-linear. This non-linearity may be caused by pressure-dependent compliances or flow resistances. Pressure transients may be influenced not only by the dynamic properties of the cochlear aqueduct but by the endolymphatic sac and duct and the compliant nature of the cochlear windows as well (Marchbanks 1990).

Prior studies have shown the varying patency of the cochlear aqueduct in man (Gopen et al. 1997; Wlodyka 1978). As the dynamic properties of the cochlear aqueduct may be related to clinical symptoms in man (such as paroxysmal vertigo, fluctuating hearing loss and tinnitus in patients with intracranial hypertension, and perilymphatic fistulae (Sismanis 1987)) further research is required, in order to obtain a better knowledge of the dynamics of inner ear pressure regulation.

References

Beentjes BIJ (1972) The cochlear aqueduct and the pressure of the cerebrospinal and endolabyrinthine fluid. Acta Otolaryngol (Stockh) 17: 112–20.

Böhmer A (1993) Hydrostatic pressure in the inner ear fluid compartments and its effects on inner ear pressure. Acta Otolaryngol (Stockh) Suppl 507: 1–24.

Carlborg BIR, Konrádsson KS, Carlborg AH, Farmer JC, Densert O (1992) Pressure transfer between the perilymph and the cerebrospinal fluid compartments in cats. Am J Otology 13: 41–8.

Carlborg B, Farmer JC (1983) Transmission of cerebrospinal fluid pressure via the cochlear aqueduct and endolymphatic sac. Am J Otolaryngol 4: 273–82.

Fox JR, Wiederhielm CA (1972) Characteristics of the servo-controlled micropipette pressure system. Microvasc Research 5: 324–35.

Gopen Q, Rosowski JJ, Merchant SN (1997) Anatomy of the normal human cochlear aqueduct with functional implications. Hear Res 107: 9–22.

Jackler RK, Hwang PH (1993) Enlargement of the cochlear aqueduct: Fact or fiction? Otolaryngol Head Neck Surg 109: 14–25.

Kerth JD, Allen GW (1963) Comparison of the perilymphatic and cerebrospinal fluid pressures. Arch Otolaryngol (Stockh) 77: 27–31.

Marchbanks RJ, Reid A (1990) Cochlear and cerebrospinal fluid pressure: their interrelationship and control mechanisms. Br J Audiol 24: 179–87.

Sismanis A (1987) Otologic manifestations of benign intracranial hypertension syndrome: Diagnosis and management. Laryngoscope 97: 1–17.

Wlodyka J (1978) Studies on the cochlear aqueduct patency. Ann Otol Rhinol Laryngol 87: 22–8.

Correspondence to EO Thalen, Department of Otorhinolaryngology, University Hospital Groningen, PO Box 30.001 NL-9700 RB Groningen, Netherlands. Tel. (+31) 503612540. Fax (+31) 503696726.

Chapter 4
CSF and posture effects on transient evoked oto-acoustic emissions – pilot study

MA FERGUSON, AC DAVIS and EA LOVELL, MRC Institute of Hearing Research, University of Nottingham, Nottingham

ABSTRACT. Change in body posture using a tilt table was used to influence cerebrospinal fluid (CSF) and intracochlear fluid pressure. The effects on transient evoked otoacoustic emissions (TEOAEs) were examined. Maximum length sequences (MLS) allowed rapid averaging of TEOAEs. MLS TEOAEs were obtained from 12 normally-hearing ears as subjects were tilted from an upright to a recumbent position, –10° below the horizontal returning to the upright position. In addition, middle-ear pressure (MEP) measurements were also obtained. Overall group results showed that tilting upwards from recumbent to upright (CSF pressure decrease) significantly increased the TEOAE amplitude. There was no overall effect seen on tilting down, although in some ears there was a significant decrease in TEOAE amplitude. Middle-ear pressure significantly increased on tilting down, and decreased on tilting up. Postural change has an effect on MEP that may explain some of the TEOAE amplitude change. However, it does not explain all the variation that may be due to CSF pressure changes.

Key words: oto-acoustic emissions, evoked emissions, posture, cerebrospinal fluid, pressure, body position, middle-ear pressure.

Introduction

A change in body posture from sitting to lying has been shown to increase cerebrospinal fluid (CSF) and intracranial fluid (ICF) pressure. Conversely, on moving from lying to sitting, CSF and ICF pressure decreases (Marchbanks and Reid 1990). Such pressure changes transfer to the intracochlear fluids as they connect directly with CSF, primarily via the cochlear aqueduct. As transient evoked otoacoustic emissions (TEOAEs) are very sensitive to changes in the status of the cochlea, indirect changes in CSF pressure due to posture may be reflected in the TEOAE's. Similarly, directly increasing or decreasing the CSF pressure during lumbar puncture procedures may alter TEOAE characteristics.

Despite a number of reports on the effects of body posture on TEOAEs, there appears to be no consensus. Antonelli and Grandori (1986) showed a reduction in TEOAE amplitude and latency when moving from sitting to lying, whereas others showed only reduced latency changes (Phillips and Farrell 1992; Buki et al. 1996). Others have reported no effect of posture on TEOAEs (Johnsen and Elberling 1982; Froelich et al. 1994). The effects of lumbar puncture on TEOAEs are also equivocal. Davis et al. (1992) reported a relationship between CSF pressure and TEOAE amplitude during lumbar puncture. However, reanalysis of the data showed that there was typically only a small reduction in TEOAE amplitude when CSF pressure was increased, compared to the much bigger increase in TEOAE amplitude when CSF pressure was subsequently reduced (for example, see Figure 4.1). However, Froelich et al. (1994) and Buki et al. (1996) did not show any effect on TEOAE amplitude when CSF pressure was increased during lumbar puncture.

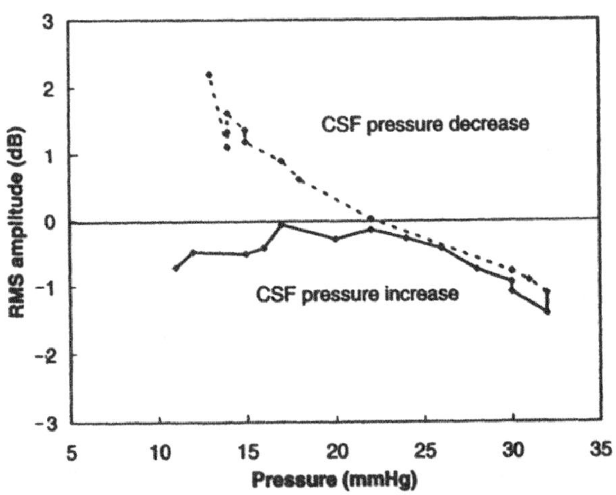

Figure 4.1: example of RMS amplitude for the derived, non-linear TEOAE as a function of CSF pressure during lumbar puncture in a single subject with normal hearing (from Davis et al. 1992).

Assuming posture change causes an almost instant transfer of pressure from CSF to intracochlear fluid (Avan et al. 1996) the time taken to obtain a TEOAE average, typically more than half a minute, may mask any immediate effect of pressure alteration on the TEOAE. This study used a technique for rapid averaging of TEOAEs by the use of maximum length sequences (MLS), described by Thornton (1993). This method enables a response to be obtained in seconds. Using change in body posture to induce change in CSF pressure and consequently intracochlear fluid pressure, we set out to:

- obtain instantaneous, rapid TEOAEs during near-continuous change in body posture;
- assess the effect of body posture on TEOAE amplitude and phase;
- examine the symmetry of the effect of posture on TEOAE amplitude;
- assess the effect of posture changes on middle-ear pressure; and
- assess the feasibility of developing a method to indirectly monitor CSF pressure using TEOAEs.

Method

Subjects

Data were collected from 12 normally-hearing ears of eight healthy volunteers, aged less than 35 years. The criteria for normal hearing was defined as average pure-tone air-conduction threshold across the octave frequencies 0.5–4 kHz less than or equal to 10 dB HL, acoustic reflex threshold (contralateral and ipsilateral) at 1 kHz less than or equal to 100 dB HL and middle-ear pressure (MEP) in the range –50 to +50 daPa.

Test parameters

Transient evoked oto-acoustic emissions were recorded using the in-house IHR POEMS (Programmable Oto-acoustic Emissions Measurement System) described by Cope and Lutman (1988), modified to obtain MLS measurements. Linear MLS emissions were obtained using click stimuli at 70 and 60 dB pe SPL. A click opportunity rate of 2000/s was used. In practice, this results in a response collection rate of around 1000/s. Each average comprised a maximum of 2000 clicks, thereby the MLS emissions could be obtained in less than 2 seconds. A reconstruction window of 5–20 ms was used. MEP measurements were made using a Grason Stadler GSI-38 screening tympanometer to allow rapid measurement of MEP. The pressure sweep was positive to negative, at a rate of 600 daPa/s for the majority of the sweep, slowing to around 200 daPa near the peak of the tympanogram.

Postural change and measurements

Each subject lay supine on a manually controlled tilt-table. This was tilted from a head-up starting position at +75° to the horizontal, through to a head-down position of –10° to the horizontal, at a rate of 2 degrees/second. Measurements of TEOAEs using the MLS technique and

MEP were made immediately when the body angle was at +75°, +60°, +30°, 0° and –10° to the horizontal. At –10° (head down), measurements continued to be taken throughout the next 3 minutes at 15 s, 30 s, 45 s, 60 s, 90 s, 120 s, 150 s and 180 s. The procedure was then reversed and the table was tilted back to the initial starting point (head-up). Measurements continued to be taken in the head-up position for a further 3 minutes using the same time intervals as for the head-down measurements.

Only one type of measurement and one repetition was made with each rotation. This allowed the tilting and subsequent measurements to be near-continuous. To prevent the probe from shifting, the TEOAE probe was held in the ear canal with earmould impression material. Subjects were given a break from testing for at least 2 minutes before obtaining the next set of measurements. Two replications of MLS emission and MEP were obtained from each ear.

In order to assess the effect of middle-ear pressure on MLS emission amplitude, independent of postural change, a second experiment was carried out on a subset of the above subjects (five ears). MLS emissions were recorded with the subjects sitting stationary in a chair. The pressure in the external ear canal was altered using a Grason Stadler GSI-33 diagnostic middle-ear analyzer coupled to the equalization tube of the TEOAE probe. MLS emissions were recorded at ear-canal pressures from +60 to –60 daPa in consecutive 10 daPa steps.

Results

Figure 4.2 shows the changes in the 70 dB MLS emission waveform which occurred at different points in the head-down rotation from the ear of the first subject. Figure 4.2a shows a significant decrease in the amplitude of the TEOAE after tilting-down to –10° compared to the amplitude in the head-up position. There is also a marked phase shift that appears to be reduced in latency on moving the head down. During the 3 minute period when the head was down, the amplitude and the phase of the TEOAE did not change substantially (Figure 4.2b). Similar effects were seen in the head-up tilt and head-up rest phase. However, at the end of the upward-tilting process the amplitude of the TEOAE 'overshot' the amplitude at the outset; returning the head to the upright position resulted in a larger amplitude than was initially recorded. This asymmetry was similar to the data described by Davis et al. (1992) and it was these initial results that encouraged us to collect data on a further 11 ears.

The data for all 12 ears showed considerable variability in terms of the relationship between the MLS emission amplitude and posture, but highly consistent results for MEP with posture change. Only MLS emission data collected at 70 dB are reported here, concentrating on the RMS amplitude for the 9–13 ms window as this tends to be the portion of the waveform that exhibits the strongest emission. Furthermore, we can be sure that the RMS amplitude values are not affected by any 'ringing' of the middle ear or ear-canal.

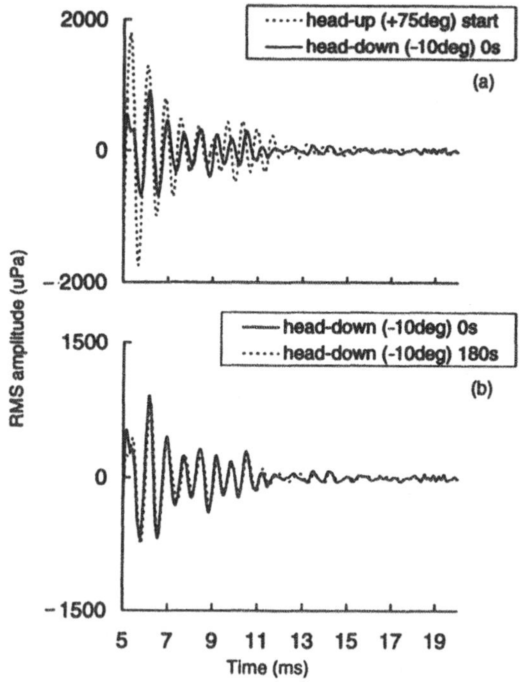

Figure 4.2: MLS emissions from subject 1 (70 dB pe SPL click) at (a) the head-up position (dashed line) and the start of the head-down (–10°) position (solid line) showing a reduction in amplitude and phase with downward tilt and (b) at the start of the head-down phase (solid line) and 180 seconds later (dashed line) showing no substantial change in the TEOAE.

Figure 4.3 shows the mean RMS amplitude of the MLS emissions and mean MEP as a function of angle of inclination and rest at the extreme angles for the whole group. There is a reduction in amplitude as the head is tilted down (and presumably CSF pressure increases) and an increase in amplitude as the head is returned to the head-up position (and presumably CSF pressure decreases). The distribution of RMS amplitude for the two replications of the MLS emissions was highly skewed and a log transform was taken to normalize the distribution. The data were analysed by a mixed model ANOVA with ear as the blocking factor. Two separate analyses were performed, one for the tilting portion of the experiment and the other for the rest period at the extreme angles. The analysis showed that there was no significant effect of angle on the RMS amplitude of the MLS emissions when tilting downwards, but this was significant when tilting upwards ($F = 2.5$; df = 4,40; $p = 0.05$). There was a symmetrical effect of posture change on MEP, which increased when the subjects were tilted down (mean = 19.41 daPa, s.d. = 9.10) and decreased on returning the subjects to the original upright position (mean = 19.31 daPa, s.d. = 8.55). This was seen in all subjects and was significant in both the tilt-down ($F = 31.9$; $df = 4,42$; $p < 0.001$) and tilt-up ($F = 55.6$; $df = 4,43$; $p < 0.001$) sections, but not during the rest periods.

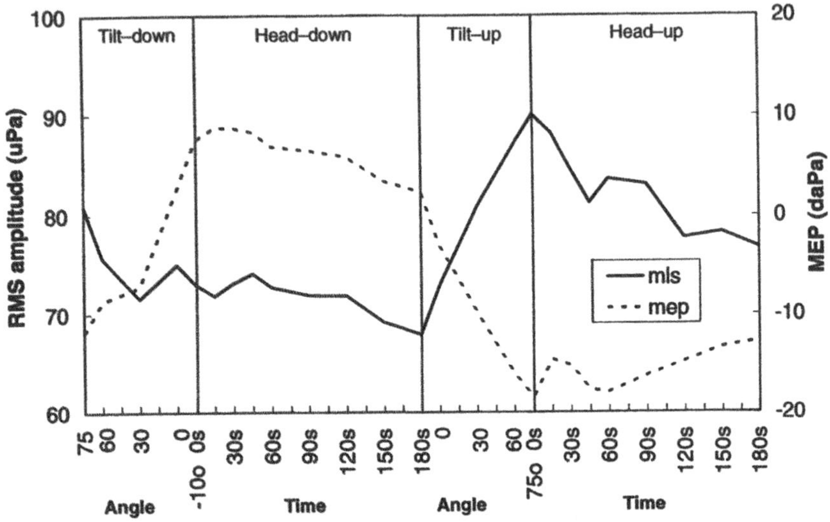

Figure 4.3: the mean middle-ear pressure and RMS amplitude of the MLS emission as a function of angle of inclination and the time at extremes, n = 12 ears (70 dB click, 9–13 ms window).

The relationship between the RMS amplitude as a function of MEP for the four phases of the tilting is shown in Figure 4.4. There is a negative correlation between the two measures ($r = -0.22$), but there appears to be a difference in this relationship depending on whether there is downward or upward rotation. For the downward tilt there is little change, but for the upward tilt there is a monotonic relationship.

Overall, the data do not show the clear trends observed in the first subject so the data for individual subjects were then analysed separately. Six ears (50%) had RMS amplitudes that were significantly increased after rotating up to +75° compared to the amplitudes at –10° (using all nine measures over 180 s at each position). Two ears showed the opposite result and four showed no difference. After rotating down, six of the ears showed a decrease in the RMS amplitude over the rest time and two showed an increase. After rotating up, most ears did not show a significant change of amplitude over the 3 minutes, although two showed an increase and two a decrease of amplitude.

In terms of angle of inclination (taking the tan of 90° (angle to the horizontal)), five ears showed an overall decrease in RMS amplitude as the angle to the horizontal decreased, in line with our prediction that increased CSF pressure would decrease emission amplitude. Taking upward and downward rotation separately (there are only five independent points per subject) six ears showed significant effects in the expected direction.

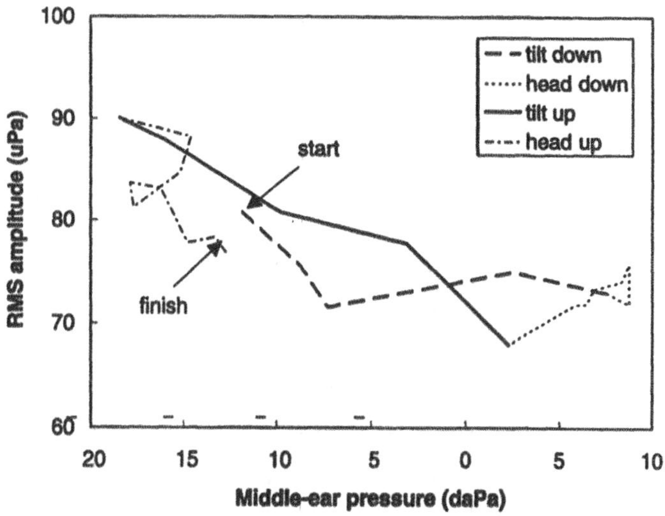

Figure 4.4: mean RMS amplitude of MLS emission as a function of MEP for the four phases of tilting, n = 12 ears (70 dB click, 9–13 ms window).

Figure 4.5 shows the mean MLS emission amplitude as a function of experimentally altered ear-canal pressure for the five ears tested. If we assume that maximum middle-ear mobility, and therefore maximum transmission of sound energy, occurs at MEP, then we can assume MEP to be at the maximum MLS emission amplitude. There is a consistent relationship, which is linear within the bounds of +30 to –30 daPa relative to MEP. Thus any deviation from initial MEP reduces the MLS emission amplitude.

Discussion

Some of the subjects, about half, showed a trend similar to that reported by Davis et al. (1992), where head-down tilting (i.e. CSF pressure increase) reduced MLS emission amplitude, and head-up tilting (CSF pressure decrease) increased the amplitude. However, the results of posture change on MLS emission amplitude showed considerable individual variation and overall group results showed that only the head-up tilting was significant. These results are broadly consistent with those reported by Antonelli and Grandori (1986). These authors also suggested that in addition to CSF pressure effects, MEP may also increase when tilting the head and body below the horizontal, and this

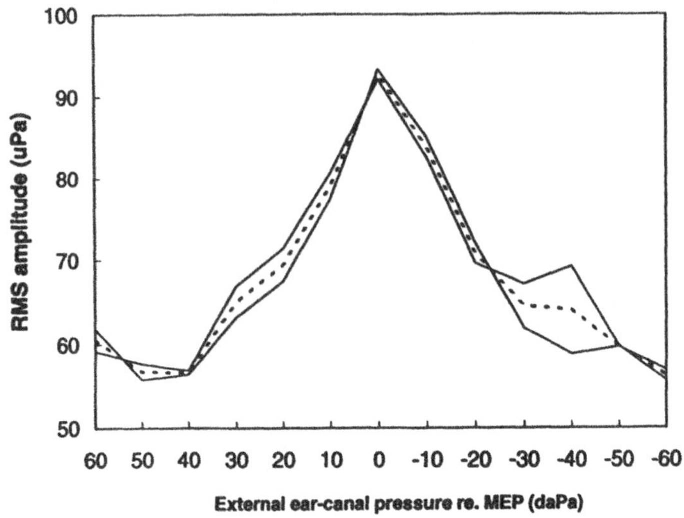

Figure 4.5: mean MLS emission amplitude as a function of experimentally altered external ear-canal pressure, n = 5 ears. (solid line) replications, (dashed line) overall mean.

may also have an additional affect on the TEOAE amplitude. There are other reports that have shown that postural changes affect MEP (Meredith et al. 1995; Buki et al. 1996). Our results are consistent with these and show a significant, symmetrical effect of posture change on MEP. This effect may be due to postural change interfering with venous drainage of blood from the capillary-rich tissues surrounding the middle ear thereby affecting volume and MEP. However, MEP changes are highly repeatable and symmetrical with postural change whereas MLS emission amplitude changes are more variable, and in some cases are asymmetrical. This suggests that although MEP may have some effect on TEOAE amplitude as a result of change in posture, the consequent CSF pressure change, or indeed some other mechanism, may also exert an effect on the cochlea and TEOAE. Interestingly, both MEP and MLS emission amplitude changed immediately with postural change, which supports our initial assumption that transmission of pressure from CSF to intracochlear fluids is instantaneous.

To conclude, it was possible to obtain MLS emissions to monitor cochlear function during changes in posture. It was found that posture and tilting do affect the RMS amplitude of MLS emissions but it was highly variable from subject to subject. The effect of posture on MLS emissions seems to be related to the direction of travel with a greater effect shown on upward tilting, which may explain differences found in the literature (Buki et al. 1996; Davis et al. 1992). There were also substantial and highly consistent changes in MEP due to posture. However, due to high inter-subject variability in the MLS emissions, the probability of using these to monitor CSF pressure is low. We were unable to obtain simultaneous, instantaneous MEP and MLS emissions measurements and examine the direct effect of MEP on MLS emission amplitude during tilting. Thus, future work will include measuring TEOAEs at constant MEP whilst tilting as well as analysis of TEOAE phase.

Acknowledgements

We would like to thank John Birchall for his helpful comments and advice.

References

Antonelli A, Grandori F (1986) Long-term stability, influence of head position and modeling considerations for evoked otoacoustic emissions. Scand Audiol Suppl 25: 97–108.

Avan P, Buki B, Lemaire J, Dordain M, Chazal J (1996) Otoacoustic emissions: a new tool for monitoring intracranial pressure. In: Ernst A, Marchbanks R, Samii M (eds) Intracranial and Intralabyrinthine Fluids. Berlin: Springer Verlag, pp. 165–73.

Buki B, Avan P, Lemaire J, Dordain M, Chazal J, Ribari O (1996) Otoacoustic emissions: a new tool for monitoring intracranial pressure changes through stapes displacement. Hear Res 94: 125–39.

Cope Y, Lutman M (1988) Oto-acoustic emissions. In: McCormick B (ed.) Paediatric Audiology 0–5 Years. London: Taylor & Francis, pp. 221–45.

Davis A, Firth J, O'Donoghue G (1992) The effect of changing intracranial pressure on click-evoked otoacoustic emissions. Abstract 75, Audiology in Europe. Cambridge, UK.

Froelich P, Ferber C, Remond J, Jabouley J, Morgan A, Duclaux R, Collet L (1994) Lack of association between transiently-evoked otoacoustic emissions amplitude and experimentation-linked factors. Hear Res 75: 184–90.

Johnsen N, Elberling C (1992) Evoked acoustic emissions from the human ear. II Normative data in young adults and influence of posture. Scand Audiol 11: 69–77.

Marchbanks R, Reid A (1990) Cochlear and cerebrospinal fluid pressure: their interrelationship and control mechanisms. Br J Audiol 24: 179–87.

Meredith R, West J, Osborne D (1995) No postural effect on middle-ear resonance frequency: A pilot study. Br J Audiol 29: 225–58.

Phillips A, Farrell G (1992) The effect of posture on three objective audiological measures. Br J Audiol 26: 339–45.

Thornton A (1993) High rate otoacoustic emissions. J Acoust Soc Am 94(1): 132–6.

Correspondence to MRC Institute of Hearing Research, University Park, University of Nottingham, Nottingham NG7 2RD, UK.

PART 2
INTRACRANIAL PATHOPHYSIOLOGY

Chapter 5
Pathophysiology of the cerebrospinal and cerebrovascular circulations

JD PICKARD, S PIECHNIK, P SMIELEWSKI, PJ KIRKPATRICK, M CZOSNYKA, Addenbrooke's Hospital, Cambridge, UK

Introduction

The brain is an expansile structure that expands and contracts with each heartbeat. As there are no valves within the venous drainage from the brain, any changes in intrathoracic pressure are transmitted to intracranial pressure. Such phenomena can readily be seen and palpated simply by examining the fontanelle of a baby, and quantified by cerebrospinal fluid (CSF) pressure recording.

Once the sutures have fused, any change in cerebral blood volume on the arterial/arteriolar side of the cerebral circulation must be compensated for either by reduction in cerebral venous volume or by phasic movement of CSF out of the skull through the foramen magnum. Expansion of the cerebral mantle with compression of the lateral ventricles during systole and movement of CSF through the aqueduct of Sylvius and to-and-fro through the foramen magnum were visualised originally on pneumoencephalography and now by dynamic magnetic resonance imaging. Inevitably there is a phase lag between the systolic increase in cerebral blood volume and the effect of the compensatory mechanisms so that CSF pressure increases to reflect in part the systolic waveform. The CSF pressure waveform is not an exact replica of the arterial waveform as it has been 'filtered' by the combined effects of the arterial wall compliance of the cerebral arteries, cerebrovascular resistance and intracranial compliance (brain + venous drainage + CSF drainage). The effect of a change in intracranial volume on intracranial pressure follows an exponential relationship (Figure 5.1) (Pickard and Czosnyka 1993 for review).

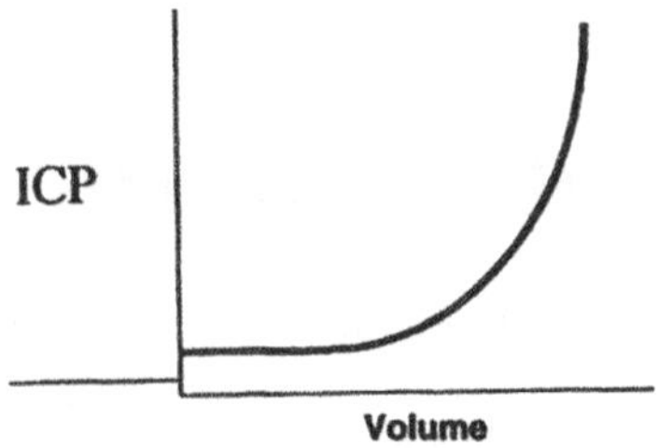

Figure 5.1: the effect of a change in intracranial volume on intracranial pressure.

Czosnyka et al. (1992) have illustrated these relationships with a hydrodynamic model and an electrical analogue as shown in Figure 5.2.

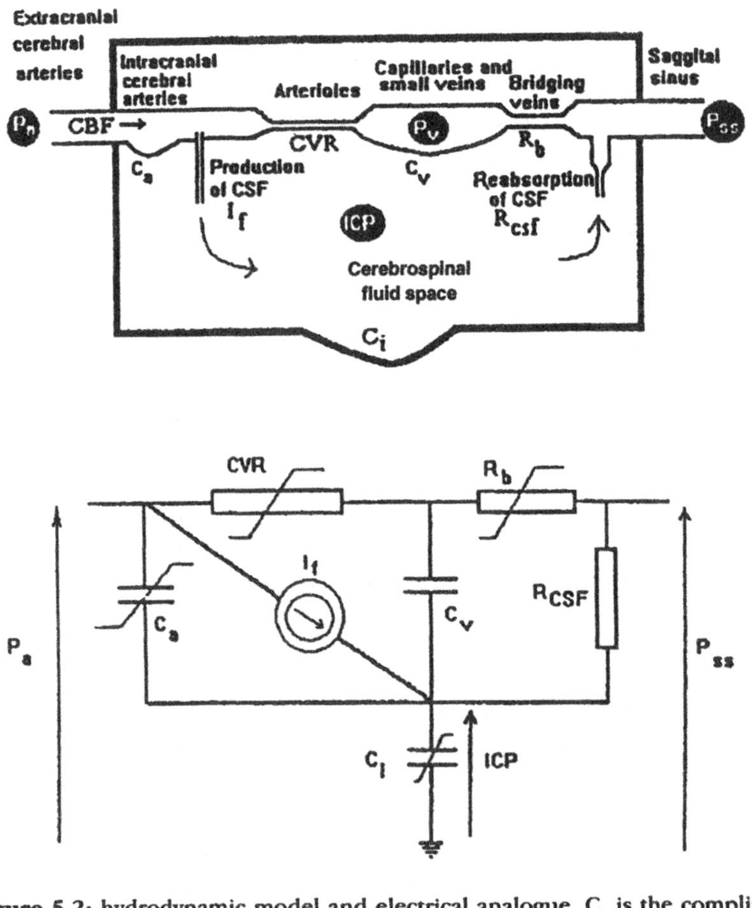

Figure 5.2: hydrodynamic model and electrical analogue. C_a is the compliance of the arterial bed; CVR is the cerebrovascular resistance; C_v is the compliance of the venous pool; R_b is the resistance of the bridging veins; C_i is the compliance of the CSF pool; R_{CSF} is the resistance to CSF outflow; If is the rate of CSF production; P_v is the intracerebral venous pressure; ICP is the intracranial pressure; and P_a is the arterial pressure in the large cerebral arteries.

There is considerable interest in modelling these relationships (Marmarou et al. 1978; Ursino and Lodi 1997) as an aid to understanding some of the complex phenomena revealed clinically by multimodality bedside monitoring (Kirkpatrick et al. 1996).

Control of CSF pressure

Eighty per cent of CSF is the product of active secretion by the choroid plexus with the remainder contributed by movement of interstitial fluid into the ventricles and subarachnoid space. Drainage is largely passive via arachnoid villi/granulations into the superior sagittal sinus and spinal root sleeve venous drainage. There is some drainage, at present unquantified, through the olfactory bulb and mucosa into the deep cervical lymphatics.

Davson's equation (Davson et al. 1987) describes the immediate relationships controlling CSF pressure:

$$Pcsf = Icsf \times Rout - Psss$$

where Icsf is the CSF production rate, Rout is the CSF outflow resistance and Psss is the superior sagittal sinus pressure.

It is possible to interrogate CSF dynamics in patients by modern, computerised adaptations of the original Katzman–Hussey CSF infusion study and measure both Rout and brain compliance. Such measurements are of considerable clinical use when assessing patients with possible CSF disturbances including benign intracranial hypertension, all forms of hydrocephalus and cerebral atrophy and shunt malfunction (Czosnyka et al. 1996).

Effects of changes in cerebral blood volume on CSF pressure

Cerebrovascular smooth muscle, and hence cerebral blood flow/volume, responds to changes in arterial carbon dioxide tension, arterial oxygen tension and cerebral perfusion pressure (CPP), where CPP is defined as arterial blood pressure (ABP) minus intracranial pressure (ICP).

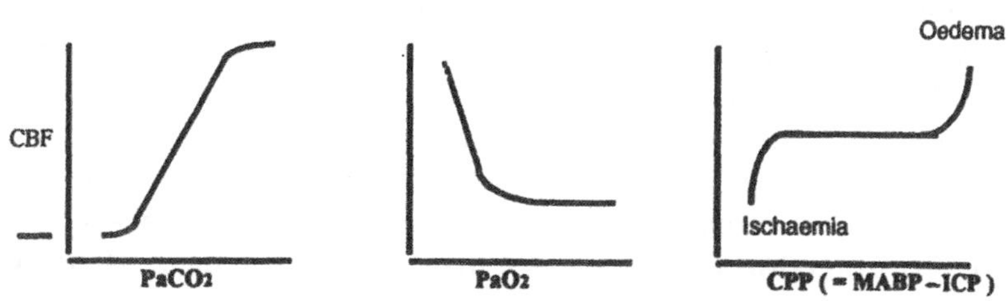

Figure 5.3: effects of changes in arterial carbon dioxide ($PaCO_2$), arterial oxygen tension (PaO_2) and cerebral perfusion pressure (CPP) on cerebral blood flow (CBF).

Cerebral vasodilatation results in a rise in ICP so that both hypercapnia and hypoxia increase ICP. Autoregulation of CBF to changes in arterial blood pressure requires active arteriolar constriction with hypertension to maintain CBF constant and ICP falls. Systemic hypotension within the autoregulatory range provokes cerebral vasodilatation and an increase in ICP. When autoregulation is defective, ICP increases and decreases with arterial blood pressure. Autoregulation of CBF may be assessed by artificial manipulation of ABP but only at infrequent intervals and with the risk that the drugs used may have a direct cerebrovascular effect. The transient hyperaemia following transient carotid compression is an all-or-none index of whether autoregulation is intact. Natural variations in ABP may be used but very precise signal processing is required. The most robust clinical method to date is to monitor the slow fluctuations in ABP, lasting from 30 seconds to a few minutes, that are almost always present in ventilated patients, and the rate of change observed is usually sufficient to provoke a noticeable vasomotor response (Czosnyka et al. 1997).

Continuous monitoring of cerebrovascular reactivity is possible with this method even during large spontaneous ICP waves ('B' and plateau). The indices derived are of prognostic significance. The concept is that defective autoregulation impairs the ability of the cerebral circulation of the acutely injured brain to withstand further insults such as periods of hypotension, hypoxia, hyperthermia or raised ICP. A CPP below 70 mmHg is associated with a worse outcome. Monitoring of ICP is invasive with a small risk of infection/haemorrhage. Where it is not available, transcranial Doppler may be used to detect the progressive changes in diastolic, mean and systolic flow velocities as autoregulation becomes exhausted.

Global lumped parameters versus regional heterogeneity: clinical trials of neuroprotective agents

It is perhaps remarkable that global measures such as ICP, CPP, jugular venous oxygen tension and indices of cerebrovascular reactivity are as successful as they are in predicting outcome in addition to the extent and severity of the initial injury compounded by secondary insults such as brief periods of hypoxia, hypotension and hyperthermia, for example. However, such 'lumped' parameters are not an adequate description of the pathophysiology of acute brain injury. There have been major disappointments with attempts to translate the enormous promise of studies of neuroprotection from the laboratory into man. Phase III studies of excitotoxic amino acid antagonists, free radical scavengers and calcium

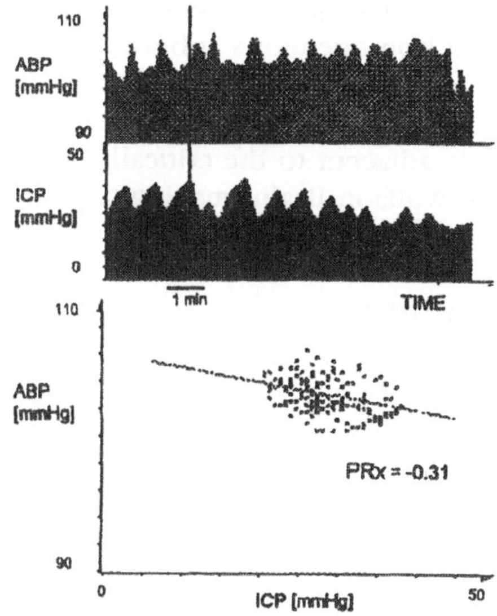

Good reactivity

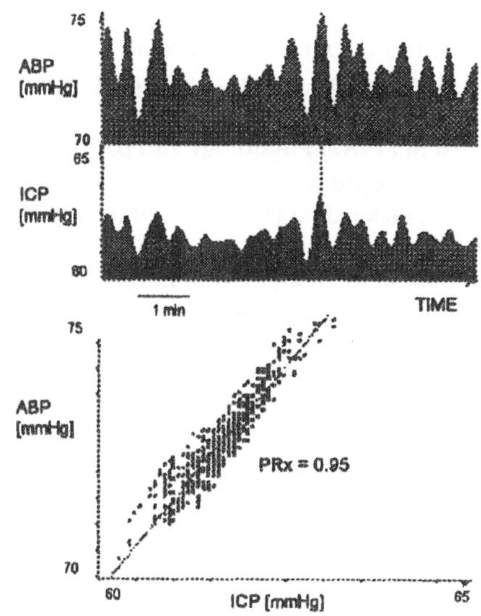

Poor reactivity

Figure 5.4: arterial blood pressure (ABP), intracranial pressure (ICP) and PRx: good and poor cerebrovascular reactivity. PRx is calculated as a linear correlation coefficient between averaged ABP and ICP from a time window of approximately 4 minutes. A positive PRx indicates impaired cerebrovascular autoregulation.

antagonists, for example, have not shown evidence of efficacy for acute brain injury except in aneurysmal subarachnoid haemorrhage. There is a need to define more homogeneous subgroups of patients who will require state-of-the-art functional brain imaging including Positron Emission Tomography and MRI/MRS. Such technology must be made available immediately adjacent to the critically ill patient, as has been accomplished in the Wolfson Brain Imaging Centre in Addenbrooke's Hospital in Cambridge.

Acknowledgements

Studies referred to have been funded by the Wolfson Foundation, Technology Foresight Challenge, MRC, Bayer, Merck Sharp and Dohme and Smithkline Beecham.

References

Czosnyka M, Pickard JD, Whitehouse H, Piechnik St (1992) The hyperaemic response to a transient reduction in cerebral perfusion pressure. Acta Neurochir (Wien) 115: 90–7.

Czosnyka M, Whitehouse H, Smielewski P, Simac S, Pickard JD (1996) Testing of cerebrospinal compensatory reserve in shunted and non-shunted patients: a guide to interpretation based on an observational study. J Neurol Neurosurg Psychiat 60: 549–58.

Czosnyka M, Smielewski P, Kirkpatrick P, Laing RJ, Menon D, Pickard JD (1997) Continuous assessment of the cerebral vasomotor reactivity in head injury. Neurosurgery 41: 11–19.

Davson H, Welch K, Segal M (1987) Physiology and pathology of cerebrospinal fluid. Edinburgh: Churchill Livingstone.

Kirkpatrick P, Czosnyka M, Pickard JD (1996) Multimodal monitoring in neurointensive care. J Neurol Neurosurg Psychiat 60:131–9.

Marmarou A, Shulman K, Rosende RM (1978) A non-linear analysis of the CSF system and intracranial pressure dynamics. J Neurosurg 48:332–4.

Pickard JD, Czosnyka M (1993) Management of raised intracranial pressure. J Neurol Neurosurg Psychiat 56:845–58.

Ursino M, Lodi CA (1997) A simple mathematical model of the interaction between intracranial pressure and cerebral hemodynamics. J Appl Physiol 82:1256–69.

Correspondence to Professor JD Pickard, Bayer Professor of Neurosurgery, Academic Neurosurgical Unit, MRC Centre for Brain Repair and Wolfson Brain Imaging Centre, Box 167, Level 4, Addenbrooke's Hospital, Cambridge CB2 2QQ, UK.

Chapter 6
Intraoperative neurophysiologic monitoring in skull base operations

AAGE R MØLLER, Department of Neurologic Surgery, University of Pittsburgh School of Medicine, Pittsburgh, PA, USA

ABSTRACT. Measurements of intracranial pressure (ICP) and various imaging techniques are the methods used for diagnosis of patients with elevated ICP. These methods measure changes in structure. Recordings of evoked potentials that can measure changes in function are seldom used in evaluating the effect of elevated ICP or in monitoring the treatment of such conditions. The use of auditory evoked potentials in intraoperative monitoring in operations in the cerebello pontine angle (CPA) illustrates the usefulness of recordings of evoked potentials in evaluating changes in the function of the ear and the auditory nerve. This chapter describes the technique of intraoperative recordings of farfield (brainstem auditory evoked potentials, BAEP) and near field auditory evoked potentials (from the exposed intracranial portion of the eighth nerve or the surface of the cochlear nucleus). The use of BAEP to detect changes in function of the brainstem caused by surgical manipulations is also discussed.

Key words: intraoperative monitoring, auditory evoked potentials, auditory nerve, cochlear nucleus, acoustic tumours.

Introduction

The diagnostic methods commonly used in patients with suspected elevated intracranial pressure (ICP) are measurements of the ICP and various imaging techniques, including ultrasound, but these methods do not provide information about abnormalities in function. This is despite the fact that readily available electrophysiologic methods can detect abnormalities or changes in function. Of the imaging techniques used, only the PET scan can, to some extent, detect change in function.

Electrophysiologic methods have been used for many years for clinical diagnosis of pathologies of the ear and the auditory nervous system (Hall 1992). The most common method is recording of brainstem auditory evoked potentials (BAEP) from electrodes placed on the scalp. Similar methods are used in the operating room to detect acute injuries of the ear and the auditory nervous system that may be caused by surgical manipulations (Raudzens 1982; Grundy 1983; Friedman et al. 1985; Fischer 1989; Linden et al. 1988; Møller and Møller 1989; Yingling and Gardi 1992; Møller 1995). Evoked potentials recorded directly from the ear or the exposed auditory nerve or from the surface of the cochlear nucleus are also frequently used for monitoring the function of the auditory system. The use of electrophysiologic methods to detect intraoperative changes in function helps to reduce the risk of permanent injuries to the nervous system (Friedman et al. 1985; Møller and Møller 1989; Radtke et al. 1989) and that is based on the assumption that changes in function occur before an injury has reached a level where postoperative permanent neurologic deficits become manifest, and that the changes are reversible.

Intraoperative monitoring of the auditory system is now in general use in operations where the auditory nerve is being manipulated, such as operations for acoustic tumours and other tumours of the cerebello-pontine angle. It is also in frequent use in vestibular nerve section (Møller et al. 1994a; Rosenberg et al. 1993) and in microvascular decompression operations to treat hemifacial spasm, trigeminal neuralgia, glossopharyngeal neuralgia and certain vestibular disorders (disabling positional vertigo, DPV) and severe tinnitus (Linden et al. 1988; Møller and Møller 1989; Radtke et al. 1989; Møller 1995).

Recording of BAEP

Brainstem auditory evoked potentials have a very small amplitude and it is therefore necessary to collect the responses to many stimuli to obtain an interpretable record. This takes time and delays detection of changes caused by surgical manipulations. Shortening that time as much as possible is important when BAEP recording is used in intraoperative monitoring. The time it takes to obtain an interpretable record depends on several factors:

- the electrical interference that reaches the recording electrodes;
- the stimulus repetition rate;
- the stimulus strength;
- filtering of the recorded potentials;
- electrode placement;
- time required for quality control.

Successful recordings of BAEP in the operating room require that the electrical interferences that reach the recording electrodes be reduced as much as possible. In the operating room, both electrical interferences from other equipment and biological potentials (such as those generated by muscles) contribute to the interference. The practical arrangements in the operating room should be focused on reducing electrical interference from other equipment. The interference from muscles is not a problem when the patient is paralysed. Muscle relaxants cannot be used, however, in operations where electromyographic potentials are recorded, as in operations for acoustic tumours where facial muscles are monitored.

Optimal processing (averaging and filtering) of the recorded BAEP can reduce the number of responses that must be collected to obtain an interpretable record (Doyle and Hyde 1981; Sgro et al. 1989; Møller 1988, 1995). Interpretation of the BAEP is based on determination of the latencies of one or two of the peaks of the BAEP and to some extent on the amplitude of these peaks. That means that not all the information in the BAEP is important for interpretation of the BAEP. Attenuating unnecessary parts of the recorded BAEP can reduce the noise and by that reduce the number of responses that must be collected to obtain an interpretable record. Usually peak V is used because it has the highest amplitude. Sometimes using peak III or the vertex negative peak that precedes peak IV and peak V is advantageous. These peaks should therefore be enhanced in the record that is to be interpreted. We have used digital filters for many years because of their greater flexibility than conventional electronic filters (Møller 1988, 1995), as have others (Doyle and Hyde 1981).

The time required to obtain an interpretable record can also be shortened by optimal choice of stimulation. The BAEP is commonly elicited by click sounds. Optimal parameters are clicks of an intensity of 100–105 dB PeSPL, corresponding to 60–65 dB hearing level when presented at a rate of 20 clicks per second. This is the rate that is often used. However, a higher stimulus rate will produce an interpretable record in a shorter time. The optimal rate is 40–60 pps (Møller 1995). It will take 25 seconds to obtain an interpretable record if 1000 responses must be averaged and the stimulus rate is 40 pps. This is a practical situation in a patient with normal hearing if optimal filtering is used and the electrical interference is low. If the interference is not well controlled, more responses need to be averaged, perhaps 2000, and that means it will take nearly one minute to obtain an interpretable record. If the patient has impaired hearing, the amplitude of the BAEP may be lower than normal and more responses must therefore be averaged. Hearing loss of cochlear origin affects the BAEP less than hearing impairment caused by injury to the auditory nerve. The BAEP in patients with acoustic neuromas often have low amplitudes besides prolonged latencies because of injury to the auditory nerve.

When BAEPs are recorded in the clinic, it is customary to replicate recordings to ensure validity. Using that method in the operating room means that the time it takes to obtain an interpretable response will be twice as long as it takes to obtain one record. Several other ways have been described for quality control that do not require two identical records to be obtained. We have used the '+/–' averaging methods (Schimmel 1967; Hoke et al. 1984; Møller 1995) for that purpose. However, commercial equipment seldom includes any of these methods for quality control.

Recordings of near field auditory evoked potentials

Recording of click evoked auditory evoked potentials directly from the exposed CNVIII yields potentials that are sufficiently large to be observed directly on an oscilloscope (Møller and Jannetta 1981; Hashimoto et al. 1981; Spire et al. 1982), or after only a few responses are averaged. A silver wire to which uninsulated tip a cotton wick is sutured is a suitable recording electrode (Møller and Jannetta 1981, 1983a; Møller 1995). Typically, the amplitude of the recorded potentials is several microvolts. Monitoring of such compound action potentials (CAP) from the exposed CNVIII has been used in operations for acoustic tumours (Silverstein et al. 1985; Linden 1988) and in MVD operations (Møller and Jannetta 1984; Møller and Møller 1989; Colletti et al. 1994).

Keeping a recording electrode in place on the proximal portion of the CNVIII in operations for acoustic tumours may be difficult and the electrode easily becomes dislocated by surgical manipulations. A better way to monitor the function of the ear and the auditory nerve in operations in the CP-angle is to record auditory evoked potentials from the cochlear nucleus. The cochlear nucleus is the floor of the lateral recess of the fourth ventricle and placing the electrode through the foramen of Luschka to reach the lateral recess (Møller and Jannetta 1983b; Møller et al. 1994b,c; Kuroki and Møller 1995; Møller 1995) yields potentials of a large amplitude. The same wick electrode as used to record from the CNVIII can be used for that purpose. The recording electrode will not easily be dislocated by surgical manipulations when placed in the foramen of Luschka and the electrode wire is tucked under the dura sutures on the caudal wall of the wound. The recorded potentials consist of a sharp positive deflection that marks the termination of the auditory nerve in the cochlear nucleus, followed by a sharp negative deflection, probably generated by firings of cochlear nucleus cells. A large slow potential follows these sharp peaks. Placing the recording electrode on the CNIX and CNX where these nerves emerge from the brainstem will yield evoked potentials from the cochlear nucleus with a similar waveform but of slightly lower amplitude.

Recording auditory evoked potentials directly from the cochlear nucleus is suitable for monitoring the function of the auditory nerve and the ear in operations of acoustic tumours where the goal is to preserve hearing. It gives an almost instantaneous feedback to the surgeon regarding injuries to the auditory nerve. The drawback is that it cannot be used before the region of the foramen of Luschka is exposed. It is therefore necessary to rely on recordings of the BAEP before these structures are exposed.

Bipolar recordings from the CNVIII

Recording auditory evoked potentials from the CNVIII using a bipolar recording electrode (Møller et al. 1994a) is useful for finding the cleavage plane between the inferior vestibular and the superior vestibular nerves for vestibular nerve section. A bipolar recording electrode is more spatially selective than a monopolar electrode. It has been reported that using clicks of a relatively low intensity gives greater spatial selectivity than using the conventional high intensity clicks as stimulation for that purpose (Rosenberg et al. 1993)

Recordings of evoked potentials from the ear

Recordings of electrocochleographic (ECoG) potentials have also been promoted as a method for monitoring surgically induced injuries during removal of acoustic tumours (Levin et al. 1984; Sabin et al. 1987). Such potentials also have a large amplitude and can be observed after adding only a few responses. However, the ECoG potentials are not affected by injuries to the intracranial portion of the auditory nerve. The ECoG potentials will change if the blood supply to the cochlea is impaired, which may occur if the labyrinthine artery is injured. Such injuries are unfortunately seldom reversible and monitoring of ECoG potentials can therefore not decrease the risk of permanent postoperative hearing loss.

Recordings of contralateral elicited BAEP

Recordings of BAEP elicited from the contralateral ear can be used to detect when surgical manipulations of the brainstem cause changes in function such as may occur in operations of large CPA tumours. Using changes in cardiovascular parameters for that purpose has been customary but Angelo and Møller (1996) showed that contralateral elicited BAEP can be a more sensitive indicator of manipulations of the brainstem. Thus changes in the BAEP could be detected before changes in cardiovascular parameters occurred. Recordings of the BAEP elicited from the contralateral ear can be made using the same methods as used for recordings of the ipsilateral BAEP. Such monitoring is useful in operations for large acoustic tumours and other space occupying lesions, the removal of which may cause displacement of the brainstem.

Practical aspects of intraoperative neurophysiologic monitoring

When is the correct time to inform the surgeon about changes? The information that should be conveyed to the surgeon has been extensively debated. Some surgeons only want information about the results of monitoring when the changes have reached a level where the results of monitoring indicate a noticeable risk of postoperative deficits. Definite problems are associated with that approach. Thus it is not known how large changes in evoked potentials such as the BAEP can be allowed before a noticeable risk of postoperative deficits become manifest. The change in the recorded potentials that indicates a noticeable risk of postoperative hearing loss is probably different for different patients and it is most probably also related to the duration of the changes in the evoked potentials. Waiting to inform the surgeon until the changes have become large also deprives the surgeon of important information about the cause of the changes and that makes it more difficult to reverse the changes.

These problems can be solved by keeping the surgeon informed about any change that is larger than the small changes that occur spontaneously without any known reason. In that way the surgeon has a choice of immediately reversing the manipulation that caused the change or waiting to see how it develops. We prefer the latter method and it is our experience that surgeons who always follow these guidelines have a much lower rate of postoperative neurologic deficits. If the changes in the recorded potential keep increasing, the surgeon knows what has caused the change and can reverse the changes at any time.

How can the value of monitoring be determined?

Only a few studies have shown a decrease in neurologic deficits because of the use of intraoperative monitoring. This is because carrying out studies in patients who were randomly selected for monitoring has been difficult. Surgeons have been reluctant to exclude patients from monitoring because they felt monitoring was beneficial to the patients.

The success of hearing preservation during operations where the auditory nerve is at risk of being injured must be documented by comparison of preoperative and postoperative hearing tests (Møller and Møller 1989). Such hearing tests must include speech discrimination tests using recorded speech material besides pure tone audiograms.

Is preservation of hearing always beneficial to the patient?

Preservation of postoperative neural function is usually desirable without doubt. Yet preserving hearing can have its negative aspects.

Many patients with hearing loss caused by injury to the auditory nerve have tinnitus. If the tinnitus is severe and the patients have useable hearing on the other side, it might have been better for the patient if hearing had not been preserved. This is because the quality of the preserved hearing after injury to the auditory nerve is often poor and many patients cannot understand speech on the affected side despite preserved hearing of pure tones. Thus when people with an acoustic tumour lose some hearing because of surgical manipulation of the auditory nerve, the tinnitus may be so severe that it would have been better if the auditory nerve had been severed during the operation.

Other benefits of intraoperative neurophysiologic monitoring

Intraoperative neurophysiologic monitoring not only decreases the risk of permanent neurologic deficits but it has other benefits. It gives the surgeon an increased feeling of confidence, it can often help the surgeon in carrying out the operation and it may reduce the operating time in some operations.

Intraoperative monitoring also provides an opportunity to do human research. Such studies have already provided valuable information both regarding the normal system and the pathological system.

References

Angelo R, Møller AR (1996) Contralateral evoked brainstem auditory potentials as an indicator of intraoperative brainstem manipulation in cerebellopontine angle tumors. Neurol Res 18: 528–40.

Colletti V, Bricolo A, Fiorino FG, Bruni L (1994) Changes in directly recorded cochlear nerve compound action potentials during acoustic tumor surgery. Skull Base Surg 4: 1–9.

Doyle DJ, Hyde ML (1981) Analogue and digital filtering of auditory brainstem responses. Scand Audiol 10: 81–9.

Fischer C (1989) Brainstem auditory evoked potential (BAEP) monitoring in posterior fossa surgery. In: Desmedt JE (ed.) Neuromonitoring in Surgery. Amsterdam: Elsevier Science Publishers, pp. 191–218.

Friedman WA, Kaplan BJ, Gravenstein D, Rhoton AL (1985) Intraoperative brainstem auditory evoked potentials during posterior fossa microvascular decompression. J Neurosurg 62: 552–7.

Grundy BL (1983) Intraoperative monitoring of sensory evoked potentials. Anesthesiology 58: 72–87.

Hall JW (1992) Handbook of Auditory Evoked Potentials. Boston: Allyn & Bacon.

Hoke M, Ross B, Wickesberg R, Luetkenhoener B (1984) Weighted averaging – theory and application to electrical response audiometry. Electroencephalogr Clin Neurophysiol 57: 484–9.

Kuroki A, Møller AR (1995) Microsurgical anatomy around the foramen of Luschka with reference to intraoperative recording of auditory evoked potentials from the cochlear nuclei. J Neurosurg 82: 933–9.

Levine RA, Ojemann RG, Montgomery WW, McGaffigan PM (1984) Monitoring auditory evoked potentials during acoustic neuroma surgery. Insights into the mechanism of the hearing loss. Ann Otol Rhinol Laryngol 93: 116–23.

Linden RD, Tator CH, Benedict C, Charles D, Mraz V, Bell I (1988) Electrophysiological monitoring during acoustic neuroma and other posterior fossa surgery. Le Journal Canadien des Sciences Neurologiques 15: 73–81.

Møller AR (1995) Intraoperative Neurophysiologic Monitoring. Luxembourg: Harwood Academic Publishers.

Møller AR, Colletti V, Fiorino F (1994a) Click evoked responses from the exposed intracranial portion of the eighth nerve during vestibular nerve section: Bipolar and monopolar recordings. Electroencephalogr Clin Neurophysiol 92: 17–29.

Møller AR, Jannetta PJ (1981) Compound action potentials recorded intracranially from the auditory nerve in man. Exp Neurol 74: 862–74.

Møller AR, Jannetta PJ (1983a) Monitoring auditory functions during cranial nerve microvascular decompression operations by direct recording from the eighth nerve. J Neurosurg 59: 493–9.

Møller AR, Jannetta PJ (1983b) Auditory evoked potentials recorded from the cochlear nucleus and its vicinity in man. J Neurosurg 59: 1013–18.

Møller AR, Jho HD, Jannetta PJ (1994b) Preservation of hearing in operations on acoustic tumors: an alternative to recording BAEP. Neurosurgery 34: 688–93.

Møller AR, Jannetta PJ, Jho HD (1994c) Click-evoked responses from the cochlear nucleus: a study in human. Electroencephalogr Clin Neurophysiol 92: 215–24.

Møller AR, Møller MB (1989) Does intraoperative monitoring of auditory evoked potentials reduce incidence of hearing loss as a complication of microvascular decompression of cranial nerves? Neurosurgery 24: 257–63.

Radtke RA, Erwin W, Wilkins RH (1989) Intraoperative brainstem auditory evoked potentials: Significant decrease in post-operative morbidity. Neurology 39: 187–91.

Raudzens PA (1982) Intraoperative monitoring of evoked potentials. Ann NY Acad Sci 388: 308–26.

Rosenberg SI, Martin WH, Pratt H, Schwegler JW, Silverstein H (1993) Bipolar cochlear nerve recording technique: a preliminary report. Am J Otol 14: 362–8.

Sabin HI, Bentivoglio P, Symon L, Cheesman AD, Prasher D, Momma F (1987) Intraoperative electrocochleography to monitor cochlear potentials during acoustic neuroma excision. Acta Neurochir (Wien) 85: 110–16.

Schimmel H (1967) The (+/–) reference: Accuracy of estimated mean components in average response studies. Science 157: 92–4.

Sgro JA, Emerson RG, Pedley TA (1989) Methods for steadily updating the averaged responses during neuromonitoring. In: Desmedt JE (ed.) Neuromonitoring in Surgery. Amsterdam: Elsevier Science Publishers pp. 49–60.

Silverstein H, Norrell H, Hyman S (1984) Simultaneous use of CO_2 laser with continuous monitoring of eighth cranial nerve action potential during acoustic neuroma surgery. Otolaryngol Head Neck Surg 92: 80–4.

Yingling CD, Gardi JN (1992) Intraoperative monitoring of facial and cochlear nerves during acoustic neuroma surgery. Otolaryngol Clin North Am 25: 413–48.

Correspondence to Professor Aage R Møller, University of Texas at Dallas, Callier Center for Communication Disorders, 1966 Inwood Rd, Dallas TX 75235-7298, USA.

Chapter 7
Neurotologic manifestations of intracranial hypertension

ARISTIDES SISMANIS, Department of Otolaryngology – Head and Neck Surgery, Medical College of Virginia/Virginia Commonwealth University, Richmond, Virginia, USA

ABSTRACT. At our institution, between August 1981 and December 1996, 64 patients with neurotologic manifestations secondary to intracranial hypertension were studied. The most common presenting symptom by far was pulsatile tinnitus, followed by hearing loss, dizziness/vertigo and aural fullness. The most common aetiology found was benign intracranial hypertension syndrome (pseudotumour cerebri). The pathophysiology of the neurotologic manifestations in intracranial hypertension is discussed and modalities of treatment for the various aetiologies are presented.

Key words: pathophysiology, neurotologic, manifestations, intracranial hypertension, tinnitus, hearing loss, vertigo, treatment.

Introduction

In 1981 it was first realized at our institution that intracranial hypertension could manifest itself with neurotologic symptoms such as pulsatile tinnitus (PT), hearing loss, dizziness and aural fullness (Sismanis et al. 1985). The purpose of this chapter is to present the author's 15 years of experience with intracranial hypertension patients presenting with neurotologic manifestations.

Material and methods

Between August 1981 and December 1996, 64 patients with neurotologic manifestations secondary to intracranial hypertension were evaluated at our institution. The evaluation protocol has been described previously (Sismanis and Smoker, 1994) and is summarized as follows.

All patients underwent thorough head and neck and neurotologic evaluation including auscultation of the ear canals, head and neck. Pure-tone (air and bone conduction) and speech audiometry was obtained in all patients. When a hearing loss of 20 dB or more was present, a repeat audiogram was performed while the patient was applying light digital pressure over the internal jugular vein (IJV) ipsilateral to the tinnitus.

Radiologic evaluation was performed in all patients prior to lumbar puncture. During the first four years of the study, temporal bone and brain computed tomography (CT) as well as carotid angiography were obtained in all patients. After 1985, when benign intracranial hypertension (BIH) syndrome was identified as a common aetiology, carotid angiography was reserved for selected cases only. Magnetic resonance imaging (MRI) and magnetic resonance angiography (MRA) were initiated in 1987 and 1992 respectively.

Auditory evoked responses (ABR), electrocochleography (ECochG) and electronystagmography (ENG) were obtained in selected cases. All patients were evaluated by a neuro-ophthalmologist. Intracranial hypertension was confirmed by lumbar puncture (LP) and measurement of cerebrospinal fluid (CSF) pressure.

Results

Intracranial hypertension was confirmed by LP with CSF pressure measurement (>200 mm of water) in 58 patients. In five patients with typical manifestations of BIH syndrome, LP was not feasible due to technical problems arising from their obesity. In one patient with obstructive hydrocephalus, LP was not carried out.

The most common neurotologic manifestation present was PT (62 patients) followed by hearing loss (25 patients), dizziness/vertigo (24 patients), aural fullness (17 patients), recurrent facial paralysis/trigeminal neuralgia (1 patient), and CSF otorrhea (1 patient). Table 7.1 summarizes the neurotologic manifestations.

Table 7.1: neurotologic manifestations

Pulsatile tinnitus	62
Hearing loss	25
Dizziness/disequilibrium	21
Vertigo	3
Aural fullness	17
Recurrent facial paralysis/trigeminal neuralgia	1
CSF otorrhea	1

Pulsatile tinnitus involved the right ear in 36 patients, the left in 14, and was bilateral in 14. In all patients PT was of the venous type (PT that subsides when light pressure is applied over the ipsilateral IJV).

A mild low frequency hearing loss with excellent discrimination was present in 25 patients. Improvement of hearing resulted immediately following application of digital neck pressure over the ipsilateral IJV and elimination of the masking effect of PT.

Dizziness, present in 21 patients, was described mainly as a sensation of lightheadedness. True vertigo was present in three patients, one of these patients had symptoms of Ménière's disease.

One BIH syndrome patient had associated recurrent bilateral facial paralysis and episodic left trigeminal neuralgia. Papilloedema was present in 21 patients with BIH syndrome and in one patient with Arnold–Chiari type I malformation.

Table 7.2 summarizes the associated neurologic manifestations. Radiologic findings are depicted in Table 7.3 and Table 7.4 summarizes the various aetiologies of intracranial hypertension.

Table 7.2: associated neurologic manifestations

Headaches	46
Visual disturbances	19
CSF rhinorrhea	1

Table 7.3: radiologic findings

Empty sella	15
Small ventricles	4
Arnold–Chiari (type I) deformity	2
Hydrocephalus-stenosis of Sylvius aqueduct	1
Cerebellar atrophy	1

Table 7.4: aetiologies of intracranial hypertension

Benign intracranial hypertension (BIH) syndrome	55
Possible BIH syndrome	5
Lupus erythematosus	1
Arnold-Chiari (type I) malformation	2
Hydrocephalus-stenosis of Sylvius aqueduct	1

The majority of these patients (60) were diagnosed with BIH syndrome. Fifty-eight of the BIH syndrome patients had PT, one presented with CSF otorrhea and one with episodic vertigo. One patient of this group had associated CSF rhinorrhea. There were 58 women and two men with a mean age of 35.3 years (range 20 to 52 years). Forty patients were black and 20 were white. There was right-ear involvement in 31 patients, left-ear involvement in 14 and bilateral involvement in 13. Tinnitus was objective in 20 patients and subjective in 38. Most of the BIH syndrome patients were overweight and 16 of them were morbidly obese (100 lb above ideal body weight).

Auditory evoked responses, obtained in 38 BIH patients, were abnormal in nine patients. The most common abnormality found was prolongation of the interpeak latencies.

Electrocochleography was abnormal in four patients.

Discussion

Fisher and Wolfson (1943) first introduced the term 'congestive inner-ear' in patients with increased intracranial pressure secondary to tumours. Saxena et al. (1969) studied the auditory function of 25 patients with intracranial hypertension arising from supratentorial tumours. The majority of these patients were found to have hearing loss. The same authors speculated that transmission of CSF pressure to the perilymph via the cochlear aqueduct was the aetiology of hearing loss in these patients. Meador et al. (1982) presented a case of BIH syndrome associated with PT, which was treated with a lumbar subarachnoid-to-peritoneal shunt. Tinnitus disappeared following surgery. Weider and Saunders (1996) presented three patients with recurrent perilymphatic fistulae as the initial manifestation of benign intracranial hypertension syndrome.

It is evident from the literature and this study that intracranial hypertension can present with neurotologic manifestations. The most common symptom found in this study was PT, which is believed to have the following pathophysiology: systolic pulsations of the CSF, mainly originating from the circle of Willis, are transmitted to the exposed medial aspect of the dural venous sinuses (transverse and sigmoid), resulting in periodic compression of their walls. These periodic compressions convert the laminar blood flow into turbulent flow resulting in vibrations of the vessel walls and production of PT (Sismanis 1987). The associated low-frequency sensorineural hearing loss is most probably secondary to the masking effect of the PT. This is supported by the normalization of hearing by eliminating the masking effect of the PT when light digital pressure is applied over the ipsilateral IJV. Auditory evoked response abnormalities, present in nine BIH syndrome patients, could be secondary to interstitial oedema and/or stretching-compression of the cochlear nerve brainstem. The same mechanism(s) could be responsible for the dizziness, which was present in 24 patients. Abnormal electrocochleography findings, present in four patients, are suggestive of endolymphatic hypertension, which could be secondary to transmission of increased CSF pressure to the perilymph through the cochlear aqueduct. Aural fullness, present in 17 patients, could arise from increased perilymphatic pressure as well. Trigeminal neuralgia and recurring facial paralysis, present in one BIH patient, could be secondary to interstitial oedema and/or stretching of the V and VII cranial nerves respectively.

A high index of suspicion and a thorough clinical evaluation are of utmost importance in evaluating these patients. Pulsatile tinnitus in young, obese, female patients is highly suggestive of BIH syndrome. Associated symptoms of headaches, visual disturbances (blurred vision, loss of vision) hearing loss and dizziness further support diagnosis of this entity. BIH syndrome (pseudotumour cerebri or idiopathic intracranial hypertension) was the most common diagnosis made in this group of patients. Fishman has described this condition as increased intracranial pressure (ICP) without focal neurologic signs, except for occasional V or VII cranial nerve palsy (Fishman 1980). Increased CSF pressure should be documented by lumbar puncture following radiologic evaluation to rule out other causes of intracranial hypertension. The non-invasive tympanic membrane displacement (TMD) test for CSF pressure measurement has been found useful in these patients (Reid et al. 1990). Recent research has revealed that obesity raises intraabdominal pressure, which increases pleural pressure and cardiac filling pressures leading to increased intracranial venous pressure, decreased CSF absorption and increased ICP (by Felton et al. 1997). This study also supports that papilloedema, present in one-third of the patients, is not a prerequisite to establish diagnosis of BIH syndrome. This has been reported by others in the literature as well (Marcelis and Silberstein 1991). Auscultation of the ear canals, skull and neck as well as funduscopy are crucial in the physical examination. For better detection of PT, it is highly recommended that auscultation be performed in a sound-treated room with a modified electronic stethoscope (Sismanis and Butts 1994). Radiologic evaluation is very important and should be individualized according to the clinical presentation. Head MRI is required for the majority of these patients. Head MRA or carotid angiography should be considered for patients suspected of intracranial vascular abnormalities (Sismanis and Smoker 1994). Management of this syndrome consists of weight reduction and administration of acetazolamide (Sismanis 1987). This treatment was effective in the majority of our patients. For morbidly obese patients who have failed conservative treatment, gastric bypass surgery should be considered (Sugerman et al. 1995).

One of the patients with Arnold–Chiari malformation was successfully treated with posterior fossa decompression. The other patient refused treatment.

Conclusions

The experience gained from this study has revealed the following:

- intracranial hypertension can present with neurotologic manifestations alone or in association with other 'traditional' neurologic symptoms and signs such as headaches and papilloedema;
- the most common presenting symptom was, by far, pulsatile tinnitus, followed by hearing loss, dizziness/vertigo and aural fullness;
- papilloedema was present in approximately one third of these patients. Absence of papilloedema does not rule out intracranial hypertension;

- the most common underlying aetiology found in this study was BIH syndrome (pseudotumour cerebri);
- otorhinolaryngologists, neurologists, neuro-ophthalmologists, and audiologists should be familiar with the neurotologic manifestations of intracranial hypertension in order to avoid misdiagnosis and inappropriate treatment.

References

Felton WL, Sugerman HJ, DeMaria EJ, Nakatsuka M, Sismanis A (1997) Increased intra-abdominal pressure and cardiac filling pressures in obesity associated idiopathic intracranial hypertension. Presented at the 1997 Meeting of the North American Neuro-Ophthalmology Society.

Fisher J, Wolfson LE (1943) In the Inner Ear. London: Heineman, p. 320.

Fishman RA (1980) Benign intracranial hypertension. In: Cerebrospinal Fluid in Disease of the Nervous System. Philadelphia: WB Saunders Co., pp. 128–39.

Marcelis J, Silberstein SD (1991) Idiopathic intracranial hypertension without papilledema. Arch Neurol 48: 392–9.

Meador KJ, Stefadouros M, Malik AJ (1982) Self heard venous bruit due to increased intracranial pressure. Lancet 1: 391.

Reid A, Marchbanks RJ, Burge DM, Martin AM, Bateman DE, Pickard JD, Brightwell AP (1990) The relationship between intracranial pressure and tympanic membrane displacement. Br J Audiol 24: 123–9.

Saxena RK, Tandon PN, Sinha A, Kacker RK, Singh K (1969) Auditory function in raised intracranial pressure. Acta Otolaryngol 68: 402–10.

Sismanis A (1987) Otologic manifestations of benign intracranial hypertension syndrome: diagnosis and management. Laryngoscope 97 (Suppl 42): 1–17.

Sismanis A, Butts FM (1994) A practical device for detection and recording of objective tinnitus. Otolaryngol Head Neck Surg 110: 459–62.

Sismanis A, Hughes GB, Abedi E (1985) Otologic symptoms and findings of the Pseudotumor Cerebri syndrome: a preliminary report. Otolaryngol Head Neck Surg 93: 398–402.

Sismanis A, Smoker WRK (1994) Pulsatile tinnitus: recent advances in diagnosis. Laryngoscope 104: 681–8.

Sugerman HJ, Felton WF, Salvant JB Jr, Sismanis A, Kellum JM (1995) Effects of surgically induced weight loss on idiopathic intracranial hypertension in morbid obesity. Neurology 45: 1655–9.

Weider DJ, Saunders RL (1996) Recurrent perilymphatic fistula as the initial and prime symptom of pseudotumor cerebri: Diagnosis and management with lumbar-peritoneal shunt – report of three cases. In: Ernst A, Marchbanks R, Samii M (eds) Intracranial and Intralabyrinthine Fluids. Berlin/Heidelberg: Springer Verlag, pp. 293–8.

Correspondence to Professor Aristides Sismanis, Department of Otolaryngology – Head and Neck Surgery, Medical College of Virginia/Virginia Commonwealth University, PO Box 980146, Richmond, Virginia 23298-0146 USA, Tel. (804) 828-3965, Fax (804) 828-5779.

Chapter 8
Idiopathic normal pressure hydrocephalus: primary disorder of CSF absorption or parenchymal cerebral disease? The relevance of vascular risk factors and cerebral white matter lesions

JOACHIM K KRAUSS, UNIVERSITY OF BERNE, SWITZERLAND

ABSTRACT. Idiopathic normal pressure hydrocephalus (NPH) of the elderly remains an enigmatic disease more than 30 years after its original description. Despite intensive research and effort the aetiology, pathogenesis and pathophysiology are still poorly understood. Altered dynamics of CSF homeostasis have been studied extensively in NPH, but the occurrence of cerebral parenchymal pathology has received much less attention. Since the early description of the clinical picture a frequent co-occurrence of idiopathic NPH and subcortical vascular encephalopathy has been noted and occasionally a high frequency of vascular risk factors has been reported. The differential diagnosis and possible associations between idiopathic NPH and subcortical vascular encephalopathy have been puzzling.

We prospectively studied the prevalence of vascular risk factors and of white matter lesions as assessed by MR in a cohort of 65 patients with idiopathic NPH. The findings were compared with those of age-matched controls. There were significant associations between idiopathic NPH and arterial hypertension (83% versus 36%, $p < 0.001$) and diabetes mellitus (49% versus 29%, $p = 0.015$). The association with hypertension remained significant after multivariate analysis ($p < 0.0001$). Periventricular as well as deep white matter lesions were significantly more frequent in a subgroup of 37 patients who all benefited from shunt surgery than in controls ($p < 0.001$). A negative correlation of the width of the anterior horns and the severity of white matter lesions was found. The degree of overall clinical improvement in 41 patients at a mean follow-up of 16 months after surgery was negatively correlated with the extension of periventricular ($p = 0.04$)

and deep white matter lesions ($p = 0.02$). Nevertheless, single patients with more widespread white matter disease still derived benefit from the operation. Our findings support the concept that NPH is a parenchymal compliance problem rather than a primary disease of CSF dynamics. The frequent co-occurrence of idiopathic NPH and subcortical vascular encephalopathy may not represent a true causal relationship. This co-occurrence rather indicates that both entities in many patients may be linked to common underlying mechanisms in which arterial hypertension may play a pivotal role. Further studies are needed to elucidate these complex interactions.

Key words: arterial hypertension, magnetic resonance imaging, normal pressure hydrocephalus, risk factors, shunt, vascular encephalopathy, white matter lesions.

Introduction

More than 30 years after its original description many aspects of idiopathic normal pressure hydrocephalus (NPH) in the elderly remain unclear. Since Adams and colleagues characterised this clinical entity in 1965 there has been considerable research and effort to help understand this intriguingly complex disease. However, until today little has been known about its aetiology, pathogenesis and pathophysiology (Vanneste 1994). In particular, the aetiology of idiopathic NPH is the subject of controversial debate (Bradley et al. 1991; Roman 1991; George 1991). There is increasing evidence that changes in the cerebral parenchyma may be responsible for the altered fluid dynamics seen in NPH patients (Bech et al. 1997; Krauss et al. 1997c).

The benefit of shunt surgery is contingent primarily upon the preoperative selection of appropriate patients. This selection continues to be a major challenge in clinical neurology and neurosurgery. Nevertheless, postoperative improvement may be seen in as many as in 60% to 90% of appropriately selected patients (Black et al. 1985; Benzel et al. 1990; Raftopoulos et al. 1996; Krauss et al. 1997b).

One of the most important differential diagnoses of idiopathic NPH is subcortical vascular encephalopathy. On the other hand, occasional co-occurrence of the two conditions was noted already in early reports (Earnest et al. 1974; Koto et al. 1977). After the introduction of MR imaging, correlates of subcortical vascular encephalopathy were observed more frequently in patients with idiopathic NPH (Jack et al. 1987; Bradley et al. 1991; Tanaka et al. 1993). Clinical symptoms can be very similar in both conditions. Cognitive impairment is more pronounced in 'pure' subcortical vascular encephalopathy, while gait disturbance is usually the main feature of NPH (Graff-Radford and Godersky 1986; Fisher 1989).

The possible associations between idiopathic NPH and subcortical vascular encephalopathy are puzzling. Whereas some studies attributed treatment failures to concomitant vascular disease, excellent improvement has been shown in other patients with neuroradiological signs of accompanying Binswanger's disease (Bradley et al. 1991; Yoshida et al. 1991). Are these always two completely different entities? Is there an overlap in some patients? Does concomitant subcortical vascular encephalopathy compromise postoperative outcome?

Diagnostic investigation of patients with suspected NPH

In order to better understand possible interactions between both entities we have studied prospectively the prevalence of vascular risk factors and of white matter disease as assessed by MR in patients with idiopathic NPH (Krauss et al. 1996a, 1996b, 1997c). Within the five-year period of this study, 118 patients with suspected NPH underwent comprehensive clinical and diagnostic investigation. The diagnosis of idiopathic NPH was confirmed in 65 patients. Diagnostic and clinical studies included analysis of clinical symptoms with the help of a videotaped standard protocol, CSF pressure overnight monitoring, hydrodynamic studies, removal of lumbar and ventricular CSF, transcranial Doppler monitoring and polysomnography, and morphological and dynamic MR imaging. Criteria thought to confirm the diagnosis of idiopathic NPH were: minimum age of 50 years; evidence of gait disturbance with or without cognitive deficits and urinary problems; chronic hydrocephalus without atrophy of the frontoparietal parasagittal cortex; and improvement of the gait disorder after CSF removal. Findings thought to indicate NPH, although they were not mandatory, were: pressure-volume index </= 13 ml; resistance to outflow > 12 mmHg/ml/minute; B-waves present during 70% of the recording time or more, and occurrence of high-amplitude B-waves. Patients with a history of subarachnoid haemorrhage, meningitis and severe head trauma were classified as having secondary NPH.

The mean age of the 65 patients who were diagnosed with idiopathic NPH was 70.8 ±7.4 years, with a range from 50 to 87 years. Thirty patients were men (47%) and 35 were women (54%). A gait disturbance was present in all patients, and urinary problems occurred in 54 of 65 patients (83%). Cognitive impairment was found in 64 of 65 patients (98%) although severe cognitive deficits were evident only in a minor proportion of patients. Fifty-three of the 65 patients presented with the complete triad (82%). The preoperative severity of the cardinal symptoms was rated with the help of four-point scales. Interestingly, in many cases we found akinetic symptoms such as hypokinesia or bradykinesia in addition to the well-known classical symptoms (Krauss et al. 1997a).

Vascular risk factors and idiopathic NPH

Vascular risk factors such as arterial hypertension or diabetes are common in subcortical vascular encephalopathy (Fisher 1989). If there were any association between idiopathic NPH and vascular encephalopathy it would be relevant to know whether and how frequently vascular risk factors are found in patients with idiopathic NPH. Few studies in the past have investigated this subject. Graff-Radford and Godersky (1987) reported that arterial hypertension was prevalent in 14 of 19 patients with idiopathic NPH (74%) and in 38 of 142 control subjects (27%). This difference was highly significant. Casmiro et al. (1989) noted a statistically significant difference for arterial hypertension, diabetes and ischaemic heart disease in a group of 17 patients with idiopathic NPH compared with controls. Jacobs (1977) reported that diabetes was present in 52% of NPH patients but only in 12% of age-matched controls.

We have screened the presence of vascular risk factors and cardiac and arteriosclerotic disease with nine variables in our group of idiopathic NPH patients and an age-matched control group (Krauss et al. 1996a). The control group consisted of 70 preoperative neurosurgical patients older than 60 years who were selected at random from three groups with a diagnosis of pituitary adenoma, cerebellopontine angle tumour or cervical disc disease. The demographic features were comparable with those of the NPH patients.

The relative frequencies of vascular risk factors and cardiac and arteriosclerotic diseases in the patients with idiopathic NPH and the control group are summarised in Tables 8.1 and 8.2. Arterial hypertension was found in 83% of NPH patients but only in 36% of the controls. There was also a marked difference for the prevalence of diabetes (49% versus 29%) and cardiac and cerebrovascular and other arteriosclerotic diseases. The difference in the prevalence of vascular risk factors between the two groups was then compared by univariate and multivariate analyses. Univariate analysis showed a highly significant association between arterial hypertension and idiopathic NPH ($p < 0.001$) and a weaker association with diabetes ($p = 0.015$). With multivariate analysis the strong association between hypertension and NPH remained highly significant ($p < 0.0001$), however, the adjusted odds ratio for diabetes did not show a significant association ($p = 0.12$).

We also looked for possible relationships between the clinical presentation and hypertension and diabetes. A positive association was recognised between severe degrees of impairment of each cardinal symptom and hypertension, whereas the associations with diabetes were much weaker. In particular, the strongest association was noted for severe gait disturbance and hypertension. This finding allows several interpretations. It is possible that these patients had a more severe coexistent vascular encephalopathy. On the other hand, it is conceivable that hypertension might have speeded up the progression of the disease.

Table 8.1: vascular risk factors, cardiac and arteriosclerotic diseases in 65 patients with idiopathic normal pressure hydrocephalus

	Number of patients with item present	Total number of patients*	Percentage
Arterial hypertension	54	65	83
antihypertensive medication	39	65	60
high blood pressure at study	39	65	60
Diabetes mellitus	31	63	49
medication with oral agents	11	63	17
application of insulin	8	63	13
Hypercholesterolemia	31	38	82
Hyperlipidemia	14	32	44
Smoking	12	49	24
Obesity	19	55	35
Cardiac diseases	46	65	71
ischaemic heart disease	37	65	57
myorcardial infarction	10	65	15
cardiac insufficiency	27	65	42
valvular disease	4	65	6
symptomatic arrhythmia	15	65	23
Cerebrovascular arteriosclerotic disease	21	65	32
history of TIAs	6	65	9
history of ischaemic stroke/PRIND	9	65	14
stenosis of extracranial cerebral arteries	14	65	22
Other arteriosclerotic diseases	23	65	35

*Total number of patients corresponds to all patients in the study population with idiopathic NPH or to subpopulations for whom the presence or absence of the variables or subcategories could be definitively established.

(from Krauss et al. 1996b)

The question remained whether duration of NPH had an impact on the presence of hypertension. When the patients were assigned to three groups according to the duration of the disease after onset of the first symptoms no relationship was seen. Thus it was possible to exclude the chance that hypertension actually might have been caused by hydrocephalus.

Table 8.2: vascular risk factors, cardiac and arteriosclerotic diseases in 70 controls with comparable age distribution

	Number of patients with item present	Total number of patients*	Percentage
Arterial hypertension	25	70	36
Diabetes mellitus	20	70	29
Hypercholesterolemia	25	39	64
Hyperlipidemia	16	38	42
Smoking	11	68	16
Obesity	24	69	35
Cardiac diseases	27	70	39
Cerebrovascular arteriosclerotic disease	9	70	13
Other arteriosclerotic diseases	8	70	11

*Total number of patients refers to all patients of the control group or to subpopulations for whom the presence or absence of the variables or subcategories could be established definitively.

(from Krauss et al. 1996b)

Our analysis lends further support to the possibility that hypertension might be an independent risk factor for idiopathic NPH. Whether accompanying vascular diseases are only secondary to the high prevalence of vascular risk factors or whether they might additionally contribute to the development of NPH in some cases awaits further clarification. There are various ways in which hypertension could interfere with the development of hydrocephalus. The most popular explanation has been the occurrence of white matter lesions secondary to hypertension. It has been postulated that such lesions could reduce periventricular tissue strength and alter elastic properties, thus predisposing the ventricles to dilate under the CSF pulse pressure (Earnest et al. 1974; Bradley et al. 1991). Defective CSF absorption in this model rather would be a consequence of secondary obliteration of the subarachnoid space over the convexity. It was even postulated that NPH might be caused by white matter lesions. However, although tempting, this hypothesis has several flaws. If the presence of white matter lesions would be sufficient to induce the dynamic changes of NPH, we should see this entity much more often in patients with diseases such as multiple sclerosis or other processes affecting the white matter.

White matter lesions and idiopathic NPH

The relevance of white matter lesions in idiopathic NPH patients has been a matter of debate. Correlates of subcortical vascular encephalopathy have been described repeatedly since the introduction of MR imaging (Figure 8.1). The occurrence of white matter lesions in the elderly and in people with various diseases, as demonstrated by MR imaging, has been the focus of considerable attention within the past decade. Due to methodological differences in the evaluation and assessment of such lesions, the incidence in healthy elderly people has been described as ranging from 11 to 89% (Ylikoski et al. 1993). Most often correlations between white matter lesions and arterial hypertension and increasing age have been found (Kertesz et al. 1988; Schmidt et al. 1995). Lesions of the deep white matter have most frequently been described as secondary to chronic ischaemia (Van Swieten et al. 1991; Munoz et al. 1993), while punctuate lesions were also found to represent dilated Virchow–Robin spaces (Munoz et al. 1993). It is assumed that periventricular lucencies (rims and caps) are secondary to transependymal passage of CSF in NPH patients, analogous to obstructive hydrocephalus, but similar lesions are also observed frequently in patients without NPH. Three previous MR studies have compared the occurrence of white matter lesions in NPH patients and controls yielding partially contradictory results (Jack et al. 1987; Bradley et al. 1991; Kristensen et al. 1996).

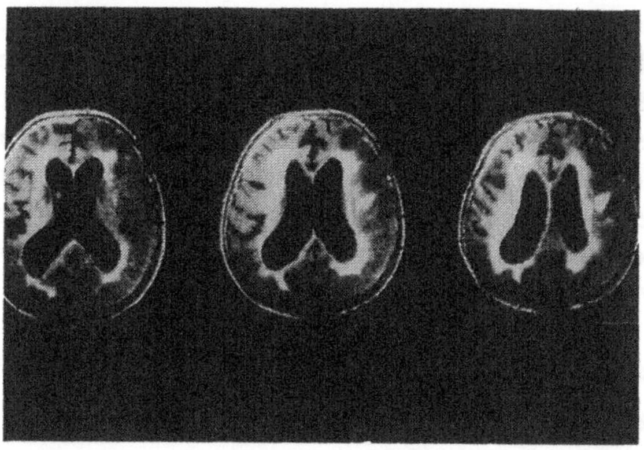

Figure 8.1: axial inversion recovery MR scans of a 74-year-old woman with coexistent idiopathic normal pressure hydrocephalus and subcortical vascular encephalopathy. The patient had moderate improvement of all three cardinal symptoms after CSF shunting.

We evaluated the extent and the severity of white matter lesions according to T2-weighted MR scans in a subgroup of 37 patients with idiopathic NPH who all showed definite improvement after shunting and compared them to the findings of 35 age-matched controls (Krauss et al. 1996b). The mean age of patients of the control group was 71.5 years, ranging from 50 to 89 years. There were nine men and 26 women. Periventricular and deep white matter lesions were assessed separately according to a numerical indexing system, scoring three distinct anatomical domains for each hemisphere. The maximal possible score for periventricular as well as for deep white matter lesions was 24, respectively, and a value of 0 indicated no lesion at all.

The distribution of white matter lesion scores in patients with idiopathic NPH and the control group were remarkably different. The mean total periventricular score was 12.1 in the NPH group, while it was 3.9 in the controls. The mean deep white matter lesion score was 12.9 in the NPH group as compared to 4.5 in the controls. For both parameters this difference was highly significant ($p < 0.001$). To investigate possible relationships between the different variables we calculated corresponding correlation coefficients. Most remarkably a significant negative correlation was found between the width of the anterior horns and the severity of white matter lesions in the NPH group ($p < 0.001$), whereas a positive correlation was seen in the control group. This finding might be partially related to underrating the severity of white matter lesions in more marked hydrocephalus and reduced white matter volume. More likely, however, it indicates a co-occurrence of idiopathic NPH and vascular encephalopathy rather than a causal relationship. In the case of a direct causal relationship we would have expected a clearly positive correlation between the severity of white matter lesions and the ventricular width.

White matter lesions are much more frequent in idiopathic NPH patients. Although it is most probable that the deep white matter lesions are secondary to chronic ischaemia, the significance of the periventricular lesions remains unclear. It is possible that these lesions also indicate ischaemic demyelination in some patients. The common link between the frequent coexistence of deep and periventricular white matter lesions and idiopathic NPH might be an underlying condition that is most likely to be arterial hypertension.

White matter lesions and outcome after CSF shunting

The relevance of different degrees of white matter lesions with regard to outcome has not been clearly established for patients with idiopathic NPH previously. While several investigators advocate CSF diversion in patients with coexistent NPH and subcortical vascular encephalopathy, others advise that such patients be excluded from shunting.

We investigated the effect of periventricular and deep white matter lesions on outcome in a prospective series of 41 patients who underwent preoperative MR examination according to a standard protocol (Krauss et al. 1997c). In this study patients were included whether they benefited or not from the shunt operation. To better quantify and compare clinical improvement, indices were calculated. For each of the three cardinal symptoms, outcome was assessed separately and marked as follows: 0 = no or only poor improvement; 1 = fair or good improvement; and 2 = excellent improvement. The overall symptomatic outcome of each patient was obtained from these measurements by the calculation of a fraction representing the total improvement index. This fraction ranged between 0 (no meaningful improvement of any cardinal symptom) and 1 (excellent improvement of all preoperative cardinal symptoms). The improvement indices were then correlated with the white matter lesion scores.

After a mean follow-up of 16 months, postoperative improvement was seen in 37 of the 41 patients (90%). Excellent improvement with regard to all cardinal symptoms was found in six patients, while four patients achieved no benefit at all. Both deep and periventricular white matter lesion scores correlated with the preoperative severity of symptoms. When the total improvement index was plotted against the periventricular lesion score a significant negative correlation was found ($p = 0.04$). A similar correlation was also seen for the deep white matter lesion score ($p = 0.02$). All six patients who had an improvement score of 1 at follow-up had deep white matter lesion scores of less than 15. Patients with white matter lesion scores of more than 20, on the other hand, did not rate higher than 0.5 for the total improvement index, with the exception of one patient. The four patients who failed to respond to shunting did not have any consistent similarities in their white matter lesion scores. Overall, although the negative correlation between outcome and white matter lesion scores was significant for the whole cohort, there was no absolute possibility to predict the degree of the total improvement index for an individual patient.

These findings show that patients with NPH and concomitant correlates of subcortical vascular encephalopathy may benefit from shunting, although the degree of improvement in general correlates with the extent and severity of the accompanying white matter changes. Usually, excellent postoperative improvement may not be expected in patients with severe white matter lesions. The occurrence of such lesions in NPH patients, however, should not engender therapeutic nihilism. Fair or good improvement can be achieved in these patients rendering worthwhile functional benefit. Once again, however, it should be emphasised that CSF diversion should only be offered to appropriately selected candidates.

Conclusions

Idiopathic NPH: primary disorder of CSF absorption or parenchymal disease?

In conclusion, one of the pivotal questions today concerning the aetiology of idiopathic NPH is whether it is a disorder primarily of intracranial fluids or a parenchymal disease. The common concept that idiopathic NPH is secondary to defective CSF absorption through the arachnoid villi has even been called a myth in a recent review (Vanneste 1994). Our data also suggest that there is more evidence for a parenchymal disease or for a more dynamic interaction. None of the hypothetical models that have been constructed in the past are capable of explaining the development and pathophysiology of idiopathic NPH.

This is not the time to propose a new model. There are still too many missing links to comprehensively explain the pathophysiological chain in the development of idiopathic NPH. Arterial hypertension may be involved with the promotion of NPH in many different ways. Besides predisposing patients for structural white matter lesions and altering elastic properties it might induce functional disturbance at several levels, particularly in the presence of impaired autoregulation and altered cerebral compliance (Table 8.3). Also, idiopathic NPH may be an inhomogeneous entity. It may have several underlying causes. Although it is often seen with arterial hypertension and vascular diseases, some patients have no such associations. NPH in these patients might actually be some kind of secondary NPH. Furthermore, it is probably not always appropriate to ask whether a patient has idiopathic NPH or subcortical vascular encephalopathy. There may be patients at one end of the spectrum with 'pure' NPH and patients at the other end with 'pure' vascular encephalopathy. The latter patients definitely should be excluded from CSF shunting. Many patients, however, will range in-between. Further clinical and experimental research is needed.

Table 8.3: arterial hypertension in idiopathic NPH: possible interference with pathomechanisms

- structural white matter lesions
- decreased periventricular blood flow
- impaired autoregulation
- interference with brainstem 'pacemakers'
- increased brain and CSF pulsatility
- homeostasis of CSF pressure and CSF flow
- impaired parenchymal compliance

References

Adams RD, Fisher CM, Hakim S, Ojemann RG, Sweet WH (1965) Symptomatic occult hydrocephalus with 'normal' cerebrospinal fluid pressure. A treatable syndrome. N Engl J Med 273: 117–26.

Bech RA, Juhler M, Waldemar G, Klinken L, Gjerris F (1997) Frontal brain and leptomeningeal biopsy specimens correlated with cerebrospinal fluid outflow resistance and B-wave activity in patients suspected of normal-pressure hydrocephalus. Neurosurgery 40: 497–502.

Benzel EC, Pelletier AL, Levy PG (1990) Communicating hydrocephalus in adults: prediction of outcome after ventricular shunting procedures. Neurosurgery 26: 655–60.

Black PM, Ojemann RG, Tzouras A (1985) CSF shunts for dementia, incontinence, and gait disturbance. Clin Neurosurg 32: 632–51.

Bradley WG, Whittemore AR, Watanabe AS, Davis SJ, Teresi LM, Homyak M (1991) Association of deep white matter infarction with chronic communicating hydrocephalus: implications regarding the possible origin of normal-pressure hydrocephalus. AJNR 12: 31–9.

Casmiro M, d'Allesandro R, Cacciatore FM, Daidone R, Calbucci F, Lugaresi E (1989) Risk factors for the syndrome of ventricular enlargement with gait apraxia (idiopathic normal pressure hydrocephalus): a case-control study. J Neurol Neurosurg Psychiatry 52: 847–52.

Earnest MP, Fahn S, Karp JH, Rowland LP (1974) Normal pressure hydrocephalus and hypertensive cerebrovascular disease. Arch Neurol 31: 262–6.

Fisher CM (1989) Binswanger's encephalopathy: a review. J Neurol 236: 65–79.

George AJ (1991) Chronic communicating hydrocephalus and periventricular white matter disease: a debate with regard to cause and effect. AJNR 12: 42–4.

Graff-Radford NR, Godersky JC (1986) Normal pressure hydrocephalus: onset of gait abnormality before dementia predicts a good surgical outcome. Arch Neurol 43: 940–2.

Graff-Radford NR, Godersky JC (1987) Idiopathic normal pressure hydrocephalus and systemic hypertension. Neurology 37: 868–71.

Jack CR, Mokri B, Laws ER, Houser OW, Baker HL, Petersen RC (1987) MR findings in normal-pressure hydrocephalus: significance and comparison with other forms of dementia. J Comput Assist Tomogr 11: 923–31.

Jacobs L (1977) Diabetes mellitus in normal pressure hydrocephalus. J Neurol Neurosurg Psychiatry 40: 331–5.

Kertesz A, Black SE, Tokar G, Benke T, Carr T, Nicholson L (1988) Periventricular and subcortical hyperintensities on magnetic resonance imaging. Arch Neurol 45: 404–8.

Koto A, Rosenberg G, Zingesser LH, Horoupian D, Katzman R (1977) Syndrome of normal pressure hydrocephalus: possible relation to hypertensive and arteriosclerotic vasculopathy. J Neurol Neurosurg Psychiatry 40: 73–9.

Krauss JK, Droste DW, Vach W, Regel JP, Orszagh M, Borremans JJ, Tietz A, Seeger W (1996a) Cerebrospinal fluid shunting in idiopathic normal pressure hydrocephalus of the elderly: effect of periventricular and deep white matter lesions. Neurosurgery 39: 292–300.

Krauss JK, Regel JP, Vach W, Droste DW, Borremans JJ, Mergner T (1996b) Vascular risk factors and arteriosclerotic disease in idiopathic normal pressure hydrocephalus of the elderly. Stroke 27: 24–9.

Krauss JK, Regel JP, Droste DW, Orszagh M, Borremans JJ, Vach W (1997a) Movement disorders in adult hydrocephalus. Mov Disord 12: 53–60.

Krauss JK, Regel JP, Vach W, Juengling FD, Droste DW, Wakhloo AK (1997b) Flow void of cerebrospinal fluid in idiopathic normal pressure hydrocephalus of the elderly: can it predict outcome after shunting? Neurosurgery 40: 67–74.

Krauss JK, Regel JP, Vach W, Orszagh M, Juengling FD, Bohus M, Droste DW (1997c) White matter lesions in patients with idiopathic normal pressure hydrocephalus and in an age-matched control group: a comparative study. Neurosurgery 40: 491–6.

Kristensen B, Malm J, Fagerlund M, Hietala SO, Johansson B, Ekstedt J, Karlsson T (1996) Regional cerebral blood flow, white matter abnormalities, and cerebrospinal fluid hydrodynamics in patients with idiopathic adult hydrocephalus syndrome. J Neurol Neurosurg Psychiatry 60: 282–8.

Munoz DG, Hastak SM, Harper B, Lee D, Hachinski VC (1993) Pathologic correlates of increased signals of the centrum ovale on magnetic resonance imaging. Arch Neurol 50: 492–7.

Raftopoulos C, Massager N, Baleriaux D, Deleval J, Clarysse S, Brotchi J (1996) Prospective analysis by computed tomography and long-term outcome of 23 adult patients with chronic idiopathic hydrocephalus. Neurosurgery 38: 51–9.

Roman GC (1991) White matter lesions and normal-pressure hydrocephalus: Binswanger disease or Hakim syndrome? AJNR 12: 40–1.

Schmidt R, Fazekas F, Koch M, Kapeller P, Augustin M, Offenbacher H, Fazekas G, Lechner H (1995) Magnetic resonance imaging cerebral abnormalities and neuropsychologic test performance in elderly hypertensive subjects. Arch Neurol 52: 905–10.

Tanaka K, Yonekawa Y, Miyake H, Kobayashi A, Tsukahara T, Niijima K, Ogata N, Kaku Y, Kazekawa K, Takeda M, Komuro T, Fukao S (1993) Idiopathic normal pressure hydrocephalus in elderly patients: its pathophysiology and diagnosis. Neurol Surg 21: 403–9. (Japanese)

Vanneste J (1994) Three decades of normal pressure hydrocephalus: are we wiser now? J Neurol Neurosurg Psychiatry 57: 1021–5.

Van Swieten JC, Van den Hout JHW, Van Ketel BA, Hijdra A, Wokke JHJ, Van Gijn J (1991) Periventricular lesions in the white matter on magnetic resonance imaging in the elderly. Brain 114: 761–74.

Ylikoski R, Ylikoski A, Erkinjuntti T, Sulkava R, Raininko R, Tilvis R (1993) White matter changes in healthy elderly persons correlate with attention and speed mental processing. Arch Neurol 50: 818–24.

Yoshida Y, Nogaki H, Shose Y (1991) Binswanger's disease with normal pressure hydrocephalus. Neurologia Medico-Chirurgica 1: 1008–11. (Japanese)

Correspondence to Privatdozent Dr. JK Krauss, Department Neurochirurgie Inselspital, University of Berne, 3010 Berne, Switzerland. Tel. (+41) 31-6323444. Fax: (+41) 31-3822414.

Chapter 9
Influence on the auditory system of cerebrospinal fluid loss: clinical and experimental results

ALICE WALSTED, Glostrup Hospital, Denmark

ABSTRACT. Hearing loss has been reported following leakage of cerebrospinal fluid (CSF). In patients exposed to either spinal anaesthesia, surgery for acoustic neuroma or other neurosurgery, hearing was tested before and several times after the surgery (in the acoustic neuroma group only the untouched ear). A control group having surgery performed under general anaesthesia without a loss of CSF was also examined. Hearing was reversibly affected only in the CSF-patients; the degree of hearing loss, the number of affected patients and the number of frequencies involved increased with increasing amount of CSF lost. Analysis pointed towards a sensorineural hearing loss with a cochlear aetiology induced by the CSF loss. In an animal model the morphological and functional changes were investigated after surgery with aspiration of CSF. An increased fluid exchange between the CSF and perilymph was found, indicated by the transmission of erythrocytes via the cochlear aqueduct into the perilymph. By electrocochleography immediately after aspiration of CSF a threshold increase of the compound action potential was demonstrated. Change in the cochlear fluid volumes after CSF aspiration, with a transitory reduction in the perilymphatic volume and resultant endolymphatic expansion, is suggested to be responsible for the observed findings.

Key words: cerebrospinal fluid loss, auditory system, hearing loss, neurosurgery, endolymph, perilymph.

Introduction

Hearing impairment has been described after procedures involving puncture or opening of the dura with leakage of cerebrospinal fluid

(CSF). The majority of cases of this impairment has been reported after spinal anaesthesia (Vandam and Dripps 1956; Panning et al. 1984; Walsted et al. 1991), but it has also occurred after myelography, lumbar puncture (Michel and Brusis 1992), after neurosurgery (Walsted et al. 1994), and in the non-operated ear after surgery for acoustic neuroma (Walsted et al. 1996a).

Clinical studies

In 126 patients exposed to surgical procedures that involved loss of CSF, hearing was measured before and after surgery. A control group of 32 patients was also investigated. Hearing changes were subsequently analysed in order to delineate the size, incidence, frequency range and time course (in some groups several postoperative audiograms were performed). The study also investigated whether age, gender, middle ear pressure or the surgical procedure influenced the outcome.

The patients participating in the studies were exposed to lumbar puncture in connection with spinal anaesthesia (Walsted et al. 1991), neurosurgery under general anaesthesia, either for unilateral acoustic neuroma (AN-group) (Walsted et al. 1996a) or other intracranial diseases (NS-group) (Walsted et al. 1994); the control group had surgery performed under general anaesthesia but this did not involve loss of CSF (Walsted et al. 1994).

The audiology tests were performed under standardised conditions and included, besides pure tone audiometry, speech audiometry, impedance tympanometry and in some patients (NS) measurement of the stapedial reflex threshold. In cases with any hearing decrease a bone-conduction test was performed to help distinguish sensorineural hearing loss (HL) from conductive HL.

To assess the extent of the hearing changes, the threshold changes from preoperative to postoperative audiograms were calculated at each frequency for each of the attending patients and the mean changes were calculated. Moreover each of the patients has been classified into arbitrarily selected classes, according to their maximal threshold increase, in one or both ears:

No HL: threshold change of less than 15 dB at any frequency.
Small HL: 15–20 dB at a single frequency.
Medium HL: 15–20 dB at more frequencies or 25 dB at a single frequency.
Large HL: 25 dB at more frequencies or 30 dB at a single frequency.

Results

In each of the groups of patients exposed to a loss of CSF a significant increase in the mean threshold could be demonstrated postoperatively at one or more frequencies. The mean thresholds in the control group remained unchanged at all frequencies. The size of the mean threshold increase was smallest in the spinal group, where a minor but statistical significant threshold increase of 2.3 dB was found only at 500 Hz. There was no mean threshold increase at the other frequencies. In the NS group a mean threshold increase of up to 7.8 dB was demonstrated at 125 Hz; apart from at 4 and 8 kHz in the right ear and at 8 kHz in the left ear there was also a statistical significant threshold increase at all the other tested frequencies. In the AN group a mean threshold increase of 9.4 dB was found at 125 Hz, and statistical significant increases were found at all frequencies other than at 2 and 4 kHz. In the NS group as well as the AN group the mean increases were most pronounced in the lower frequencies. The largest threshold increase was observed in a female patient in the AN group, who developed a postoperative hearing loss of 60 dB in the low frequencies.

In total, 26% of the spinal patients, 48% of the NS patients and 57% of the AN patients had developed a small, medium or large HL at the first postoperative test performed between days one and three. Four patients (13%) of the control group also had a threshold increase that in no cases exceeded 15 dB. In the AN group the hearing loss was classified as large in 20% of the patients while it was classified as large in 16% of the NS group and in only 3% of the spinal group.

In the AN study four postoperative measurements were performed to delineate the time course. The low frequencies were affected immediately whereas the treble became involved during the first week. The overall hearing seemed most affected after one week. After three months there was no significant mean threshold change compared with the preoperative value at any of the frequencies. In the NS group analysis of the stapedial reflex thresholds showed recruitment indicating a cochlear aetiology of the HL. The HL seemed only slightly influenced by reductions in middle ear pressure, as middle-ear pressure was found to be equally reduced in the NS group and a control group, but only the NS group developed significant HL. Furthermore the spinal group had no general anaesthesia, and accordingly no influence on middle ear pressure. The AN patients had significantly reduced middle-ear pressure at the first postoperative test, implying a possible contribution to the HL, but the most pronounced HL was found after one week, when the middle-ear pressure was within preoperative limits.

The influence of sex, age, tumour size, surgical approach and duration of surgery was investigated in the AN study. Only age seemed to exert an influence, for patients with and without a threshold increase showed a different age composition, with more individuals over 50 years of age among the affected patients.

The postoperative HL seemed to be predominantly related to the amount of CSF lost, as the hearing changes were smallest in the spinal patients who were considered to suffer the smallest amount of CSF loss, and most pronounced in the AN patients who were considered to have suffered the biggest loss of CSF.

Comments

The HL seems to be a result of the loss of CSF and is suggested to be caused by a direct effect on the cochlea. Changes in the inner ear fluids after a leak of perilymph through the cochlear aqueduct simultaneously with the CSF loss could result in a transitory endolymphatic hydrops by a mechanism similar to the one that is supposed to act in perilymphatic fistulas. When the CSF volume and perilymph volume are restored, hearing will be normalised.

Animal experiments

Twelve anaesthetised guinea pigs were subjected to loss of CSF produced by opening of the cranial cavity and puncture with drainage of CSF. The inner ear changes were evaluated by light microscopy after one or 24 hours and compared to a control group of 6 animals not exposed to a leak of CSF (Walsted et al. 1994). In another experiment 18 anaesthetised guinea pigs had electrocochleography with recordings of the thresholds performed before and after CSF aspiration while a control group exposed to anaesthesia and surgery (skin incision only) without CSF aspiration had two measurements performed at the same time intervals (Walsted et al. 1996b).

Results

The histological examination showed that red blood cells were dispersed in the perilymph of the animals that underwent operations but not in the controls. In the animals examined after one hour the erythrocytes were primarily found in the scala tympani near the round window or in the basal part of the cochlea. In the animals examined after 24 hours, red blood cells were more abundant in both the scala vestibuli scale and the scala tympani. Gross bleeding had occurred in all the animals that underwent operations at the puncture site and was microscopically also demonstrated in the subarachnoid space and in the cochlear aqueduct. No displacement of the membranes was observed. In the control group no blood was found.

The electrocochleographic evaluations demonstrated a mild but significant increase of 9.2 dB in the click evoked thresholds of the compound action potential after CSF aspiration, but the thresholds of the control group remained unchanged.

Comments

After the surgery and CSF aspiration red blood cells were shown to pass from the CSF through the cochlear aqueduct to the perilymph. Since the cochlea is a hard-walled bony compartment, only the windows being compliant, an increase in volume of one of its containing components will result in a decrease in another or a change in the position of the cochlear windows or intracochlear membranes. In our experiment, the inflow of CSF into the perilymph did not result in detectable changes of some of the other compartments of the cochlea. An initial outflow of perilymph in connection with the CSF aspiration with a simultaneous compensatory expansion of the endolymphatic volume could explain this finding. The enhanced inflow of CSF carrying the red blood cells represents a return to normal conditions with the restoration of the fluid volumes. The threshold increase demonstrated immediately after CSF aspiration evaluated by electrocochleography could also be explained by a transitory endolymphatic expansion.

Conclusion

The clinical studies showed that CSF aspiration resulted in a transitory threshold increase. The animal experiments, evaluated by electrocochleography, supported this observation. Histological examination showed an enhanced fluid exchange between the CSF and perilymph through the cochlear aqueduct after the loss of CSF.

Change in the cochlear fluid volumes after CSF aspiration, with a transitory reduction in the perilymphatic volume and resultant endolymphatic expansion, is suggested to be responsible for the observed findings.

References

Michel O, Brusis T (1992) Hearing loss as a sequel of lumbar puncture. Ann Otol Laryngol 101: 390–4.

Panning B, Lehnhardt E, Mehler D (1984) Zur passageren Tieftonschwerhoerigkeit nach Spinalanaesthesie. Anaethesist 33: 593–5.

Vandam LD, Dripps RD (1956) Long-term follow-up of patients who received 10,098 spinal anesthetics JAMA 161: 586–91.

Walsted A, Salomon G, Olsen KS (1991) Low-frequency hearing loss after spinal anesthesia. Perilymphatic hypotonia? Scand Audiol 20: 211–15.

Walsted A, Garbarsch C, Michaels L (1994) Effect of craniotomy and cerebrospinal fluid loss on the inner ear. An experimental study. Acta Otolaryngol (Stockh) 114: 626–31.

Walsted A, Nielsen OA, Borum P (1994) Hearing loss after neurosurgery. The influence of low cerebrospinal fluid pressure. J Laryngol Otology 108: 637–41.

Walsted A, Salomon G, Thomsen J, Tos M (1996a) Cerebrospinal fluid loss and threshold changes. Hearing loss in the contralateral ear after operation for acoustic neuroma: An analysis of the incidence, time course, frequency range, size and pathophysiological considerations. Audiol Neurootol 1: 247–55.
Walsted A, Nilsson P, Gerlif J (1996b) Cerebrospinal fluid loss and threshold changes. Electrocochleographic changes of the compound action potential after CSF aspiration: An experimental study. Audiol Neurootol 1: 256–64.

Correspondence to Alice Walsted, Melchiorsvej 6, DK-2920 Charlottenlund, Denmark. Tel. and Fax (+45) 39631926.

Chapter 10
OAE and TMD recordings in ICP changes

A ERNST, Department of Otolaryngology at UUB, Berlin, Germany, T LENARZ, Department of Otolaryngology, Hannover Medical School, Germany, and K WECKBECKER, Department of Neurology, Hannover Medical School, Germany

ABSTRACT. The cerebrospinal fluid (CSF) filled spaces and the labyrinth are connected via the so-called cochlear aqueduct. Changes of the CSF pressure, i.e. the intracranial pressure (ICP), can therefore influence audiovestibular function. It was the aim of the present paper to apply audiological techniques (transient otoacoustic emission (TEOAE) and tympanic membrane displacement (TMD) recordings) to indirectly assess ICP. A total of 52 patients (average age 43) with intracranial hypotension (6) and hypertension (46) were included. They presented at the hospital with the following diseases: benign intracranial hypertension (BIH), hydrocephalus, shunt complication in hydrocephalus (blockade or overdrainage), narrow spinal canal, subarachnoid haemorrhage, myelopathy, neurolues, sinus thrombosis, and CSF overdrainage after lumbar puncture. Trauma and tumour patients were excluded. Retrospectively 13 (different) patients were excluded for a pure-tone hearing loss exceeding 30 dB HL or a lack in cochlear aqueduct patency, which were tested after the patient had recovered from the disease.

The patients were tested by lumbar puncture or permanent insertion of an epidural pressure probe. At the same time, 1 hour later and 24 h later, TEOAE and TMD recordings were performed (by the otolaryngologists) for follow-up while the treatment was started. An ICP of below 6 mm Hg (82 daPa) or above 15 mm Hg (204 daPa) was considered as hypotensive and hypertensive respectively. Transient evoked otoacoustic emissions are – at least presently – not easily applicable for non-invasive ICP recording. TMD, however, is closely correlated to ICP changes, although inter-individual variations have to be considered. Both audiological techniques are, however, well suited to diagnosing hearing loss or tinnitus as the result of underlying ICP changes.

Key words: CSF, ICP monitoring, OAE, TMD recording.

Introduction

The cerebrospinal fluid (CSF) filled spaces and the labyrinth are connected via the so-called cochlear aqueduct. Changes of the CSF pressure, i.e. the intracranial pressure (ICP), can therefore influence audiovestibular function (Densert et al. 1981). The ICP transfer leads to an increase in pressure load on the cochlear partition, which, in excess, can impair cochlear micromechanics (Marchbanks and Reid 1990). This is reflected by a depression of cochlear microphonics (Yoshida and Uemura 1991) and a change in auditory evoked responses (Matsuura et al. 1986; Sismanis 1990). In addition to these animal experimental findings there are reports on a sensorineural hearing loss (SNHL) after CSF drainage on one hand, and about vertigo and low-frequency tinnitus in hydrocephalic patients with CSF overproduction and shunt blockade on the other hand (Sismanis 1987; Reid et al. 1990). It was therefore the aim of the present paper to apply audiological techniques (TEOAE and TMD recordings) to indirectly assess ICP.

Material and methods

A total of 52 patients (average age 43) with intracranial hypotension (6) and hypertension (46) were included. They arrived at the hospital with the following diseases: benign intracranial hypertension (BIH), hydrocephalus, shunt complication in hydrocephalus (blockade or overdrainage), narrow spinal canal, subarachnoid haemorrhage, myelopathy, neurolues, sinus thrombosis, CSF overdrainage after lumbar puncture. Trauma and tumour patients were excluded. Retrospectively, 13 (different) patients were excluded for a pure-tone hearing loss exceeding 30 dB (HL) or a lack in cochlear aqueduct patency which were tested after the patient had recovered from the disease.

Upon delivery, the patients were regularly seen by a neurologist or neurosurgeon who confirmed the diagnosis. Thereafter, the otolaryngologists could perform otoscopy and start testing (see below). The patients were tested (neurologically or neurosurgically) by lumbar puncture or permanent insertion of an epidural pressure probe. At the same time, 1 hour later and 24 hours later, TEOAE and TMD recordings were performed (by the otolaryngologists) for follow-up while the treatment was started. Intracranial pressures of below 6 mm Hg (82 daPa) or above 15 mm Hg (204 daPa) were considered as hypotensive and hypertensive, respectively.

TMD recordings (10 runs) were performed by ipsilateral acoustic stimuli with sound bursts of 1 kHz, 500 ms, 20 dB above stapes reflex threshold in supine. The mean tympanic membrane displacement (Vm) was taken for each ear.

TEOAE recordings were performed with standard ILO-92 equipment, also using the analysis software. The click amplitude amounted to be 65 dB SPL, 260 runs (of 20 ms each) were averaged for each ear. The TEOAE amplitudes at 1, 2, 3 and 4 kHz as well as the latency of TEOAE onset were evaluated. Only emissions above 60% reproducibility were taken into account.

Statistical analysis (see Winer 1970) was undertaken using the chi-square test ($p < 0.05$) with ICP being related to the TMD and TEOAE values of each ear.

Results

Intracranial hypotension

A total of six patients were available for evaluation. Four patients could be evaluated only where all the recordings were complete and in sufficient quality (see above).

The initial ICP was 4.1 mmHg (± 3.9) on average. The ICP was normalized in three of the cases within the next 24 hours (10.6 ± 5.1). TMD recordings showed an initial Vm of (+ 146 ± 44.3 nl) which was nearly unchanged after 1 h, but had reversed after 24 h (–127 ± 33.1 nl). Transient evoked otoacoustic emission amplitudes and latencies remained unchanged on each of the three tests. No statistical analysis was performed because of the small number of samples.

Intracranial hypertension

For evaluation, a total of 46 patients were available. Forty-one of these patients could be evaluated. The ICP was at 19.5 (15.2 – 28.6) mmHg on average. In 93% of the cases, the Vm was largely negative with an age-related decrease. The mean Vm was –288 (± 56; –188 to – 645) nl. After 1 h, it was decreased to –196 (± 44) nl. It had nearly normalized after 24 hours (–112 ± 55 nl) of treatment. All these changes are statistically significantly correlated to the ICP changes.

Transient evoked otoacoustic emission recordings showed a significant change over treatment at 1 kHz in amplitude. The latency of emissions at 1 kHz was also shortened significantly but not at 2 kHz. The same holds true for the amplitude change at 2 kHz.

The amplitude at 1 kHz was reduced on average initially by 2.3 dB compared to the controls after 24 h. At all other frequencies, no changes could be found (Table 10.1).

Table 10.1: ICP, TMD and TEOAE (at 1 kHz) changes in 41 patients with intracranial hypertension (the recordings were taken initially 1 hour and 24 hours after starting the treatment). The data are presented as (mean ± SD). A correlation between ICP and TMD or TEOAE was considered significant at p = 0.05 (*).

Time of recording	ICP (mmHg)	TMD (Vm, nl)	TEOEA (at 1 kHz) amplitude (dB)	latency (ms)
Initial	19·5 (±5·1)	288 (56)	2 (1·2)	6·2 (1·3)
After 1 hour	17·3 (4·0)	196 (44)	3 (3·1)	7·2 (3·6)
After 24 hours	12·1 (7·1)*	112 (55)	5 (2·1)*	9·6 (2·2)*

Discussion

The intracranial pressure (ICP) is closely correlated to the intra-labyrinthine pressure because the cochlear aqueduct directly connects the CSF-filled spaces and the scala tympani of the cochlea which is filled with perilymph. Under normal conditions, there is no evidence for a flow of CSF into the cochlear perilymph even with a patent cochlear aqueduct (Carlborg et al. 1982). In contrast to this, under conditions with increased or decreased ICP, CSF can interfere with labyrinthine fluids and change primarily cochlear micromechanics as evidenced by ABR and CM recordings (Matsuura et al. 1986; Yoshida and Uemura 1991).

The present study was therefore aimed at searching for non-invasive tools to evaluate ICP as reported earlier and to correlate TMD and TEOAE recordings with the ICP at the same time. It could be shown that the TMD recordings significantly reflect ICP changes in case of a patent cochlear aqueduct. This is in line with previous findings (Ernst et al. 1995). TEOAE was seemingly another promising tool in non-invasive ICP monitoring since Avan et al. (1996) reported a significant phase shift correlated to the ICP. In general, frequencies, amplitudes and envelopes in this study also remained largely unchanged.

Froehlich et al. (1994) also demonstrated that physiological changes (positional changes, alertness etc.) hardly influence TEOAE amplitude. A 'stiffening effect' at the cochlear partition resulting from a rise in ICP should have its maximum effect in OAE phase lead at around 1 kHz, but should become negligible at 1.5–2 kHz (Avan et al. 1996). This is reflected by the significant TEOAE latency and amplitude changes at 1 kHz, which could be described in the present paper.

In essence, TEOAE are – at least presently – not a clinical tool that is easy to apply for non-invasive ICP recording. TMD, however, is closely correlated with ICP changes, although inter-individual variations have to be considered (Marchbanks and Reid 1990). Both audiological techniques are, on the other hand, well suited to diagnose hearing loss or tinnitus as the result of underlying ICP changes (Sismanis 1987; Walsted et al. 1991).

References

Avan P, Buki B, Lemaire JJ, Dordain M, Chazal J (1996) Otoacoustic emissions: a new tool for monitoring intracranial pressure. In Ernst A, Marchbanks R, Samii M (eds) Intracranial and Intralabyrinthine Fluids. Berlin: Springer-Verlag, pp. 165–75.

Carlborg B, Densert B, Densert O (1982) Functional patency of the cochlear aqueduct. Ann Otol 91: 209–15.

Densert O, Carlborg B, Stagg J (1981) Transmission of low-frequency pressure steps to the perilymphatic fluid. Acta Otolaryngol (Stockh) 91: 55–64.

Ernst A, Snik A, Mylanus I, Cremers CWRJ (1995) Non-invasive assessment of the intralabyrinthine pressure. Arch Otolaryngol Head Neck Surg 121: 926–9.

Froehlich P, Ferber C, Remond J, Jaboulay JM, Morgon A, Duclaux R, Collet L (1994) Lack of association between transiently evoked otoacoustic emission amplitude and experimentation linked-factors (repeated acoustic stimulation, cerebrospinal fluid pressure, supine and sitting positions, alertness level). Hear Res 75: 184–90.

Harada N, Ernst A, Zenner HP (1993) Hyposmotic activation hyperpolarizes outer hair cells of guinea pig cochlea. Brain Res 614: 205–11.

Marchbanks RJ, Reid A (1990) Cochlear and cerebrospinal fluid pressure: their interrelationship and control mechanisms. Br J Audiol 24: 179–87.

Matsuura S, Kuno M, Nakamura, T (1986) Intracranial pressure and auditory evoked responses of the cat. Acta Otolaryngol (Stockh) 102: 12–9.

Reid A, Marchbanks RJ, Burge DM, Martin, AM, Bateman DE, Pickard JD, Brightwell AP (1990) The relationship between intracranial pressure and tympanic membrane displacement. Br J Audiol 24: 123–9.

Sismanis A (1987) Otologic manifestations of benign intracranial hypertension syndrome: Diagnosis and management. Laryngoscope 97 (Suppl 42): 1–17.

Sismanis A, Callari RH, Slomka WS, Butts FM (1990) Auditory-evoked responses in benign intracranial hypertension syndrome. Laryngoscope 100: 1152–5.

Walsted A , Salomon G, Thomsen J, Tos M (1991) Hearing decrease after loss of cerebrospinal fluid. A new hydrops model? Acta Otolaryngol (Stockh) 111: 468–76.

Winer BJ (1970) Statistical principles in experimental design. London: McGraw-Hill.

Wunderlich K, Brinker T, Ernst A (1995) The correlation of TMD measurements to direct ICP recordings in cat. Midwinter Meeting of the Assoc Res Otolaryngology, Clearwater Beach, FL, Abstracts, 28.

Yoshida M, Uemura T (1991) Transmission of cerebrospinal fluid pressure changes to the inner ear and its effect on cochlear microphonics. Eur Arch Otorhinolaryngol 248: 139–43.

Zenner HP, Ernst A (1993) Cochlear-motor, transduction and signal-transfer tinnitus: models for three types of cochlear tinnitus. Eur Arch Otorhinolaryngol 249: 447–54.

Correspondence to Professor Arne Ernst, Direktor, HNO-Klinik, Unfall Krankenhaus Berlin, Rapsweg 55, 12683 BERLIN, Germany. Tel. (+49) 30 5681 2901. Fax (+49) 30 5681 2703.

Chapter 11
A review of the management and current methods of investigating intracranial pressure in shunted hydrocephalics and an introduction to the tympanic membrane displacement test

MADAN SAMUEL, ROBERT J MARCHBANKS and DAVID M BURGE, Department of Paediatric Surgery, Southampton General Hospital, Southampton, UK

ABSTRACT. The grotesque appearance of a hydrocephalic prompted early recognition of this condition and recommendations for potential effective management. The middle of the nineteenth century saw the renaissance of innovative surgical techniques for the management of hydrocephalus with poor results. During the 20th century various effective valve systems were introduced in combination with ventriculo-atrial and ventriculoperitoneal shunts. Despite improved results the treatment of hydrocephalus by shunting remains problematic and assessment of shunt function post-shunt surgery can be difficult. There is a need for a reliable non-invasive method of shunt assessment. The tympanic membrane displacement (TMD) test as a non-invasive technique in the assessment of shunt malfunction may be an ideal diagnostic test, though not a complete answer.

Introduction

Hippocrates (460–377 BC) was the first to describe hydrocephalus (he recognised that fluid was responsible for the enlargement of the head) and recommended decompression-trepanation as an effective method of treatment of this condition (Walker 1967). Since then the management of hydrocephalus has been revolutionised with advancements in technology, equipment and surgical skills. The introduction of a combination of valves incorporated in ventricular shunts has improved the results of shunt surgery. However, the evaluation of intracranial pressure (ICP) changes can be difficult at times in shunted hydrocephalic children and adolescents. There is considerable variability in the performance characteristics of individual shunts and they do not, in practice, function in accordance with the manufacturers' specifications. A shunt system can malfunction after a prolonged quiescent period and regular assessment of these patients is essential to prevent morbidity and mortality. There is the need for a reliable non-invasive means of assessing shunt function. The tympanic membrane displacement (TMD) test, which is an audiometric technique that assesses ICP indirectly by measuring perilymphatic pressure, is described. The purpose of this study was to summarise the surgical progress in the field of management of hydrocephalus and the investigative modalities currently available in the assessment of ICP, and also to introduce the TMD test as a diagnostic tool in the evaluation of ICP changes in shunted hydrocephalic patients.

Progress over the centuries in the management of hydrocephalus

Table 11.1 summarises the progress made in the management of hydrocephalus following the first description by Hippocrates in 460 BC.

Table 11.1: surgical treatment of hydrocephalus

Ventricular and lumbar punctures

	Contributor	Year	Contribution
1	Hippocrates 460 BC	Walker 1967	Frontal decompression trephination
2	The Arabs	Rhazes 1511	Cautery for trepanation at the suture lines and temporal veins
3	Fabricius Hildanus	1646	Puncture of the lateral ventricles by trocar
4	Monro	1783	Incision and drainage to avoid injury to vessels
5	Keen Broca	1888 1891	Posterior temporal area for ventricular drainage
6	Quinke	1891	Lumbar puncture
7	Chipault	1894	Differentiated subarachnoid and ventricular spaces. Performed drainage from varied regions on the head and also through the nose
8	Bilhaut	1894	Trepanopuncture with circular craniotomy and compression
9	Nölke Cassel	1897 1897	Combination of lumbar and ventricular drainage
10	Chater	1845	Seton Grantham (1854) and Ricketts (1904) applied this procedure following failure of repeated ventricular punctures

(contd)

Table 11.1: (contd)

Ventricular and lumbar punctures

	Contributor	Year	Contribution
Open ventricular drains			
1	Wernicke	1881	Permanent open drainage High mortality from use of large tubes
2	Keen	1891	Thin cord of horse hairs Thicker bundles of horse hairs and later rubber tubing replaced this
3	Illingworth	1891	Introduced Southey trocar and cannula Trephination and drainage
4	Ingraham	1941	The Southey trocar and cannula was the precursor of the Frazier needle and the technique of ventricular drainage by catheter as recommended by Ingraham and Campbell
Closed ventricular drains			
Ventriculo-subcutaneous drains			
1	Von Miculicz	1896	Ventricles to the subcutaneous space Inserted nail-shaped mass of glass wool in the lateral ventricle to the positioned periostal flap Later used gold and glass tubes Poor results due to infection and inadequate drainage and blockage
Ventriculo-subdural drains			
1	Von Miculicz	1896	Ventricles to the subarachnoid space and subdural spaces
	Sutherland and Cheyne	1898	Failed due to infection, blockage and inadequate drainage
2	Senn	1903	Perforated rubber tube
3	Taylor	1909	Catgut suture covered Cargile tube
4	Krause and Heymann	1917	Flanged silver tube extending to the pericranium
5	Forrest	1957	Achieved arrest of hydrocephalus in 50% of patients. The 50% mortality reported was from progressive hydrocephalus, intracranial infections, septicaemia, haemorrhage, blockage and inadequate drainage
Ventriculo-subgaleal drains			
1	Perret	1977	Transient treatment of hydrocephalus and drainage of subdural haematomas in 173 patients

Table 11.1: (contd)

Ventricular and lumbar punctures

	Contributor	Year	Contribution
Ventriculo-vascular drains and ventriculo-atrial shunts			
1	Gärtner	1895	First suggested drainage of ventricles into the lymphatic and venous system
2	Payr	1908	Lateral ventricular drainage into the lateral sinus by a saphenous vein graft. All patients succumbed to blockage and infection within four months post-operatively
3	McClure	1909	Lateral ventricles to cervical veins
4	Payr	1911	Used heterologous denatured anterior and posterior tibial arteries (dogs and calves) covered by autogenous saphenous veins
5	Haynes	1913	Cisterna magna to cranial sinuses
6	Payr	1919	Drainage into the internal jugular vein (8/15 patients survived, improvement in four patients)
7	Nulsen	1949	First shunt system with ball-valve mechanism. Successfully treated a 14 month old with hydrocephalus
8	Gupta	1950	Slit valves distal end-drainage into proximal segment of an obstructed facial vein and external jugular vein
9	Pudenz	1955	Animal experiments. Resulted in use of silicon as tubing material. Demonstrated good functional results with ventriculo-atrial shunts. Shunted a two year old with stenosis of the aqueduct of Sylvius. Died two years later from acute shunt malfunction
10	Holter	1955	Holter valve. Precision toolmaker commenced work on a valve following the birth of a baby girl with myelomeningocele.
11	Sikkens	1957	Lateral ventricles to the caval vein with an integrated ball valve imbedded in methyl methacrylate. Good results

(contd)

Table 11.1: (contd)

Ventricular and lumbar punctures

	Contributor	Year	Contribution
Ventriculo-lymphatic shunts			
1	Yokoyama	1959	Ventricles to thoracic duct 100% (3/3) mortality due to obstruction of the duct
2	Kempe	1977	Twelve-year experience of ventriculo-lymphangiostomy Used simple silicon tube without valves and relayed on competence of the valves of the thoracic duct Good results in 62% (16 patients)
Ventriculo-peritoneal and other abdominal shunts			
1	Ferguson	1898	Laminectomy of the fifth lumbar vertebra and opened the vertebral body to the peritoneal cavity. Used a sling of silver wire directed to the cauda equina to maintain patency. 100% mortality (3/3)
2	Nicoll	1905	Omental strip to maintain patency between peritoneal cavity and spinal cavity via a paravertebral incision
3	Cushing	1905	Anastomosed the spinal subarachnoid space to the retroperitoneal space. Laparotomy and laminectomy of the 4th lumbar vertebra and insertion of silver cannula
4	Heile	1908	Used silk threads (procedure similar to Cushing) The child succumbed to infection Later, described saphenous vein graft that was replaced by a rubber tube due to occlusion. The shunt functioned for 12 years
5	Kausch	1908	First rubber catheter ventriculoperitoneal shunt Patient died from over-drainage within 20 h post-operative Fell into disuse due to high mortality from infection, occlusion, septicaemia and over-drainage
6	Payr	1919	First lumbar peritoneal shunt
7	Heile	1925	Anastomosis of subarachnoid space to renal pelvis
	Matson	1956	following nephrectomy 95% success rate Matson-Connected lumbar subarachnoid

Table 11.1: (contd)

Ventricular and lumbar punctures

	Contributor	Year	Contribution
			space and the ureter by means of a moulded polyethylene drain-Ureteral drain. No mortality in 108 patients Complications: mechanical obstruction, recurrent infection, electrolyte and fluid imbalance, meningitis and renal failure (Matson 1956)
8	Cone	1949	Revived ventriculoperitoneal shunt
9	Ziemnowicz Schulze	1950 1956	Subarachnoid intravertebral shunting for communicating hydrocephalus Connected the subarachnoid space with the vertebral bone marrow by means of a hollow metal screw (14-year-old-female) with good results Schulze – unsatisfactory results in five patients
10	Harsh	1954	Salpingothecal and ventriculosalping-ostomies
11	Hakim	1955	Spinal subarachnoid space to epidural space Excellent results in four children with communicating hydrocephalus Modified Torkildsen's ventriculocistern-ostomy by placing distal catheter tip into the epidural space Poor long-term results resulted in the development of the popular Hakim valve shunt system
12	Picaza Dodge	1956 1957	Lumbar subarachnoid space to the retro-Omental region (Omental bursa) To prevent shunt obstruction from the omentum Good results in treating hydrocephalus
13	Smith	1959	Ventriculocholecystostomy Based on experimental work on dogs Specially designed valve system Poor functional results. Blockage and infection
14	Neumann	1959	Lumbar subarachnoid space to the ileum via a diverted and sterilised ileal loop and a synthetic tube
15	Scraff	1963	Reported 230 patients with ventriculo-peritoneal shunt Hydrocephalus arrest rate 55% Mortality 13%

(contd)

Table 11.1: (contd)

Ventricular and lumbar punctures

	Contributor	Year	Contribution
16	Alther	1965	Ventricles to stomach
	Lamesch	1972	Alther constructed a gastric valve system
	Weiss	1975	Low complication and good functional results
17	Portnoy	1971	First antisiphon valve
18	Smith	1980	Ventriculoureteral shunts without nephrectomy Proximal ureter reimplantation Distal ureter for shunt insertion – good results
19	Pudenz	1981	Slit valve catheter is inserted through the Winslow's foramen into the Omental bursa, anchored to the posterior peritoneum Peritoneoscope for insertion Excellent results in five children
Ventriculo-thoracic shunts			
1	Heile	1914	First ventriculo-pleural shunt
2	Ransohoff	1954-60	Initial success rate of 65%
	Milhorat	1976	Numerous shunt revisions

The shunt systems available at present have valves with varying pressure ranges. Antisiphon devices to prevent over-drainage and distal catheters with or without terminal slit valves have reduced the risk of shunt malfunction. Millipore filters incorporated within the shunt system prevent the spread of metastases and infection. The surgical techniques of shunt insertion have improved and are simplified with the introduction of percutaneous insertion. However, peritoneal and atrial shunting have limited long-term efficiency in spite of the simple operative technique, as a shunt system is fundamentally a simple piece of tubing that is prone to malfunction. A shunt system is well known to vary its *in vivo* performance characteristics despite the *in vitro* specifications outlined by the manufacturers. There is the need for a reliable, preferably non-invasive, means of assessing shunt function and ICP changes, post-shunt surgery.

Current methods available for intracranial pressure monitoring in children and adolescents with shunted hydrocephalus

Table 11.2 summarises the available investigations for the evaluation of ICP changes and shunt function.

Table 11.2: investigations for the assessment of ICP in patients with shunted hydrocephalus

Investigations	Comments
Clinical assessment	
Symptoms and signs in shunted hydrocephalic patients are poor indicators of shunt malfunction. Pumping of shunt reservoir is unreliable method of assessment and may induce blockage in children with low ICP or slit ventricles.	
Invasive procedures	
1. Lumbar puncture (transducer attached to the needle)	Direct ICP measurement remains the most reliable procedure in the evaluation of ICP, despite being invasive. Dynamic studies (pressure-volume index, dynamic and static outflow resistance) can be carried out via the ventricular catheter. Continuous ICP monitoring is always possible. There is the risk of infection, haemorrhage and seizures.
2. Percutaneous perforation of ventriculostomy reservoir (Rickham or Ommaya)	
3. Telemetry and telethermography	A pressure-balanced radio-telemetry system can be incorporated in a ventricular shunt system or can be used separately. The telesensor is less accurate at high ICP values. The measurement is a single instantaneous value. Cosmetically the system is more obtrusive than a single shunt system. Expensive removal if infection occurs. (Zervas et al. 1977; Cosman et al. 1979)
Radiological procedures	
1. Computed axial tomography (CAT) scan	Radiological investigations are static and do not measure absolute ICP. MRI however can be used to measure CSF production rate and volume.
2. Magnetic resonance imaging (MRI) (Gadolinium-enhanced T1-weighted spin echo images)	
3. Dynamic MRI	Can monitor CSF flow in shunt tubing and ventricles. Injection of isotope confirms patency but gives no information about ICP. There is the risk of infection.
4. Sequential transcranial Doppler	Detects abnormally raised pulsatile index associated with shunt blockage

(contd)

Table 11.2: (contd)

Investigations	Comments
Clinical assessment	
	with a sensitivity of 56% and specificity of 97% (Pople et al. 1990). Does not provide an absolute measure of pressure.
Electrophysiological monitoring	
1. Somatosensory evoked potentials Brain-stem auditory evoked potentials Visual evoked potentials	Somatosensory and visual evoked potentials are poorly correlated with changes in raised ICP. Moreover elevated ICP may not affect neural function. Only elevated ICP affecting neural function resulting in ischaemia and injury to neural tissue will induce consistent changes in specific evoked potentials (Moulton et al. 1994).
Audiometric techniques	
1. Evoked otoacoustic emissions	Measures otoacoustic emissions from the cochlear outer hair cells. The test depends on middle-ear acoustic impedance. Qualitatively assess ICP via the cochlea. The patients must have detectable evoked otoacoustic emissions in the low frequency range (Kemp 1978; Avan et al. 1996).
2. Tympanic membrane displacement	CSF and perilymph communicate through the cochlear aqueduct. ICP changes in the resting position of the stapes footplate Analysis of perilymphatic pressure and indirectly ICP by measuring volume displacement of tympanic membrane to acoustic stimulation of the stapedius (Kast 1985; Marchbanks et al. 1987; Reid et al. 1989; Samuel et al. 1996) Normal middle-ear function, patent cochlear aqueduct and an intact stapedial reflex are essential for conducting the test. Semi-quantitative.

Tympanic membrane displacement test for assessing intracranial pressure

The CSF fluid communicates with the labyrinthine fluid through the cochlear aqueduct, which is the main route of pressure transmission between the ventricles and the labyrinth (Beentjes 1972). Intracranial pressure (ICP) changes are reflected in the intralabyrinthine fluid pressure, resulting in alterations in the dynamics of the middle ear. An indirect assessment of the ICP is possible by studying the movements of the tympanic membrane (TM). The TM movement reflects the contraction of the stapedius as well as the kinetics of the ossicles of the middle ear, and in particular the resting position of the stapes footplate.

Extratympanic manometry is probably the most basic method for detecting small middle- and inner-ear pressure changes in terms of induced pressure changes in the sealed external ear canal resulting from TM movement. The three types of displacement of the TM on stapedius contraction are inward, outward and inward/outward (Møller 1964; Yonowitz et al. 1976). The magnitude of this pressure will depend on the volume of the external ear canal. Microflowmeters that use air as a medium can accurately determine this volume displacement of the TM on stapedius contraction and are largely independent of the volume of the enclosed ear canal. However these devices vent to the atmosphere and are therefore very susceptible to ambient pressure changes and noise. Also the problem of backpressure persists as flowmeters depend upon a pressure differential with the ambient. (Backpressure is the pressure change that occurs in the enclosed ear canal and may tend to limit or modify the TM displacement. The smaller the space enclosed, larger the pressure changes.) Ambient noise is the main problem with microflowmeters and backpressure may not be relevant with reference to measuring gross TM displacements.

The above effects were excluded with the introduction of tympanic membrane volume displacement with a virtual free field of movement (Marchbanks 1980). This device, developed by Marchbanks, also incorporated the use of ensemble averaging to improve the signal-to-noise ratio as described by Weiss (Weiss et al. 1962). The equipment contains a transducer based upon a servo-system, which is connected to a small probe. The probe is sealed into the entrance of the external auditory meatus. The servo-system ensures a constant external auditory meatal pressure during the measurement of the volume displacement of the TM, by a microphone concealed in the headset. Since the pressure is kept constant, the volume displacement of the TM is directly proportional to the displacement of a servo-driven reference diaphragm.

Figure 11.1 demonstrates these measurements made under virtual free field conditions. The figure shows a probe hermetically sealed in the external auditory canal and connected to a cavity containing a reference diaphragm that is servo-controlled to keep the pressure in the closed system constant. The movement of the diaphragm is identical to the TM as the diaphragm instantaneously compensates the TM displacement. The measurement technique uses a closed volume and does not require a pressure chamber. This enables its usage in a much wider environment irrespective of external noise and low frequency pressure variations.

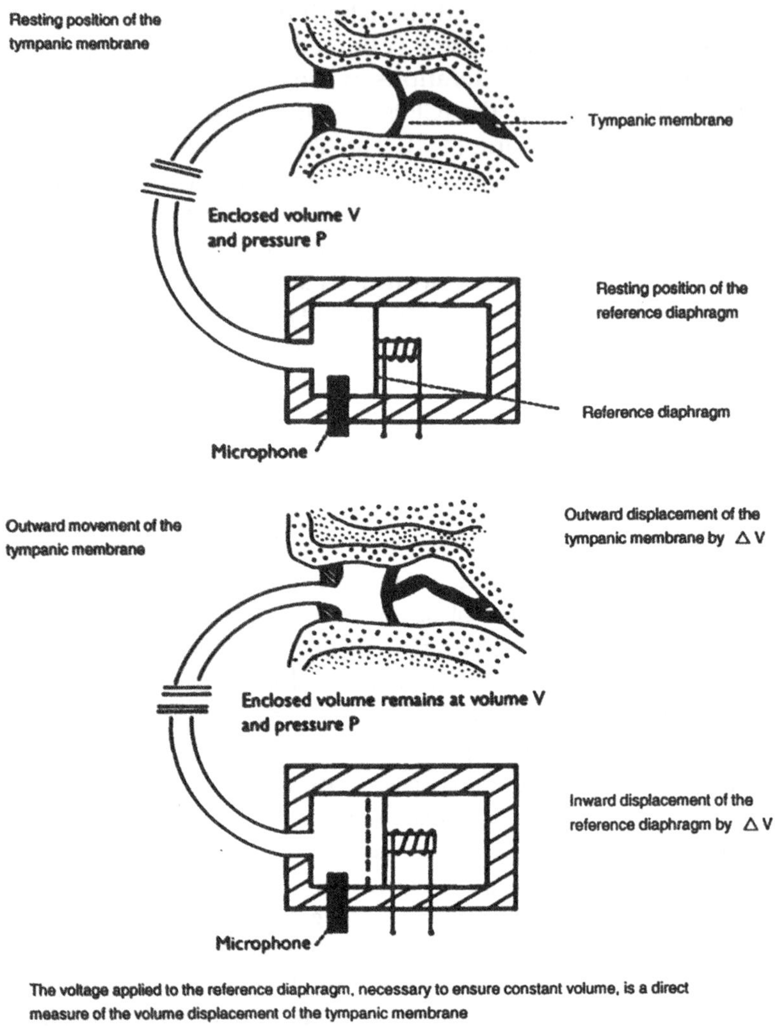

Figure 11.1: principle of operation of the tympanic membrane displacement measurement technique.

The TMD technique developed by Marchbanks in 1976 combines the advantages of pressure and flowmeter measurement devices to give a direct determination of the displacement of the TM (Marchbanks 1984).

The main advantages are:

- the absence of a probe tone. The technique depends on stimulus frequency and intensity of the acoustic reflex threshold and the probe tone is excluded which is inherent in impedance measurements. This eliminates the frequency and intensity of a probe tone, which may affect reflex response. The interference between the probe tone and the stimulus, which results in artefacts, was also eliminated;

- it involves a closed measurement system. The measurements are isolated from ambient pressure or temperature changes to a certain degree;
- the elimination of backpressure, which is attenuated within the operating bandwidth by 40% and at low frequencies by 99%;
- the response is signal averaged. The signal-to-noise ratio level is reduced depending on the number of averages taken for each response. This facilitates the differentiation of stapedius and tensor tympani reflex responses as they differ in latencies with different graph pattern for the displacement of the TM;
- it is a computerised audiometric measurement system. The integrated system allows recording and analysis of reflex responses in terms of TMD and can be linked to a multi-purpose microcomputer for more in-depth analysis;
- closed measurement system providing a degree of free field test environment.

The labyrinthine fluid pressure mirrors the CSF pressure changes through the main transmission pathway, the patent cochlear aqueduct. The hydrodynamic pressure interactions of the intracranial and intra-labyrinthine fluids are reflected in the dynamics of the middle ear and its influence on TM movement on stapedius contraction (Carlborg et al. 1981; Carlborg et al. 1992; Densert et al. 1977). The TMD technique was initially used to demonstrate that the cochlear fluid pressure is an indirect measure of the intracranial pressure (ICP) and in the assessment of the patency of the cochlear aqueduct in the normal population non-invasively (Brask 1978; Marchbanks 1980; Kast 1985; Phillips et al. 1989). This infrasonic procedure was used for the first time clinically in the assessment of ICP in three hydrocephalic patients in 1986 (Marchbanks et al. 1987). The first pilot study of 58 patients with an age range of 5 to 77 years demonstrated that the most consistent results were seen in young individuals with benign intracranial hypertension. (Reid et al. 1989). In a subsequent study, 43 hydrocephalic children with shunted hydrocephalus aged between four to 17 years were assessed when asymptomatic by TMD. Eleven (26%) of these children were later admitted with symptoms suggestive of acute shunt blockage. The TMD assessment correlated with clinical and/or operative finding in 10 patients, with one false positive result (Moss et al. 1991). The TMD technique was then compared with direct ICP measurements to procure a quantitative measure of tympanic ICP measurements and the reproducibility of the test was also evaluated (Samuel et al. 1996). The TMD and direct ICP measurements were shown to have linear correlation in children with shunted hydrocephalus and this has been corroborated by experimental studies conducted on cats, despite intersubject variation (Samuel et al. 1996, Wunderlich et al. 1996).

Conclusion

The TMD test as a non-invasive technique in the assessment of shunt malfunction may be an ideal diagnostic test, although not a complete answer.

References

Alther E (1965) Das Magenventil: eine neue Operationsmethode zur Behandlung des kindlichen Hydrocephalus. Schweiz Med Wochenschr 95: 234–9.

Avan P, Buki B, Lemaire JJ, Dordain M, Chazal J (1996) Otoacoustic emissions: a new tool for monitoring intracranial pressure. In Ernst A, Marchbanks RJ and Samii M (eds) Intracranial and Intralabyrinthine Fluids. Basic Aspects and Clinical Applications. Berlin: Springer Verlag, pp. 165–73.

Beentjes BIJ (1972) The cochlear aqueduct and the pressure of cerebrospinal and endolabyrinthine fluids. Acta Otolaryngologica 73: 112–20.

Bilhaut EA (1894) Traitement chirurgical de l'hydrocéphalie. Rev Chir (Paris) 14: 920–7.

Brask T (1978) Extratympanic manometry in man. Scandinavian Audiology (suppl): 7–11.

Broca A (1891) Drainage des ventricules cérébraux pour hydrocéphalie. Rev Chir (Paris)11: 37–40.

Carlborg B (1981) On physiological and experimental variation of the perilymphatic pressure in the cat. Acta Otolaryngologica 91: 19–28.

Carlborg BIR, Konrádsson KS, Carlborg AH, Farmer JC, Densert O (1992) Pressure transfer between the perilymph and the cerebrospinal fluid compartments in cats. Am J Otology 13: 41–8.

Cassel D (1897) Discussion remark. Dtsch Med Wochenschr 23: 231.

Chater G (1845) Case of hydrocephalus; treatment by puncture and seton; autopsy. Prov Med Surg J 599–602.

Chipault A (1894) Chirurgie operative du système nerveux. Paris: Rueff, vol 1, p. 734, vol 2, pp. 1–799.

Cosman ER, Zervas NT, Chapman PH, Cosman BJ, Arnold MA (1979) A telemetric pressure sensor for ventricular shunt system. Surg Neurology 11: 287–94.

Cushing H (1905) The special field of neurological surgery. Cleve Med J 4: 1–11.

Davidoff LM (1929) Treatment of hydrocephalus: historical review and description of a new method. Arch Surg 18: 1737–41.

Davidson RJ (1976) Peritoneal bypass in the treatment of hydrocephalus: historical review and abdominal complications. J Neurol Neurosurg Psychiatry 39: 640–8.

Densert O, Ivarrson A, Pederson K (1977) The influence of perilymphatic pressure on the displacement of the tympanic membrane. Acta Otolaryngol 84: 220–6.

Dodge HW, Remine WH, Leavens MD (1957) The treatment of hydrocephalus by spinal subarachnoid-omental bursa shunt. Minn Med 40: 227–9.

Duff TA, Rosenthal J, Jane JA (1977) Evaluation and revision of ventriculo-gastric shunts by gastrostomy. Neurochirurgia (Stuttg) 20: 28–32.

El Shafei IL (1975) Ventriculo-venous shunt to the proximal segment of an occluded neck vein: a new method for the shunting of cerebrospinal fluid to the venous circulation. Surg Neurol 3: 237–9.

Fabricius Hildanus G (1646) Observationum et curationum medico-chirurgicarum. Operaquae extant omnia. Frankfurt: Beueri, pp. 112, 1044, 1045.

Ferguson AH (1898) Intraperitoneal diversion of the cerebrospinal fluid in cases of hydrocephalus. NY Med J I 67: 902–7.

Fincher EF, Strewler GJ, Swanson HSb (1948) The Torkildsen procedure: a report of 19 cases. J Neurosurg 5: 213–16.

Forrest DM, Laurence KM, Macnab GH (1957) Ventriculosubdural drainage in infantile hydrocephalus. Selection of patients by subdural dye excretion test. Lancet II: 827–9.

Forrest DM, Laurence KM, Macnab GH (1957) Ventriculosubdural drainage in infantile hydrocephalus: analysis of early results. Lancet II: 1274–7.

Gartner G (1895) Discussionsbemerkung. Naturforscherversammlung, Lubeck, 1895; part 2, sect 2, pp. 146–63.

Grantham J (1854) A case of hydrocephalus externus treated with repeated puncturing and the seton. Med Times Gaz 9: 110–12.

Gupta SM (1950) New surgical treatment for idiopathic hydrocephalus. Indian J Med Sci 4: 85–90.

Hakim S, Jimenez A, Rosas F (1955) Drainage of cerebrospinal fluid into the spinal epidural space. Acta Neurochir (Wien) 4: 224–8.

Hakim S, Adams RD (1965) The special clinical problem of symptomatic hydrocephalus. J Neurol Sci 2: 307–12.

Harsh GR (1954) Peritoneal shunt for hydrocephalus utilizing the fimbria of the fallopian tube for entrance to the peritoneal cavity. J Neurosurg 11: 284–7.

Haynes JS (1913) Congenital internal hydrocephalus: its treatment by drainage of the cisterna magna into thc cranial sinuses. Ann Surg 57: 449–51.

Haynes JS (1913) Further experiences in the treatment of hydrocephalus by cisterna-sinus drainage (author's operation). NY State J Med 16: 174–8.

Heile B (1910) Zur chirurgischen Behandlung der Spina bifida mit Hydrocephalus. Berl Klin Wochenschr 47: 2298–304.

Heile B (1908) Zur Behandlung des Hydrocephalus. Dtsch Med Wochenschr 34: 1468–72.

Heile B (1914) Zur chirurgischen Behandlung des Hydrocephalus internus durch Ableitung der Cerebrospinalflussigkeit nach der Bauchhühle und nach der Pleurakuppe. Arch Klin Chir 105: 501–9.

Heile B (1925) Uber neue operative Wege zur Druckentlastung bei angeborellem Hydrocephalus (Ureter-Duraanastomose). Zentralbl Chir 52: 2229–31.

Heile B (1927) Zur Behandlung des Hydrocephalus mit Ureter-Duraanastomose. Zentralbl Chir 54: 1859–63.

Hippocrate. Oeuvres complètes d'Hippocrate. Baillière, Paris, 1953, vol 8: 40, p. 676. Cited in Walker AE 1967.

Hoffman HJ, Hendrick EB, Humphreys RP (1976) Metastasis via ventriculoperitoneal shunt in patients with hydrocephalus. J Neurosurg 44: 562–6.

Illingworth CR (1891) Tapping the ventricles. Br Med J 1: 755–8.

Immerwol. FD Beitrag zur Heilung des Hydrocephalus (1901) Arch Kinderheilkd 32: 329–34.

Ingraham FD, Campbell JB (1941) An apparatus for closed drainage of the ventricular system. Ann Surg 114: 1096–8.

Ingraham FD, Matson DD, Alexander E Jr, Woods RP (1948) Studies in the treatment of experimental hydrocephalus. J Neuropathol Exp Neurol 7:123–31.

Ingraham FD, Sears RA, Woods RP, Bailey OT (1949) Further studies on the treatment of experimental hydrocephalus: attempts to drain the cerebrospinal fluid into the pleural cavity and the thoracic duct. J Neurosurg 6: 207–10.

Jackson JJ (1951) A review of the surgical treatment of internal hydrocephalus. J Pediatr 38: 1–11.

Jackson JJ, Snodgrass SR (1955) Peritoneal shunts in the treatment of hydrocephalus and increased intracranial pressure: a 4-year-survey of 62 patients. J Neurosurg 12: 216–21.

Jackson JJ, Thompson RK (1959) Pediatric Neurosurgery. Springfield IL: Thomas, pp 143–65.

Kast R (1985) A new method for non-invasive measurement of short-term cerebrospinal fluid pressure changes in humans. J Neurology 232: 260–1.

Kausch W (1908) Die Behandlung des Hydrocephalus der kleinen Kinder. Arch Klin Chir 87: 709–15.

Keen WW (1888) Exploratory trephining and puncture of the brain almost to the lateral ventricle, for intracranial pressure supposed to be due to an abscess in the temporosphenoidal lobe: temporary improvement; death on the fifth day; autopsy; meningitis with effusion into the ventricles, with a description of a proposed operation to tap and drain the ventricles as a definite surgical procedure. Med News 53: 603–11.

Keen WW (1891) Tapping the ventricles. Br Med J 1: 486–7.

Kemp DT (1978) Stimulated acoustic emissions from within the human auditory system. J Acoust Soc Am 64: 1386–91.

Kempe LG, Blaylock R (1977) Ventriculolymphatic shunt. J Neurosurg 47: 86–91.

Lamesch AJ (1972) Ventriculogastrostomy by means of a gastric tube for the treatment of hydrocephalus: a preliminary report. J Pediatr Surg 7: 55–61.

Marchbanks RJ (1980) A study of tympanic membrane displacement. PhD thesis 1980. Br Lib Ref D34215/80.

Marchbanks RJ (1984) Measurement of tympanic membrane displacement arising from aural cardiovascular activity, swallowing and intra-aural muscle reflex. Acta Otolaryngol (Stockh) 98:119–29.

Marchbanks RJ, Reid A, Martin AM, Brightwell AP, Baheman D (1987) The effect of raised intracranial pressure on intracochlear fluid pressure: three case studies. Br J Audiology 21: 127–30.

Matson DD (1949) New operation for the treatment of communicating hydrocephalus: report of a case secondary to generalized meningitis. J Neurosurg 6: 238–49.

Matson DD (1956) Current treatment of infantile hydrocephalus. N Engl J Med 255: 933–5.

McCarty CS, Pratt JH, Ewert JC (1957) Congenital communicating hydrocephalus: treatment by subarachnoid-fallopian tube peritoneal shunt for a patient with unusually high intelligence. Proc Staff Meet Mayo Clin 32: 23–7.

McClure RD (1909) Hydrocephalus treated by drainage into a vein of the neck. Johns Hopkins Hosp Bull 1909; 20: 110–16.

McCullogh DC, Fox JL (1974) Negative intracranial pressure hydrocephalus in adults with shunts and its relationship to the production of subdural hematoma. J Neurosurg 40: 372–81.

Miculicz J (1896) Beitrag zur Pathologie und Therapie des Hydrocephalus. Mitt Grenzgeb Med Chir 1: 264–9.

Milhorat TH (1976) Structures and function of the choroid plexus and other sites of cerebrospinal fluid formation. In: Bourne GH, Danielli JF (eds) International review of cytology. New York: Academic, vol 47, pp. 225–88.

Møller AR (1964) Effect of tympanic muscle activity on measurement of the ear drum, acoustic impedance and cochlear microphonics. Acta Otolaryngol 58: 525–9.

Moss SM, Marchbanks RJ, Burge DM (1991) Long-term assessment of intracranial pressure using the tympanic membrane displacement measurement technique. Eur J Pediatr Surg I (suppl) 25–6.

Moulton RJ, Shedden PM, Tucker WS, Muller PJ (1994) Somatosensory evoked potential monitoring following severe closed head injury. Clin Invest Med 17:187–95.

Neumann CG, Hoen TL, Davis DA (1959) The adaptation of ileo-eutectrophy to the control of congenital communicating hydrocephalus. Plast Reconstr Surg 23: 159–68.

Neumann CG, Richman H, Aridizone R (1959) Reconstructive surgery for hydrocephalus: anastomosis of an isolated ileal segment to the lumbar subarachnoid space (modified ileoeutectropy). Surg Clin North Am 39: 491–9.

Nicoll JH (1905) Case of hydrocephalus in which peritoneo-meningeal drainage has been carried out. Glasgow Med J 63: 187–90.

Nolke H (1897) Beobachtungen zur Pathologie des Hirndruckes. Dtsch Med Wochenschr 23: 618–23.

Nulsen FE, Becker DP (1967) Control of hydrocephalus by valve-regulated shunt: infections and their prevention. Clin Neurosurg 14: 256–9.

Nulsen FE, Spitz EB (1952) Treatment of hydrocephalus by direct shunt from ventricle to jugular vein. Surg Forum 2: 399–402.

Olds JM (1962) Surgical drainage of hydrocephalus via the thoracic duct and subclavian vein. Can Med Assoc J 86: 696–702.

Payr E (1908) Drainage der Hirnventrikel mittels frei transplantierter Blutgefäße: Bemerkungen uber Hydrocephalus. Arch Klin Chir 87: 801–6.

Payr E (1911) Uber Ventrikeldrainage bei Hydrocephalus. Verh Dtsch Ges Chir 40: 515–21.

Payr E (1919) Elfjahriger Dauererfolg einer Ventrikeldrainage bei Hydrocephalus. Med Klin 49: 1247–9.

Perret GE, Graf CJ (1977) Subgaleal shunt for temporary ventricle decompression and subdural drainage. J Neurosurg 47: 590–7.

Phillips AJ, Marchbanks RJ (1989) Effects of posture and age on tympanic membrane displacement. Br J Audiology 23: 279–84.

Picaza JA (1956) The posterior-peritoneal shunt technique for the treatment of internal hydrocephalus in infants. J Neurosurg 13: 289–97.

Pilcz A (1899) Die Behandlung des Hydrocephalus. Zentralbl Grenzgeb Med Chir 2: 681–8.

Pople IK, Quinn MW, Bayston R, Hayward RD (1991) The Doppler pulsatility index as a screening test for blocked ventriculoperitoneal shunts. Eur J Pediatr Surg 1 (suppl I): 27–9.

Portnoy HD, Schulte RR, Fox JL, Croisaant P, Tripp L (1973) Antisiphon and reversible occlusion valves for shunting in hydrocephalus and preventing post-shunt subdural hematomas. J Neurosurg 38: 729–36.

Pudenz RH (1981) The surgical treatment of hydrocephalus – an historical review. Surg Neurol 15:15–19.

Pudenz RH (1958) Experimental and clinical observations on the shunting of cerebrospinal fluid into the circulatory system. In Pudenz RH, Clinical Neurosurgery. Congress of Neurological Surgeons. Baltimore: Williams & Wilkins, pp 98–115.

Raimondi AJ, Matsumoto S (1967) A simplified technique for performing the ventriculo-peritoneal shunt: technical note. J Neurosurg 26: 357–9.

Ransohoff J, Shulman K, Fishman RA (1906) Hydrocephalus. A review of etiology and treatment. J Pediatr 56: 399–402.

Ransohoff J (1954) Ventriculo-pleural anastomosis in treatment of midline obstructional neoplasms. J Neurosurg 11: 295–307.

Reid A, Marchbanks RJ, Martin AM, Burge DM, Batemen DE (1988) Raised intracranial pressure and the perilymph. Br J Audiology 22: 148–53.

Reid A, Marchbanks RJ, Batemen DE, Martin AM, Brightwell AP, Pickard JD (1989) Mean intracranial pressure monitoring by a non-invasive audiological technique: a pilot study. J Neurol Neurosurg Psychiatry 52: 610–12.

Ricketts BM (1904) Surgery of hydrocephalus: a historic review. Am Med 7: 983–95.

Ritter F, Lawerence M (1965) A histological and experimental study of cochlear aqueduct patency in the adult human. Laryngoscope 75: 1224–33.

Samuel M, Marchbanks RJ, Burge DM (1996) Tympanic membrane displacement test in regular assessment of children with shunted hydrocephalus. Presented at the 40th Annual Meeting of the Society for Research into Hydrocephalus and Spina Bifida, Utrecht, Netherlands, July 1996. Eur J Pediatr Surg I (suppl): 47–51.

Samuel M, Burge DM, Marchbanks RJ (1996) The accuracy of the tympanic membrane displacement test in detecting low intracranial pressure in children with shunted hydrocephalus. Presented at the meeting of the British Association of Paediatric Surgeons, XLIII Annual International Congress, St Helier, Jersey, July 1996

Scarff JE (1963) Treatment of hydrocephalus: an historical and critical review of methods and results. J Neurol Neurosurg Psychiatr 26: 1–10.

Schaltenbrand G (1938) Neuere Anschauungen zur Pathophysiologie der Liquorzirkulation. Zentralbl Neurochir 3: 290–300.

Schaltenbrand G (1940) Die akute Aliquorrhoe. Verh Dtsch Ges Inn Med 52: 473–81.

Schaltenbrand G (1953) Normal and pathological physiology of the cerebrospinal fluid circulation. Lancet 1: 805–8.

Schulze HE (1956) Die Behandlungsmoglichkeit des Hydrocephalus communis durch intervertebrale Schraubendrainage im Lumbarbcreich. Chirurg 27: 420–9.

Senn N (1903) Subcutaneous drainage in the surgical treatment of hydrocephalus internus. Alienist Neurol 24: 317–22.

Senn N (1903) Internal hydrocephalus. Int Clinics 13(1): 131–4.

Sikkens TB (1957) Traitement de l'hydrocéphalie du nourrison par ventriculo-jugulo-stomie. Neurochirurgie 3: 65–73.

Smith GW, Moretz WH, Pritchard WL (1959) Ventriculobiliary shunt. A new treatment for hydrocephalus. Surg Forum 9: 701–10.

Smith JA, Lee RE, Middleton RG (1980) Ventriculoureteral shunt for hydrocephalus without nephrectomy. J Urol 123: 224–31.

Sutherland GA, Cheyne WW (1898) The treatment of hydrocephalus by intracranial drainage. Br Med 2: 1155–7.

Sutherland GA (1898) Case of chronic hydrocephalus treated by intracranial drainage. Lancet I: 791–4.

Taylor AS (1909) Subdural autodrainage in internal hydrocephalus. J Nerv Ment Dis 36: 422–9.

Taylor AS (1904) The treatment of chronic internal hydrocephalus by autodrainage. Am J Med Sci 2: 255–9.

Walker AE (1967) A History of Neurological Surgery. New York: Hafner, pp. 334–47. [Facsimile edn (Williams & Wilkins, Baltimore, 1951].

Weiss MH, Jane JA, Apuzzo MLJ, Heiden JS, Kurze T (1975) Ventriculogastrostomy, an alternate means for CSF diversion: a preliminary study. Bull Los Angeles Neurol Soc 40: 140–7.

Weiss HS, Mundie JR, Cashin JL, Shinabarger EW (1962) The normal human intra-aural muscle reflex in response to sound. Acta Otolaryngol 55: 505–11.

Wernicke C (1881) Lehrbuch der Gehirnkrankheiten fur Arzte und Studierende. Berlin: Fischer, vol 3, pp. 253–572.

Wunderlich K, Brinker T, Ernst A (1996) Correlation of direct and non-invasive ICP measurements in cat. In Ernst A, Marchbanks RJ, Samii M (eds) Intracranial and Intralabyrinthine Fluids. Basic Aspects and Clinical Applications. Berlin: Springer Verlag, pp 93–103.

Yokoyama J, Aoki H, Takebayashi K, Hirai T, Matsumoto T, Fukushima T (1959) Ventriculolymphangiostomy: a shunting operation for hydrocephalus to drain cerebrospinal fluid into the thoracic duct. Folia Psychiatr Neurol Jpn 13: 305–19.

Yonowitz A, Harris JD (1976) Ear drum displacement following stapedius muscle contraction. Acta Otolaryngol 81: 1–7.

Zervas NT, Cosman ER, Cosman BJ (1977) A pressure balanced radio-telemetry system for the measurement of intracranial pressure. A preliminary design report. J Neurosurg 47: 899–911.

Ziemnowicz S (1950) A new trail of operative treatment of hydrocephalus communicans progressivus. Zentralbl Neurochir 10: 11–18.

Correspondence to Dr Madan Samuel, Department of Paediatric Surgery, Southampton General Hospital, Southampton SO16 6YD, UK. Tel. (+44)(0)1703 794947. Fax (+44)(0)1703 794947. E-mail: Madan.Samuel@Virgin.net.

Chapter 12
The application of the tympanic membrane displacement (TMD) test as part of the clinical service for hydrocephalus patients

SARAH FLYNN, ROBERT MARCHBANKS and DAVID BURGE, Non-Invasive Intracranial Pressure Assessment Unit (NIPA), University Child Health, Southampton University Hospital, Southampton, UK

ABSTRACT. The tympanic membrane displacement (TMD) test has been shown to be an effective ambulatory method for the non-invasive assessment of intracranial pressure (ICP) in patients with hydrocephalus. Previous studies have shown that TMD measurements correlate with direct ICP measurements. Cerebrospinal shunt dysfunction can be accurately detected with this technique in patients who have previously had baseline TMD measurements. Since there are no risks or discomfort to the patient, repeated TMD testing is possible. The use of the TMD test as part of the clinical service for patients with hydrocephalus has been evaluated in 50 patients who had 79 TMD tests over an 18-month period. Questionnaires completed by clinicians before and after TMD testing were used to compare TMD results with clinical signs and symptoms and to ascertain whether the TMD results influenced patient management or reduced the need for invasive investigations.

After TMD testing, 12 proposed invasive investigations in 10 patients did not go ahead. Moreover, TMD results were cited as the primary reason for patient management being changed in four cases. Tympanic membrane displacement results gave clinicians reassurance about patient management and supported the decision to discharge two patients from hospital. Overall, TMD was more accurate in determining ICP than clinical assessment and useful TMD results were obtained in four cases where clinical suspicions were unclear. Clinicians felt that the TMD test was helpful in 94% cases.

To illustrate the clinical potential of the TMD technique as part of a

team approach to shunt assessment three case histories are described in detail. Repeated TMD testing was helpful in diagnosing abnormal ICP, monitoring fluctuating ICP and assessing ICP levels before and after shunt revision surgery. Tympanic membrane displacement testing has considerable clinical potential and, if adopted as part of the routine management of shunted patients, TMD measurements will result in more precise monitoring of ICP levels and shunt function and is likely to reduce the need for invasive investigations. Tympanic membrane displacement testing could also be used to evaluate the effectiveness of new shunts.

Key words: TMD test, intracranial pressure, hydrocephalus, cerebrospinal shunts.

Introduction

The tympanic membrane displacement (TMD) test has been used as a non-invasive measure of intracranial pressure for over 10 years (Marchbanks et al. 1987; Reid et al. 1989). An indirect measurement of cerebrospinal fluid is made by observation of the tympanic membrane during acoustic stapedial reflex contraction provided that there is communication between intracochlear and intracranial fluids. The cochlear aqueduct allows communication between these fluids and its patency can be checked by performing the TMD test in supine and sitting postures (Phillips and Marchbanks 1989; Murphy et al. 1992). The results of the TMD test have been shown to correlate with direct ICP recordings in the cat (Wunderlich et al. 1996). Tympanic membrane displacement testing in normal subjects has shown that stable TMD results are obtained during the day and after exercise (Rosingh et al. 1996). These normative data also showed that there were large intersubject differences but the correlation of test results within one individual was high. The TMD test therefore gives the most definitive results when an asymptomatic baseline is available for comparison. Previous studies with the TMD test in hydrocephalus and shunted patients have shown that TMD results vary with symptoms of shunt dysfunction and pre- and post-shunt revision surgery (Moss et al. 1990). The test has also been useful in tracking ICP changes with fluctuating symptoms over periods of days or weeks (Moss et al. 1991).

The clinical signs and symptoms associated with abnormal ICP in patients with hydrocephalus and/or cerebrospinal shunts are often subtle and hard to diagnose. Similar symptoms may occur with low or raised ICP but the treatment of each of these conditions is very different. Also signs and symptoms of abnormal ICP can be confused with other unrelated disorders. The difficulties of diagnosing shunt disorders are even greater in children who are unable to describe symptoms clearly. Prior to the development of the TMD test clinicians relied on the clinical history, presenting symptoms and by pumping the shunt chamber. Computerised tomography (CT) scanning and MR imaging have not proved conclusive for detecting ICP or shunt disorders (Watkins et al. 1994). The integrity of the shunt system and ICP status is, therefore, often unknown unless invasive measurements are carried out. Intermittent shunt blockages in a child, particularly, could go undetected for months or years but may nevertheless affect the social and educational welfare of the child.

Burge et al. (1996) reported preliminary findings on some of the benefits of the clinical use of the TMD test on a small group of patients. This study showed the potential of the TMD test in aiding the diagnosis of non-acute raised ICP, low ICP and headaches due to other causes. This larger study aims to determine whether TMD measurements can be carried out on most patients and whether TMD testing would influence patient management, especially the need for invasive investigations. The accuracy of the TMD results are compared with other assessments. The clinical value of the TMD test is demonstrated on three patients seen for TMD testing over several years with changing symptoms.

Patients and method

The patients comprised 13 hydrocephalic adults (aged 20–44 years) and 37 hydrocephalic children (aged 5–19 years). The primary diagnoses of the patients were congenital hydrocephalus (22), spina bifida (18), intraventricular haemorrhage (4), benign intracranial hypertension (BIH) (3), post-meningitis (3) and tumour (1). During the study some patients were seen more than once. Tympanic membrane displacement testing was carried out when patients were asymptomatic 27 times and when patients had symptoms 52 times.

Middle ear function and stapedial reflex thresholds were recorded using a GSI-33 middle ear analyser. Tympanic membrane displacement measurements were performed using the MMS-10 TMD Cerebral and Cochlear Pressure Analyser. An ipsilateral 500 ms stimulus tone of 1000 Hz set to 20 dB above the acoustic reflex threshold was used. Wherever possible testing was carried out in sitting and supine postures to check intracranial-cochlear pressure transmission, which is presumed to be via a patent cochlear aqueduct.

A three-part questionnaire was used to assess the clinical value of the TMD test. The answers to the first questionnaire, completed when the patient was referred for TMD testing, and two follow-up questionnaires, completed after TMD testing, were analysed to compare the results of TMD tests with other findings and to determine the effects of the TMD test on patient management.

The three patients described in detail comprise two of the adults tested on 16 and four occasions respectively and one child tested on three occasions. Tests were undertaken on these patients over a period of 7 years including the 18 months of the trial.

Results

TMD results were successfully obtained from 83% of the tests. Of these the results of the questionnaire revealed that the TMD test has considerable clinical benefit, with 94% of the tests classed as helpful by clinicians. The use of the TMD test resulted in changes in patient management with 12 invasive investigations being cancelled on receipt of the TMD results and four further cases where management was changed as a direct result of the TMD test. The TMD test gave reassurance about the management of two patients and backed up the decision to go ahead with surgery in two patients. Tympanic membrane displacement results contributed to the decision to discharge two patients from hospital. There were only two cases where the TMD test gave misleading results and in both cases there was no previous asymptomatic baseline for comparison. Overall TMD was 93% accurate and clinical assessment was 66% accurate. Additionally the TMD test gave useful information where clinical suspicions were unclear. Figure 12.1 shows a comparison of proposed patient management before TMD testing with actual management after TMD testing.

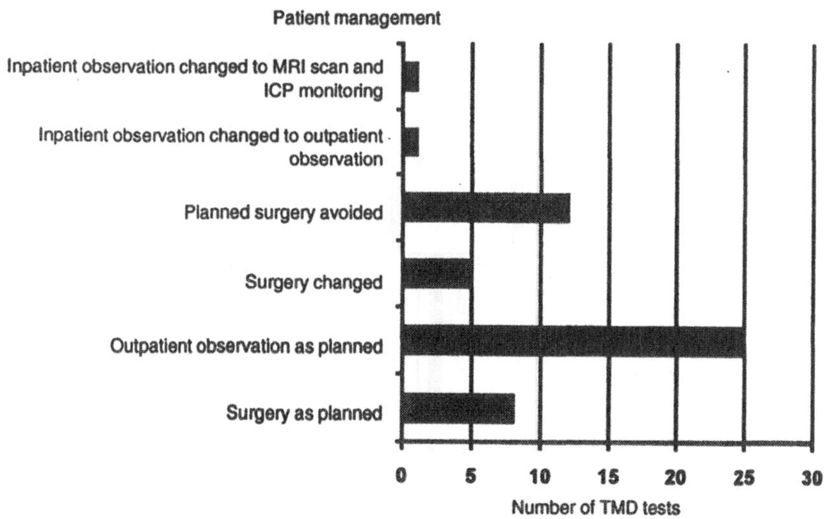

Figure 12.1: comparison of proposed patient management before TMD testing with actual patient management after TMD testing.

The key benefits of TMD testing include; long-term management of a hydrocephalic patient, monitoring shunt function and shunt revision, and differentiating raised from low ICP. These benefits are described in detail in the following three case histories.

Patient 1

This female patient was born in 1971 with spina bifida and hydrocephalus. The patient was first shunted at two weeks old. A baseline TMD measurement was carried out in 1989 when the patient was asymptomatic. Since the baseline measurement, TMD testing has been carried out on 16 occasions over 7 years (Figure 12.2). Testing occurred in the presence of symptoms seven times. On five of these occasions raised pressure was found with TMD testing and shunt revision was required before the symptoms of headache and vomiting resolved. Post-operative TMD tests revealed a reduction in ICP on all of these occasions. From 24 April 1989 to 12 October 1992, routine ICP assessments were undertaken which demonstrated normal pressure. On 8 April 1991, the symptoms were a few intermittent headaches and a sensation of pressure behind the eyes. The TMD result showed results within the normal range. These symptoms resolved without surgical intervention suggesting that the symptoms were probably due to a cause other than raised ICP.

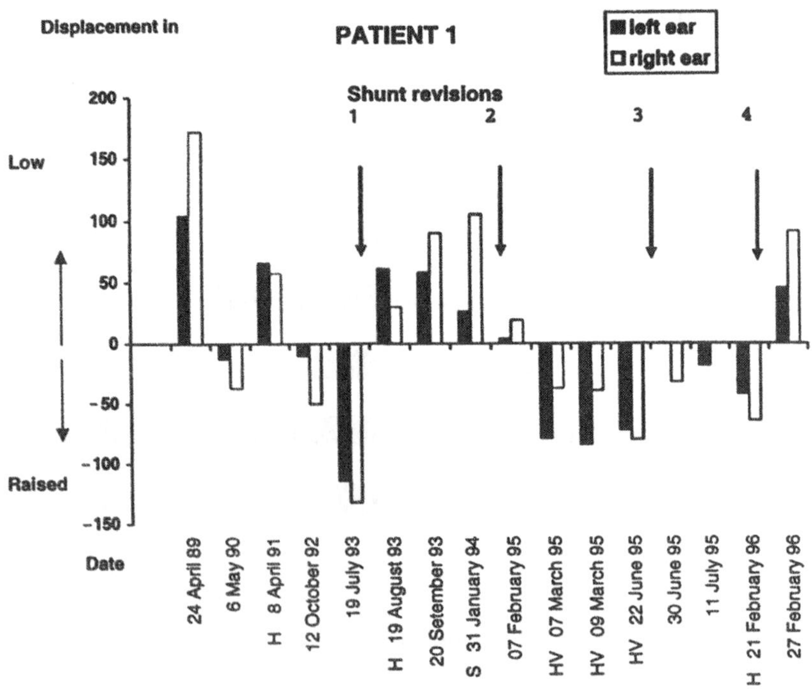

Figure 12.2: serial measurements used to assist long-term management of patient 1 – a 26-year-old female born with spina bifida and hydrocephalus. A ventricular peritoneal shunt was fitted at the age of two weeks and subsequently revised.

On 31 January 1994 TMD results showed normal pressure but the patient was experiencing seizures. Apart from these seizures, the clinical signs supported normal ICP and no surgical intervention was made. The patient reported increasing severity of dizziness, nausea and headaches over a six-month period prior to testing on 30 January 1995 when pressure monitoring showed raised ICP of 32–48 mmHg. A shunt revision was undertaken on 2 February 1995. Subsequent meningitis was treated with antibiotics. Tympanic membrane displacement testing on 7 February 1995 showed normal ICP and the patient was discharged being asymptomatic. On 2 March 1995 the patient was readmitted with a recurrence of symptoms which were previously found to be indicative of raised ICP. Tympanic membrane displacement tests on 7 March 1995 and 9 March 1995 concurred, showing an increase in ICP compared with post-shunt revision values 19 August 1993. On 14 March 1995, raised ICP was confirmed by direct measurements and the shunt was revised. Repeat episodes of shunt blockages occurred on 22 June 1995 and 21 February 1996. On these occasions TMD measurements were made pre- and post-surgery. Tympanic membrane displacement indicated a decrease in ICP with shunt revision on both occasions, which coincided with the patient becoming asymptomatic, although only on 27 February 1996 did the ICP become similar to that post-revision on 19 August 1993.

For this patient, therefore, the benefits of repeated TMD testing can be seen to help in deciding when surgery is necessary and in monitoring the outcome of surgery, thus helping to determine when the patient can leave hospital.

Patient 2

Tympanic membrane displacement testing proved useful in monitoring shunt function in this patient with benign intracranial hypertension (BIH). This patient was born in 1956 and experienced tinnitus from the age of 11. In 1977 she developed BIH with symptoms of headaches, unsteadiness and some deterioration of her eyesight. High cerebrospinal fluid (CSF) pressure could not be controlled with medication and an L-P shunt was inserted. The first TMD test was done in 1991 when the patient was well and had undergone a shuntogram that showed normal shunt function. In 1995 the patient had an increase in headaches, unsteadiness, tinnitus and deterioration of her eyesight. At this time TMD testing showed low pressure (Figure 12.3). MRI showed a Chiari type I malformation as a result of low pressure caused by the L-P shunt. The L-P shunt was removed and at surgery the ICP was indeed low (5–8 mmHg). Subsequently the ICP remained low for the duration of a CSF leak but later became raised (LP = 25 mmHg).

Removal of 15 ml of CSF temporarily improved the high pressure headaches. On 27 October 1995 a left ventricular peritoneal shunt was inserted to control the pressure. Low pressure headaches persisted and on 1 December 1995 the TMD tests again indicated low pressure which was confirmed with lumbar puncture (LP < 4 mmHg). The condition was treated conservatively and a repeat TMD test on 19 January 1996 demonstrated that the ICP had increased to a more normal level for the patient for whom all the symptoms except for the tinnitus resolved completely. On 30 August 1996 a repeat MRI confirmed that the Chiari had resolved.

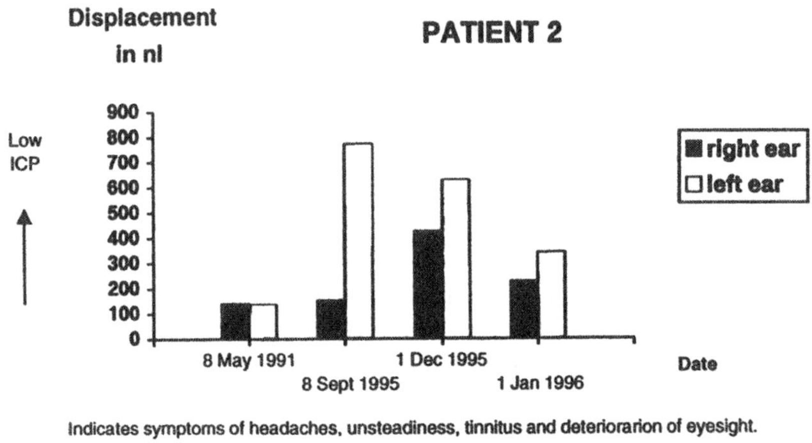

Figure 12.3: serial measurements used to assist long-term management of patient 2 – a 41-year-old female with tinnitus since the age of 11, and diagnosed as having benign intracranial hypertension at the age of 30 years. A lumbar peritoneal shunt was fitted to control the CSF pressure.

With this patient the TMD test was helpful in detecting low ICP as a result of the Chiari malformation and in demonstrating a return to more normal ICP levels with insertion of the VP shunt.

Patient 3

The TMD test was effective in distinguishing low from high ICP in this child who was experiencing symptoms but for whom clinical assessment was difficult. Patient 3 is a boy, born in 1985 at 40 weeks' gestation following an uneventful pregnancy. Hydrocephalus was diagnosed at 2 weeks of age following intraventricular haemorrhage of unknown origin. The first shunt was fitted at age 11 weeks. A baseline TMD measurement was obtained in 1990 when the patient was only 5 years old and asymptomatic. In 1995 the patient was having intermittent headaches and the clinical assessment was of raised ICP but TMD testing revealed low ICP (Figure 12.4). Investigative surgery was carried out and ICP was found to be low (3 mmHg) confirming that the TMD results were correct and the investigative surgery could have been avoided. A delta valve was fitted to overcome the low pressure. The TMD test showed an increase in ICP with the new valve although not quite to the 1990 values and the patient's symptoms resolved. Results for the left ear only are shown because there is a query over the patency of the cochlear aqueduct in the right ear. Clearly the value of the TMD test for this patient was in determining the difference between low and high ICP that may have similar presenting symptoms especially in chidren who have been shunted all their lives and may be unable to describe their symptoms. In this case investigative surgery could have been avoided had the TMD result been taken into consideration.

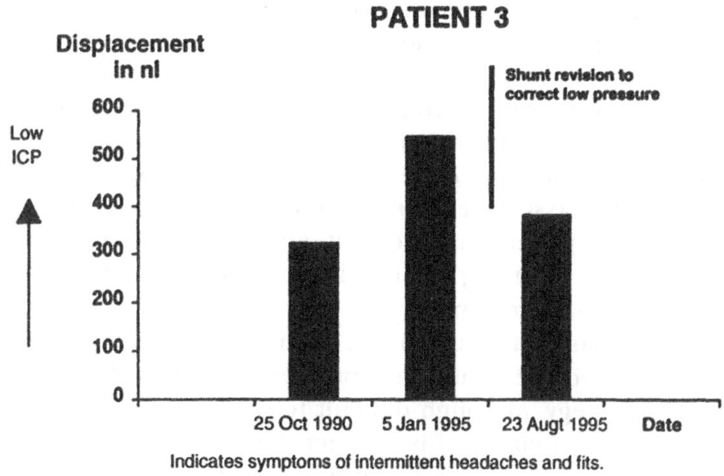

Figure 12.4: serial measurements used to assist long-term management of patient 3 – a 12-year-old male who was diagnosed as having hydrocephalus at 2 weeks of age following an intraventricular haemorrhage of unknown origin. A ventricular peritoneal shunt was fitted at age 11 weeks, which was subsequently revised.

Discussion and conclusions

To be adopted in routine clinical practice a test should be accurate and cost-effective. It should provide information that cannot be obtained by other methods or that reduces the risk or discomfort of currently used procedures. This study shows that the TMD test fulfils these criteria and is a valuable addition to the clinical service for patients with ICP disorders and shunt malfunction. Tympanic membrane displacement results can aid diagnosis, help in the decision of when to carry out surgery and monitor the effect of treatment for ICP disorders or shunt dysfunction.

TMD measurements can be obtained from most patients from the age of 4 years. The TMD test is distinct from other methods of investigating ICP in that it is non-invasive and there is no risk or discomfort to the patient, thus enabling the test to be repeated as often as necessary. The test equipment is mobile and therefore flexible to clinical circumstances. It only takes 30–60 minutes to perform a TMD test and the results can be available within 10 minutes of completing a test. The cost of TMD testing is relatively low and could be offset against savings from the reduction in the need for invasive investigations and possible earlier discharges of inpatients after treatment. The test has been carried out by audiological scientists for this study. Most hospital audiology departments already have audiological scientists who could carry out the test with only minimal additional training. Alternatively the test could be done by medical physicists with some background or training in audiology.

Both the questionnaire results and the case histories demonstrate the accuracy of the TMD test when an asymptomatic baseline is available for comparison. Tympanic membrane displacement results were more often correct than clinical suspicions. The only two cases where the results were misleading were cases with no previous asymptomatic baseline for comparison. Care should therefore be exercised when interpreting results without previous baseline results. The test supported information from clinical assessment and other investigations, thus giving reassurance about patient management, and measurably improved the clinical service by reducing the need for invasive investigations and changing patient management.

The TMD test also acted to support other diagnostic information and therefore increased the clinicians' confidence in their diagnostic and management strategy. Although difficult to quantify, the TMD test was demonstrably beneficial in this respect. For example, routine post-operative testing will help in the decision about when the patient may leave hospital and provide a new baseline with the new shunt. Testing pre-surgery could help correlate TMD results with direct ICP measurements more closely. Routine testing at outpatient appointments could help build a profile of normal versus abnormal ICP fluctuations in hydrocephalus and or shunted patients. If abnormal pressure is identified intervention may be implemented earlier. Shunt-dependent hydrocephalus patients are also prone to other disorders unrelated to the shunt system such as migraine, allergies and psychological problems as well as common flu and infections. The routine use of the TMD test could help distinguish true shunt malfunction from these other disorders which often have similar symptoms (Samuel et al. 1997, 1998).

References

Burge DM, Flynn SL, Marchbanks R (1996) Assessment of the clinical value of the Tympanic Membrane Displacement (TMD) test in children with shunted hydrocephalus. In Ernst A, Marchbanks R, Samii M (eds) Intracranial and Intralabyrinthine Fluids: Basic Aspects and Clinical Applications. Berlin/Heidelberg: Springer Verlag, pp. 211–16.

Marchbanks RJ, Reid A, Martin AM, Brightwell AP, Bateman D (1987) The effect of raised intracranial pressure on intracochlear fluid pressure: three case studies. Brit J Audiol 21: 127–30.

Moss SM, Marchbanks RJ, Burge DM (1990) Non-invasive assessment of ventricular shunt function using the Tympanic Membrane Displacement measurement technique. Z Kinderchir 45 (Suppl I): 26–8.

Moss SM, Marchbanks RJ, Burge DM (1991) Long term assessment of intracranial pressure using the Tympanic Membrane Displacement measurement technique. Eur J Pediatr Surg Suppl I: 25–6.

Murphy G, Marchbanks RJ, Watenpaugh DE, Meyer U, Eliashberg N, Hargens AR (1992) Increased intracranial pressure in humans during simulated microgravity. Physiologist 35 (Suppl 1): S-184–5.

Phillips AJ, Marchbanks RJ (1989) Effects of posture and age on tympanic membrane displacement. Brit J Audiol 23: 279–84.

Reid A, Marchbanks RJ, Bateman DE, Martin AM, Brightwell AP, Pickard JD (1989) Mean intracranial pressure monitoring by a non-invasive audiological technique: a pilot study. J Neurol Neurosurg Psychiatry 52: 610–2.

Rosingh HJ, Wit HP, Albers FWJ (1996) Non-invasive perilymphatic measurement in normal hearing subjects using the MMS-10 Tympanic Displacement Analyser. Acta Otolaryngol (Stockh) 116: 382–7.

Samuel M, Burge DM, Marchbanks RJ (in press) Tympanic membrane displacement test in regular assessment of ICP in 8 children with shunted hydrocephalus. J Neurosurg.

Samuel M, Burge DM, Marchbanks RJ (1997) Quantitative assessment of ICP by TMD audiometric technique in patients with shunted hydrocephalus. Presented at 10th International Symposium on Paediatric Surgical Research, Oct 1997, Zurich, Switzerland. Also Eur J Pediatr Surg (in press).

Watkins, L., Hayward, R., Andar, U., Harkness, W (1994) The diagnosis of blocked cerebrospinal fluid shunts: a prospective study of a paediatric neurological unit. Child Nerv Syst 10 (2): 87–90.

Wunderlich K, Brinker Th, Ernst A (1996) Correlation of direct and non-invasive ICP measurements in cat. In Ernst A, Marchbanks R, Samii M (eds) Intracranial and Intralabyrinthine Fluids: Basic Aspects and Clinical Applications. Berlin/Heidelberg: Springer Verlag, pp. 73–80.

Correspondence to Robert Marchbanks, Non-invasive Intracranial Pressure Assessment Unit (NIPA), University Child Health, Southampton University Hospital, Southampton, UK SO16 6YD. Tel. (+44) 1703 794649. Fax (+44) 1703 796378.

Chapter 13
Quantitative assessment of intracranial pressure by the tympanic membrane displacement test in patients with shunted hydrocephalus

MADAN SAMUEL, DAVID M BURGE, Wessex Regional Centre for Paediatric Surgery, Southampton General Hospital, Southampton, UK, and ROBERT J MARCHBANKS, Consultant Scientist, Department of Child Health, Southampton General Hospital, Southampton, UK

ABSTRACT. Tympanic membrane displacement, induced by reflex acoustic stapedial contraction, provided an indirect measure of the intracranial pressure through the cochlear aqueduct. A volume displacement of the tympanic membrane equal to or less than –200 nanolitres was indicative of raised cerebrospinal fluid pressure. A volume displacement of +200 nanolitres or greater was diagnostic of low intracranial pressure. This non-invasive ambulatory diagnostic investigation can be used on a regular basis in the objective assessment of intracranial pressure in patients with shunted hydrocephalus.

Key words: tympanic membrane displacement test, intracranial pressure, shunted hydrocephalus.

Introduction

The tympanic membrane displacement (TMD) test is an indirect assessment of the cochlear fluid pressure, which is determined by observing TMD during acoustic reflex stapedial contraction. The perilymphatic pressure is a direct reflection of the intracranial fluid pressure through the cochlear aqueduct, which is patent in 90% of the population less than 40 years of age (Wlodyka 1978; Phillips et al. 1989). Hence the paediatric and young adult population is a good cohort for assessment of intracranial pressure (ICP) by TMD test.

The cochlear aqueduct is the main fluid communication pathway between the intracranial and perilymphatic fluids (Figure 13.1). The endolymph indirectly communicates with the intracranial fluid through the endolymphatic sac and maintains a homeostasis with the perilymphatic pressure across the Reissner's and other intralabyrinthine membranes. Experimental studies on cats have demonstrated that the labyrinthine pressure correlates with that of the cerebrospinal fluid pressure and that a change in the intralabyrinthine pressure can be measured only if the cochlear and vestibular aqueducts are patent (Carlborg et al. 1992). The communication between the intracranial and labyrinthine spaces by perivascular and perineural routes is doubtful, and if it exists, it is secondary to that of the cochlear and vestibular aqueducts. Histopathological studies have demonstrated that the patency of the cochlear aqueduct is age-dependent and that it is patent in children and adolescents with hydrocephalus (Wlodyka 1978; Bachor et al. 1996). The patency of the cochlear aqueduct, which is the main route of pressure transmission, can also be determined non-invasively by measuring significant variation in the mean (Vm) TMD displacement with posture change from sitting to supine.

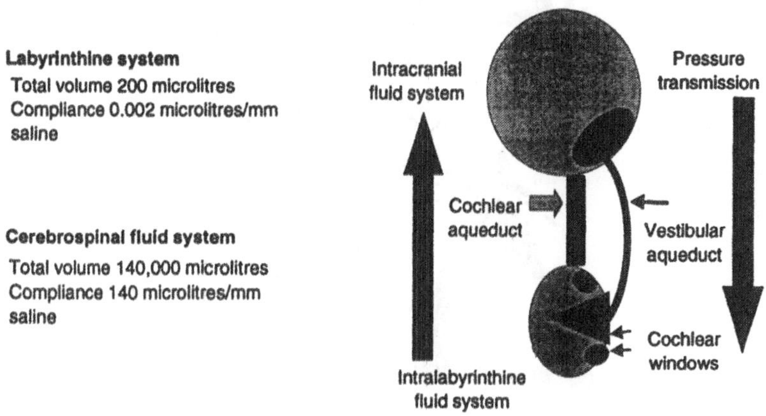

Figure 13.1: intracranial and intralabyrinthine fluid interaction.

The perilymphatic pressure is assessed indirectly by recording displacement of the tympanic membrane (TM) during reflex stapedial contractions elicited by a supra-threshold tone of 1 kHz. A raised perilymphatic pressure displaces the resting position of the stapes footplate laterally, causing a higher degree of motion in the medial direction. This results in an inward displacement of the TM on stapedial contraction. Low perilymphatic pressure results in an outward displacement of the TM (Figure 13.2). Computer-based instrumentation that resolves volume displacements (Vm) as small as a nanolitre (nl) is used to measure the movements of the TM on stapedial contraction.

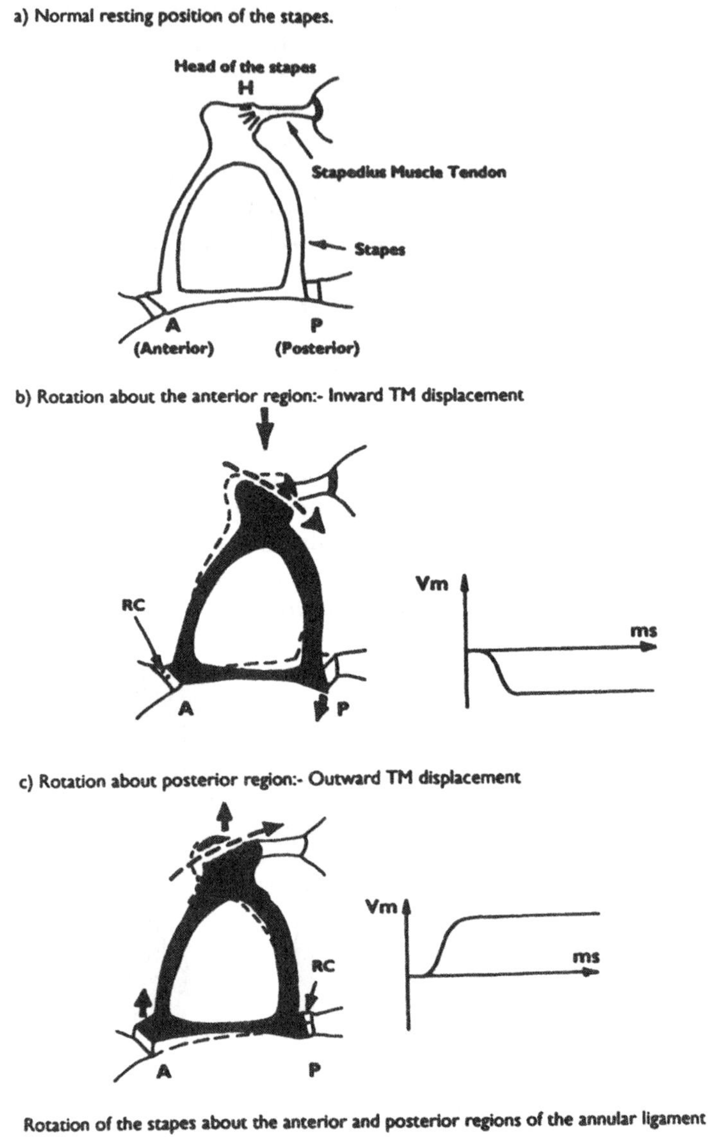

Figure 13.2: (1) the high perilymphatic pressure displaces the resting position of the stapes footplate laterally, thereby allowing a higher degree of motion in a medial direction, and results in a more inward going TMD on stapedial contraction. The volume displacement of the TM is negative (b). (2) low perilymphatic pressure demonstrating an opposite effect, with the volume displacement of the TM being positive and outwards (c).

TMD analysis has been used on an individual basis to provide a qualitative assessment whether the ICP is raised, normal or low (Reid et al. 1989). The test has been accurate in determining raised or low ICP in children with shunted hydrocephalus (Samuel et al. 1996). The aim of the current prospective study is to procure a quantitative measure of the ICP by TMD test by comparing volume displacement of the TM on stapedial contraction and direct ICP measurement.

Patients

A prospective comparative evaluation of 61 clinical episodes of shunt malfunction was assessed by volume displacement of the TM and direct ICP measurements in 40 patients with shunted hydrocephalus between January 1995 and June 1996.

In seven patients who were asymptomatic, normative data were obtained during elective admission for shunt lengthening procedure by measuring cerebrospinal fluid (CSF) pressure by TMD and direct ICP monitoring.

The mean age of the patients was 13.3 years and male to female ratio was 19:21. Table 13.1 summarizes the primary disorder.

Patients with abnormal middle ear pressure were excluded from the study.

Table 13.1: Primary aetiology of hydrocephalus in 40 patients with shunted hydrocephalus

Primary disorder	Patients (%)
Spina bifida (SB)	18 (45)
Congenital hydrocephalus (CH)	12 (30)
Post-haemorrhagic hydrocephalus (PHH)	7 (17.5)
Others – meningitis (two patients)	3 (7.5)
Holoprosencephaly (one patient)	

Percentages are shown in parentheses.

Methods

The test involves a transducer probe attached to a headset, which is hermetically sealed in the patient's external auditory meatus. A 1 kHz ipsilateral stimulus induced controlled stapedial muscle contraction, and ossicular and TM movement. The duration of the stimulus was 500 ms and the recording window was 1500 ms. The TMD was measured and analysed by a computer-based instrumentation, MMS-11, tympanic displacement analyser, which is available commercially. The records were ensemble averaged up to a maximum of 10 records to improve the signal to noise ratio. The ICP was then indirectly quantified in terms of the mean TMD (Vm), measured from the time of maximum inward displacement achieved while the acoustic stimulus is present, to the time of stimulus switch off.

The time needed initially to test both ears and the patency of the cochlear aqueduct by postural variation (comparative analysis of TMD in sitting and supine posture) was 45 to 60 minutes and subsequent, repeat tests were performed in 15 to 30 minutes.

Direct ICP monitoring was performed in all episodes in all 40 patients. The opening and the maximum sustained ICP were recorded through the Rickham ventriculostomy reservoir with the patient recumbent under a general endotracheal isoflurane anaesthesia and normocarbic condition. The measurement technique involved percutaneous puncture of the Rickham ventriculostomy shunt reservoir, which was then connected via rigid saline-filled tubing to a strain gauge transducer. The ventricular pressures were referenced to the external auditory meatus and extracted from the diastolic component of the cardiac and respiratory cycles.

Figure 13.3 demonstrates the correlation between TMD test and maximum sustained ICP measurements at each instance for individual patients.

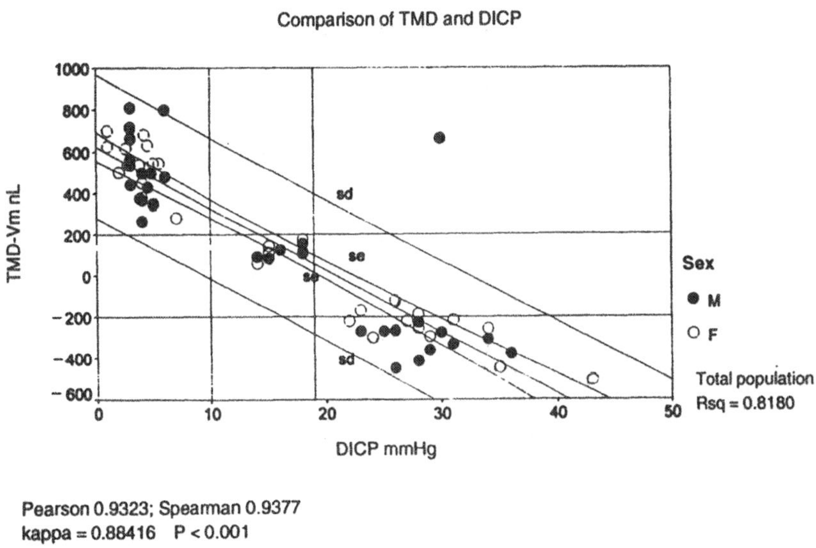

Figure 13.3: demonstrates a clear dissociation between the three categories, raised, normal and high ICP. A cut-off measure of –200 nl and less is indicative of raised ICP and + 200 nl and greater is indicative of low ICP. The Pearson coefficient correlation for the TMD test measurement and direct ICP measurement was 0.939. K = 0.88. P= < 0.001.

Results

Table 13.2 summarizes the TMD and direct ICP measurements in 40 patients. Sixty-eight TMD test measurements were used to assess 27 (40%) raised episodes of ICP, 30 (44%) low instances of ICP and 11 (16%) of normal measured ICP.

Table 13.2: comparison of TMD Vm values and direct ICP measurements

Episodes of shunt malfunction	Vm in nanolitres	Direct intracranial pressure in mmHg
Raised ICP N = 27	Mean = –250 Range = –120 to –506	Mean = 29 Range = 23 to 40
Normal ICP N = 11	Mean = 129 Range = 83 to 154	Mean = 14.7 Range = 15 to 18
Low ICP N = 30	Mean = 530 Range = 263 to 810	Mean = 3.8 Range = 1 to 6

At the time of each evaluation all patients had normal middle-ear pressure ranging from –30 and +25 daPa (normal range = –50 to +50 daPa) and compliance 0.2 to 1.5 ml (normal range = 0.2 to 1.5 ml). The normal ipsilateral acoustic reflex threshold at 1 kHz, ranged from 80 to 100 dB SPL. The TMD test measurement (Vm) was performed at a supra-reflex threshold stimulus of 15 to 25 dB (Figure 13.4(1) and (2)).

Figure 13.5(1) and (2) show the normal Vm values obtained by TMD test when the patients were asymptomatic. The test values were obtained following recovery after a shunt revision and at outpatient clinically follow up when normal and well.

Statistics

Inter-rater agreement kappa (κ)

The results demonstrated a correlation between mean Vm values and direct ICP measurements, with a linear regression (Figure 13.3). The subjects were classified into one of the three groups, which were, low ICP, normal ICP and high ICP. Comparing the three different observations made the agreement between categorical assessments. The kappa (κ) statistical analysis was used to measure the agreements between the groups (Landis et al. 1979). The strength of the agreement was very good, (k was 0.88) and the P value was < 0.001 (Figure 13.3; Table 13.3).

Sensitivity and specificity, and objective measure

The sensitivity and specificity of the test were defined with the aim of obtaining an objective measurement and for assessing TMD as a diagnostic modality. This would enable the CSF pressure changes in a patient to be predicted without the need to rely on the previous baseline test results for that particular individual. In cases of raised ICP the TMD is expected to be inward, and in low ICP outward, and of a magnitude that will tend to be larger than normal. As such the sensitivity and

(1)

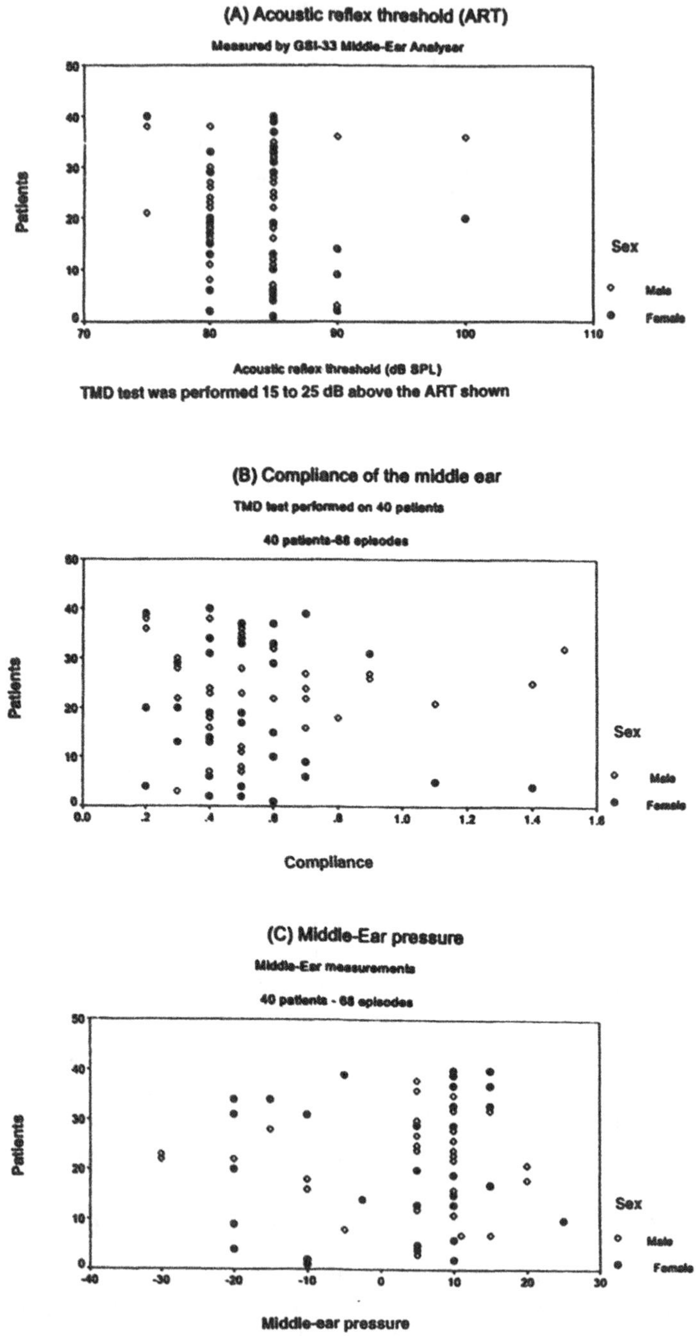

Figure 13.4: (1) shows the acoustic reflex threshold, compliance and middle-ear pressure of all 40 patients tested in 68 episodes. (2) shows the measure of middle-ear pressure and reflex threshold at 1 kHz as determined by the GSI 33 middle-ear analyser with a nominal probe frequency of 226 Hz in three patients.

(2)

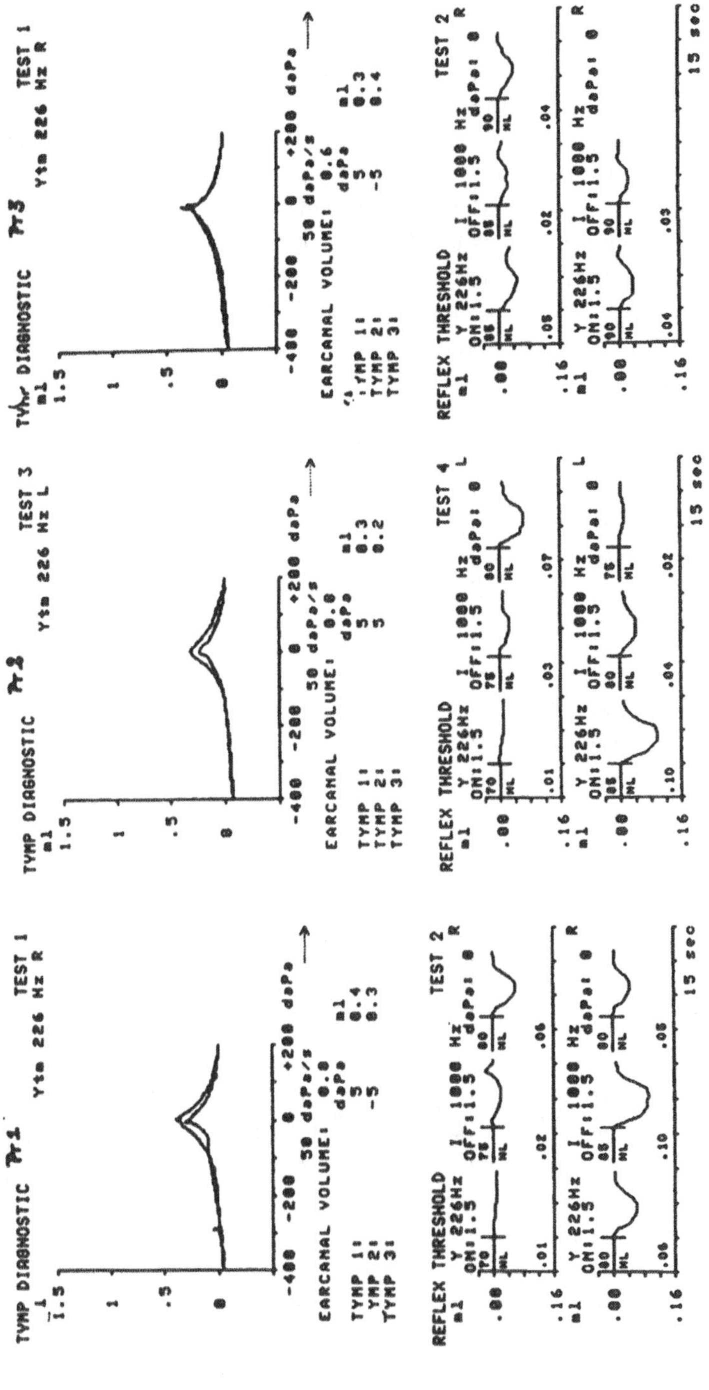

Figure 13.4: (contd)

specificity of the TMD test as a detector of raised and low ICP will vary depending on the chosen inward and outward TMD limit between what is considered to be a TMD magnitude representative of normal, raised or low ICP. Figure 13.3 demonstrates the data for TMD test (Vm) and direct ICP measurements of shunted hydrocephalus patients. These data show a clear dissociation between normal, raised and low ICP. The sensitivity was 82% and specificity was 100% for a normal-raised TMD limit set to –200 nl and for a normal-low TMD limit set to +200 nl the sensitivity was 100% and specificity was 96%. The predictive value was 100% and negative predictive value was 89% for raised ICP and for low ICP predictive value was 96% and negative predictive value was 100%. The TMD test as a non-invasive diagnostic tool in predicting changes in ICP had a sensitivity of 83% and specificity of 100% (Table 13.4).

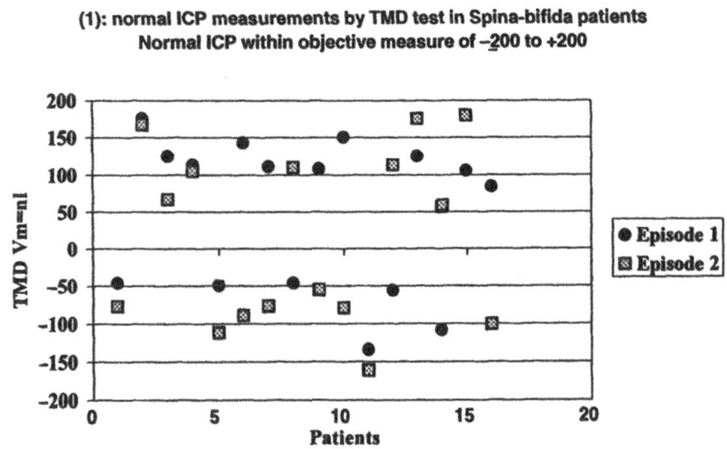

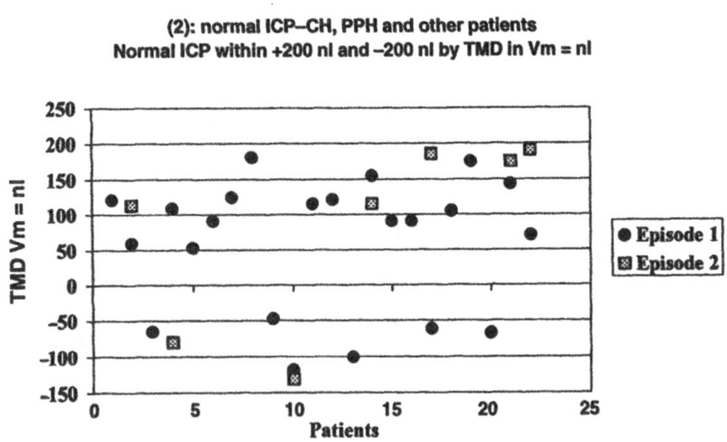

Figure 13.5: shows the normal ICP measurements by TMD in 40 patients when asymptomatic.

Table 13.3: assessment of inter-rater agreement between TMD and direct ICP measurements

Statistics	Value	Asymptotic standard error Ase 1	Value/Ase 1	P (probability value or significance level)	95% confidence interval
1. Kappa (κ)	0·8842	0·0489	10·0495	0·0000	0·8353 to 0·9330
2. Pearson's correlation coefficient (r)	0·9323	0·0359	20·9356	0·0000	0·8964 to 0·9682
3. Spearman correlation coefficient (rs)	0·9377	0·0361	21·9271	0·0000	0·9016 to 0·9738

The standard error (se) for κ was se (κ) = 0·4885 and a 95% confidence interval for κ were between 0·8353 to 0·9330.

GUIDELINES FOR STRENGTH OF AGREEMENT FOR κ

Value of κ	Strength of agreement
< 0·20	Poor
0·21–0·40	Fair
0·41–0·60	Moderate
0·61–0·80	Good
0·80–1·00	Very good

Landis and Koch (1977).

Table 13.4(1): analysis of 68 episodes of ICP measurements by TMD test and direct ICP monitoring

TMD test	Raised ICP	Direct ICP Normal ICP	Low ICP	Row total
Raised ICP (Count)	22	0	0	22
Expected value	8·7	3·6	9·7	32·4%
Row percentage	100%	0%	0%	
Column percentage	81·5%	0%	0%	
Total percentage	32·4%	0%	0%	
Normal ICP (Count)	4	11	0	15
Expected value	6	2·4	6·6	22·1%
Row percentage	26·7%	73·3%	0%	
Column percentage	14·8%	100%	0%	
Total percentage	5·9%	16·2%	0%	

(contd)

Table 13.4(1): (contd)

TMD test	Raised ICP	Direct ICP Normal ICP	Low ICP	Row total
Low ICP (Count)	1	0	30	31
Expected value	12·3%	5	13·7	45·6%
Row percentage	3·2%	0%	96·8%	
Column percentage	3·7%	0%	100%	
Total percentage	1·5%	0%	44·1%	
Column	27	11	30	68
Total	39·7%	16·2%	44·1%	100%

Table 13.4(2): analysis of the TMD test data for specificity and sensitivity (40 patients, 68 episodes)

	Raised ICP	Low ICP	TMD test
Sensitivity	81·5%	100%	93%
Specificity	100%	96%	100%
Positive predictive value	100%	96·8%	100%
Negative predictive value	89·1%	100%	73%

The objective measure of ICP by TMD test with the Vm value of –200 nl and more negative should be indicative of raised CSF pressure and a Vm value of +200nl and greater, for low ICP (Figure 13.3).

Discussion

Clinical features and current investigative modalities for evaluation ICP

In shunt-dependent hydrocephalic patients' overt neurological symptoms of relative rapid onset is usually attributed to raised ICP. At the other end of the spectrum of shunt malfunction is overdrainage resulting in low ICP, the clinical manifestation of which may be more subtle and chronic. Shunt malfunction if not detected early and appropriately managed would progress to obtundation and brainstem dysfunction. The variability of the clinical features within an individual makes it difficult to diagnose raised or low ICP due to shunt dysfunction. Clinical symptoms are not reliable predictors of changes in CSF pressure (Kirkpatrick et al. 1989). Physical examination of a shunt, especially in children and adolescents with subtle or intermittent symptomatology, by pumping of the shunt reservoir, is an unreliable method and may even induce blockage in patients with slit ventricles (Piatt 1992). Radiological imaging is a static investigation and does not invariably indicate the presence of raised or reduced ICP. The patency of the shunt system can be evaluated by the injection of an isotope. This technique carries the risk of introduction of infection and also gives no information about ICP. Dynamic magnetic resonance imaging can monitor flow in the shunt tubing and ventricles, but does not determine changes in ICP and would require general anaesthesia in children. Sequential transcranial Doppler studies sometimes help but do not provide an absolute measure of pressure (Pople et al. 1991).

Telethermography and telemetric pressure sensing techniques are not of proven value (Chapman et al. 1981). Direct measurement of ICP remains the standard to establish changes in ICP, but being invasive introduces the risk of infection (Guillaume 1951). There exists a need for an alternative non-invasive, ambulatory, reliable and a repeatable investigation for regular assessment of patients with shunted hydrocephalus. The TMD test as a diagnositic modality may be useful in the regular assessment of shunt function in these patients.

TMD test

Intracranial pressure is transmitted through the cochlear aqueduct to the perilymph of the cochlea provided the aqueduct is patent. The cochlear aqueduct was patent in all patients as determined by postural variation. The other factors essential for interpretation of the TMD test such as middle-ear pressure and an intact acoustic stapedial reflex, were normal. The high sensitivity and specificity demonstrated that the TMD test was viable despite inter-subject variation. A linear correlation was seen between direct ICP measurement and TMD test. A similar linear correlation between TMD and direct ICP measurement has been demonstrated in animal (cat) experiments (Wunderlich et al. 1996).

The technique is quick, painless and repeatable, hence easily applicable in monitoring children and adults with shunted hydrocephalus. Care was taken not to overexpose the patients to acoustic stimulation as it can result in a temporary auditory threshold shift.

Our study has shown that the sensitivity and specificity of the TMD test as a diagnostic modality is 83% and 100%. The predictive value of the test is 100% and the negative predictive value is 73%. This study shows that the TMD test is relatively accurate in predicting changes in the CSF pressure in-patients with shunted hydrocephalus.

Conclusion

- The objective measure of ICP by TMD test with the Vm value of −200 nl and more negative should be indicative of raised CSF pressure and a Vm value of +200 nl and greater, for low ICP.

- The TMD test as a non-invasive, ambulatory investigation is accurate in detecting changes in ICP in-patients with shunted hydrocephalus. It can therefore be used for regular assessment to enable correlation of ICP with symptoms in individual patients.
- TMD testing at the onset of minor symptoms may allow detection of ICP change at an earlier stage without the risks of direct monitoring. This in turn may allow earlier treatment and prevention of morbidity such as that due to chronic high or low ICP.
- It may also be used to monitor changes in ICP with different forms of treatment and shunt modification.

References

Bachor E, Karmody CS (1996) Hydrocephalus and the status of endolymphatic membranes in temporal bones of children. In Ernst A, Marchbanks RJ, Samii M (eds) Intracranial and intralabyrinthine fluids – basic aspects and clinical applications. Berlin, Heidelberg, New York: Springer Verlag, pp. 93–103.

Carlborg BIR, Konrádsson KS, Carlborg AH, Farmer JC, Densert O (1992) Pressure transfer between the perilymph and the cerebrospinal fluid compartments in cats. Am J Otol 13: 41–8.

Chapman PH, Cosman ER, Zervas NT (1981) Concepts in Pediatric Neurosurgery I. Basel, New York: Karger, pp. 120–7.

Guillaume J, Janny P (1951) Manométrie intracranienne continue intérêt de la méthode et premiers résultats. Revue Neurologique 84: 131–42.

Kirkpatrick M, Engleman H, Minns RA (1989) Symptoms and signs of progressive hydrocephalus. Arch Dis Childhood 64: 124–8.

Landis JR, Koch GG (1977) The measurement of observer agreement for categorical data. Biometrics 33: 159–74.

Phillips AJ, Marchbanks RJ (1989) Effects of posture and age on tympanic membrane displacement. Br J Audiol 23: 279–84.

Piatt JH (1992) Physical examination of patients with cerebrospinal fluid shunts: is there useful information in pumping the shunt? Pediatr 89: 470–3.

Pople IK, Quinn MW, Bayston R, Hayward RD (1991) The Doppler pulsatility index as a screening test for blocked ventriculoperitoneal shunts. Eur J Pediatr Surg Suppl I: 27–9.

Reid A, Marchbanks RJ, Bateman D, Martin AM, Brightwell AP (1989) Mean intracranial pressure monitoring by a non-invasive audiological technique: a pilot study. J Neurol Neurosurg Psychiatry 52: 610–12.

Samuel M, Marchbanks RJ, Burge DM (1996) Tympanic membrane displacement test in regular assessment of children with shunted hydrocephalus. Presented at the 40th Annual Meeting of the Society for Research into Hydrocephalus and Spina Bifida, Utrecht, Netherlands, July 1996.

Samuel M, Burge DM, Marchbanks RJ (1996) The accuracy of the tympanic membrane displacement test in detecting low intracranial pressure in children with shunted hydrocephalus. Presented at the meeting of the British Association of Paediatric Surgeons, XLIII Annual International Congress, St Helier, Jersey, July 1996.

Wlodyka J (1978) Studies on cochlear aqueduct patency. Ann Otol Rhinol Laryngol 87: 22–8.

Wunderlich K, Brinker Th, Ernst A (1996) Correlation of direct and non-invasive ICP measurements in cat. In Ernst A, Marchbanks RJ, Samii M (eds) Intracranial and Intralabyrinthine Fluids: Basic Aspects and Clinical Applications. Berlin/Heidelberg: Springer-Verlag, pp. 93–103.

Correspondence to Dr M Samuel, Wessex Regional Centre for Paediatric Surgery, Southampton General Hospital, Southampton SO16 6YD, UK. Tel. (+44)(0)1703 796677. Fax (+44)(0)1703 794750.

PART 3
INNER EAR PHYSIOLOGY

Chapter 14
Physical processes underlying homeostasis of cochlear endolymph

ALEC N SALT and JOHN E DEMOTT, Department of Otolaryngology, Washington University School of Medicine, St. Louis, USA

ABSTRACT. Cochlear endolymph has a unique chemical composition for an extracellular fluid, containing predominantly potassium instead of sodium and having extremely low calcium content. Under normal conditions, the chemical composition of endolymph is maintained by active homeostatic processes located in tissues bounding the endolymphatic space in the cochlea. Ionic movements radially and longitudinally within the endolymph are driven by prevailing concentration and electrical gradients. However, under some conditions significant solute movements can occur by other processes, including bulk longitudinal flow.

If endolymph volume is reduced by osmotic dehydration, apically directed endolymph flow occurs, which contributes to the restoration of normal volume. Similarly, when endolymph volume is increased by direct injection of artificial endolymph, basally directed flow occurs, which contributes to the removal of the excess volume. There is no detectable increase in pressure of endolymph with respect to perilymph during this volume disturbance. The dependence of endolymph movements on cochlear pressures is complex. Substantial changes of endolymphatic and perilymphatic pressure can occur without inducing longitudinal movements of endolymph, provided the cochlea remains sealed. On the other hand, when the cochlea is perforated, smaller perilymphatic pressure changes induce longitudinal endolymph movements. These observations add to our understanding of the relationships between cochlear fluid pressures, volume, flow and composition that is necessary to understand clinical pathologies which disturb inner-ear fluids.

Key words: cochlea, endolymph, perilymph, endolymphatic hydrops, homeostasis.

Introduction

The gross anatomy of the inner ear is dominated by a series of interconnecting fluid spaces running through channels in the temporal bone, as shown schematically in Figure 14.1. The perilymphatic spaces of the cochlea, scala tympani (ST) and scala vestibuli (SV) are connected to each other at the apex of the cochlea via the helicotrema, and connect respectively to the cerebrospinal fluid spaces of the cranium and to the perilymphatic space of the vestibule. The endolymphatic space of the cochlea, scala media (SM), connects to the saccule through the ductus reuniens and from there to the endolymphatic sac via the vestibular aqueduct. It is notable that the endolymph is completely bounded by tissues with no ducts or direct communications with the surrounding perilymph. The perilymph is similar in composition to other extracellular fluids. In contrast, the endolymph has a unique composition, requiring specialized homeostatic mechanisms regulated ina different way from those in other fluid-transporting tissues. Chemically, the major cation in endolymph is K^+, rather than Na^+, which predominates in other extracellular fluids. In the cochlea, endolymph is maintained at a positive potential with respect to perilymph, which is called the endocochlear potential (EP). Partially as a result of the EP, cochlear endolymph Ca^{2+} content is extremely low, and HCO_3^-, Cl^- and pH levels are somewhat higher than perilymph (reviewed recently by Wangemann and Schacht 1996). The composition of both endolymph and perilymph are relatively heterogeneous, showing substantial concentration differences in different regions of the ear. K^+, Na^+ and Ca^{2+} levels all vary systematically between the endolymphatic sac, the utricle, the saccule and the cochlea. Even within the cochlea, longitudinal gradients exist between basal and apical regions for K^+, Ca^{2+} and EP, reflecting regional differences in homeostatic processes. To understand the contributions to endolymph homeostasis of the various tissues and cell types in different regions of the ear, the processes by which different fluid regions interact under normal and pathological conditions must be considered.

Solute movements between different regions of the endolymphatic space occur by two main processes: passive diffusion and longitudinal volume flow. Diffusion of solutes along the cochlear scalae must occur but diffusion can be remarkably slow, as illustrated in Figure 14.2. In this figure the rate of diffusion has been calculated for a small ion (K^+, Diffusion coefficient = 1.34×10^{-9} m^2/s) from a site maintained at

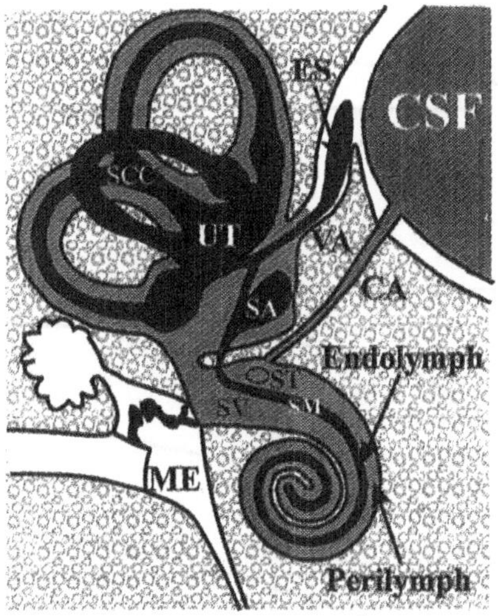

Figure 14.1: schematic representation of the fluid spaces that dominate the anatomy of the inner ear. Cerebrospinal fluid (CSF) and perilymph are shown light grey and the endolymph is shown dark grey. The abbreviations are ES: endolymphatic sac; SCC: semi-circular canals; UT: utricle; SA: saccule; VA: vestibular aqueduct; CA: cochlear aqueduct; ST: scala tympani; SV: scala vestibuli; SM: scala media; ME: middle ear.

constant concentration at one end of the tube. The length of the tube was chosen to represent the 18 mm basal-to-apical distance in the guinea pig cochlea. In humans the cochlea would be longer, approximately 32 mm from base to apex. At sites within a few millimetres of the base, concentrations approaching the source concentration are rapidly achieved. In contrast, for distances greater than 10 mm, only low concentrations are achieved – and that after a long time period. Furthermore, if the solute is larger, thereby diffusing more slowly, or if local processes affect the solute, such as clearance to blood, entry into another scala, uptake by cells or binding to the tissue (as in the case of a drug), then substantially lower concentrations will be achieved at the indicated distances. The slow rate of longitudinal diffusion means that, by diffusion alone, longitudinal solute movements or mixing will only occur slowly. This is relevant to experimental studies and clinical treatments in which drugs are applied to the round window membrane. Under these conditions it cannot be assumed that a uniform concentration of drug is generated within the perilymph within a reasonable time frame. Since radiotracer studies have shown that the major ions turn over rapidly in endolymph, with half-times for K^+ estimated to be close to 1 hour (Konishi et al. 1978), it is apparent that local homeostasis occurs more rapidly than longitudinal diffusion. This explains the observation of the lack of chemical homogeneity of the endolymph between different regions of the inner ear, as local transport is the predominant

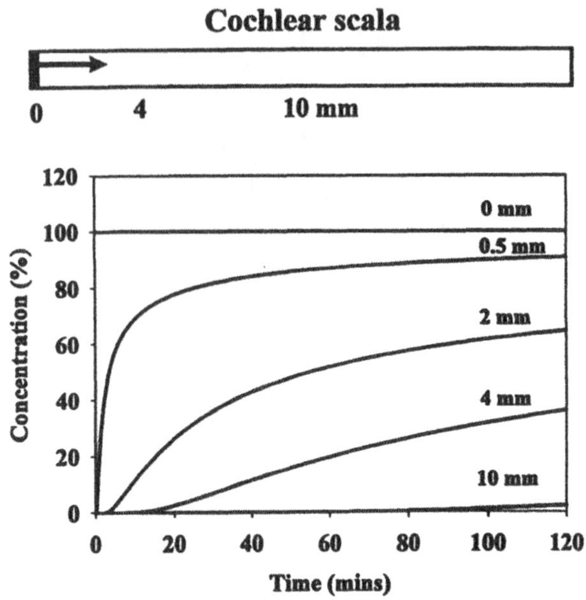

Figure 14.2: calculation of solute diffusion along a tube representing a cochlear scala. A diffusion coefficient of 1.34×10^{-9} m^2/s was used, corresponding to that of a small ion such as K^+. The lines represent the concentration time courses at varying distances along the tube from the source, which is maintained at constant concentration (100%) at one end. Note that diffusion occurs very rapidly over short distances and extremely slowly over long distances. If clearance, binding (as in the case of a drug) or loss to another scala occurs, then even lower concentrations would be reached along the scalae.

process. In contrast, it can be calculated that radial diffusion across the scala media would occur relatively rapidly, with equilibration of a concentration difference across the scala within a few minutes. Consequently, the concept of 'radial flow' of endolymph may not be meaningful since solute movements generated by radial flow will be small compared to the rate of spread by diffusion. Radial flow of volume between the different structures in the cochlea has never been shown to occur. It appears more likely that endolymph homeostasis is comparable to that of cytoplasm in a cell, which is not secreted and resorbed in different regions, but rather solutes are transported in or out of the cell as necessary, with water following passively according to the prevailing osmotic gradients and water permeabilities.

The second process by which solutes may be moved between different parts of the endolymphatic system is by longitudinal volume flow. Early studies by Guild (1927) and Lundqvist et al. (1964) showed that chemical markers injected into endolymph could subsequently be detected in the endolymphatic sac. These observations gave rise to the notion that endolymph was secreted in volume in the cochlea and flowed basally down the cochlea, passing through the ductus reuniens and saccule to be resorbed by the endolymphatic sac. Subsequently, more sensitive flow-measurement techniques were developed in which tracer amounts of a flow marker ion such as tetramethylammonium (TMA) could be injected into endolymph by iontophoresis, without any volume disturbance (Salt and Thalmann 1989). Movements of TMA along the scala were monitored with ion-selective microelectrodes. Based on the recorded concentration time courses, the rate of longitudinal flow could be calculated. In contrast to the earlier findings, results obtained with this technique gave no indication of a standing longitudinal flow of endolymph along the cochlea. On the contrary, they demonstrated that the rate of flow in the normal cochlea was exceedingly slow, to the extent that flow effects were negligible relative to local turnover and diffusion. These apparently conflicting results have now been reconciled by findings from a recent study in which endolymph flow was monitored during volume disturbances induced by microinjections into endolymph (Salt and DeMott 1997a). In this study, endolymph flow and scala media area changes were measured in the basal turn during injections of artificial endolymph into the second turn. Prior to injections, the average endolymph flow rate was close to zero, in accordance with the previous *in vivo* studies. In contrast, small injections at rates less than 60 nl/min were found to induce acute hydrops and substantial basally directed flow as summarized in Figure 14.3. The measured flow rate was found to correlate closely with the volume injected. These findings indicate that in the abnormal state, when endolymph volume is increased, flow towards the base represents a significant mechanism by which excess endolymph is removed. The results are in complete agreement with the original observations of Guild (1927) and Lundqvist et al. (1964) in demonstrating that flow towards the endolymphatic sac occurs when endolymph volume is increased by injection. However, the original interpretation of these studies is now shown to be incorrect, as the observed basally directed flow does not represent the normal physiological state but rather was induced by the tracer injection itself. The present explanation is also consistent with prior hypotheses that suggested that endolymph was maintained without longitudinal flow (Naftalin and Harrison 1958, Lawrence et al. 1961). On the basis of the data in Figure 14.3, in which the intercept of the fitted line differs significantly from zero, it can be estimated that approximately 80 nl can be injected into the second turn without inducing flow in the basal turn. This suggests that local mechanisms may exist by which small volume disturbances are corrected, although the capacity of such mechanisms appears to be extremely limited. It was concluded from these data that at least two mechanisms exist for endolymph volume regulation, consisting of a limited-capacity, local process and a flow-based system in which the endolymphatic sac functions as an 'overflow valve', removing excess endolymph under conditions that exceed the capacity of the local processes.

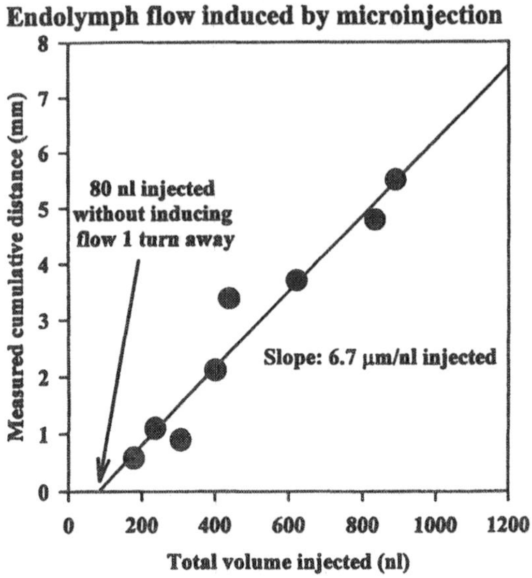

Figure 14.3: movements of endolymph measured in the basal turn of scala media during 15 mins injection of artificial endolymph into the second turn at rates varying up to 60 nl/min. The distance moved correlated with the total volume injected. The intercept of the fitted line was significantly different from zero which indicated that approximately 80 nl could be injected without inducing measurable flow in the basal turn.

Other studies have shown that under conditions where endolymph volume is reduced, an apically directed flow plays a significant role in restoring normal volume (Salt and DeMott, 1995). In this study, scala media area, flow and K^+ content were monitored during osmotic dehydration of the endolymph, produced by perfusing hypertonic solution through the perilymphatic spaces. Dehydration was found to reduce the cross-sectional area of scala media, with a concomitant apically directed endolymph movement. The longitudinal flow was estimated to contribute approximately one third of the K^+ increase seen in the endolymph as a result of dehydration. Previous studies have suggested that the endolymphatic sac may be resorptive or secretory under different conditions (Friberg et al. 1986). They examined the endolymphatic sac in animals following labyrinthectomy and found it to be dilated and filled with a stainable substance that was identified to be proteoglycan. In subsequent studies, Takumida et al. (1989, 1995) observed that the sac appeared to take on a secretory role after intravenous infusion of glycerol. These findings suggest that longitudinal flow indeed contributes to endolymph homeostasis with the endolymphatic sac acting as a 'bi-directional overflow valve', absorbing or secreting endolymph as required to maintain a constant volume.

It is apparent that the homeostatic processes that locally maintain the ionic and electrical status of endolymph are exquisitely balanced in the normal state in a manner that maintains electrochemical composition without producing volume change. This is likely to be necessary due to the adverse consequences of endolymph volume disturbances, presumably mediated by extremely small pressure differences between endolymph and perilymph. In the normal cochlea, the measured endolymph pressure does not differ significantly from perilymph pressure and both fluctuate with respiratory and cardiac cycles (Bohmer 1993). When endolymphatic hydrops is induced by ablation of the endolymphatic sac, although hydrops develops within days, no pressure elevation is detected until much later (> four weeks) when severe hydrops has developed and Reissner's membrane has lost its compliance (Bohmer 1996). Similarly, acute injections into endolymph that induce hydrops and cause longitudinal endolymph flow produce equal pressure increases in both endolymph and perilymph but no significant elevation of endolymph pressure with respect to perilymph (Takeuchi et al. 1991; DeMott and Salt 1997). These findings suggest that at least part of the endolymphatic boundary must be extremely compliant, thereby minimizing pressure differences across the sensory structures. Although this may be a necessary adaptation to maintain auditory sensitivity under varying conditions, it presents a number of technical difficulties with regard to maintaining normal endolymph volume, if volume is predominantly maintained by a single structure such as the endolymphatic sac. The structure would need to be capable of detecting extremely small pressure disturbances even though cochlear fluid pressures constantly fluctuate with respiration, heartbeat, posture and movements. It is logically possible that the stretching of a highly compliant structure could signal volume disturbance but the specific tissues involved and details of such a mechanism remain to be identified. Although the endolymphatic sac remains a likely candidate, it is still unclear whether the endolymphatic sac represents the most compliant region of the endolymph boundary. Certainly, the endolymphatic sac appear to be highly sensitive to mechanical disturbance, as evidenced by substantial changes in the endolymphatic sac potential that were observed when the surrounding tissues were perfused with control media (Couloigner et al. 1995)

The dependence of endolymph homeostasis on perilymphatic pressure changes has also been investigated (Salt and DeMott 1997b). Endolymphatic flow, scala media area and EP were measured during injections into scala tympani at 1.5 μl/min with no outlet. This procedure resulted in an elevation of intracochlear pressure by approximately 4 mmHg, but no significant endolymph movements, area changes or EP changes were observed. In comparison, injections at an identical rate were also performed in cochleae in which an outlet was made in scala vestibuli. This released the resting pressure of the cochlea and drastically reduced the measured injection-induced pressure changes in the cochlea. Under these conditions, the onset of injection produced a systematic longitudinal movement of endolymph in the basal direction and transient reduction of the EP by a few mV. The endolymph displacement and EP change were reversed when the injection ceased. The magnitude of the endolymph disturbance increased as a function of injection rate, suggesting a mechanical origin to the phenomenon. These results demonstrate that endolymph in the normal, sealed cochlea is extremely insensitive to large pressure fluctuations in the perilymph, yet in the perforated cochlea endolymph is highly sensitive to far smaller pressure disturbances. An important difference between the two conditions, and which accounts for these findings, is that in the condition with no outlet, the pressures in scala tympani and vestibuli are both equally increased since no outflow of scala vestibuli is possible. The observation that no longitudinal movement is detected in this condition indicates that pressure differences between the endolymph in the endolymphatic sac and the CSF in the cranium do not give rise to endolymph volume disturbance in the cochlea, a finding that is not in keeping with the notion of the endolymphatic sac being sensitive to small pressure changes. In contrast, when an outlet is present, a very small pressure difference is generated across the tissues between scala tympani and scala vestibuli, which produces the flow through the scalae to the outlet. The endolymph disturbance that we observe is apparently a consequence of this extremely small pressure difference. This explanation is supported by experiments in which injections were performed into scala vestibuli with an outlet in scala tympani. This condition produced the opposite result, with an endolymph displacement toward the apex and an EP increase that reversed when the injection stopped. These findings suggest the real possibility that endolymphatic volume may be more sensitive to disturbance under abnormal conditions where pressures are not adequately balanced between the perilymphatic scalae. These findings may be relevant to clinical conditions such as perilymph fistula or large vestibular aqueduct syndrome.

Acknowledgement

These studies were supported by research grant number 5 RO1 DC01368 from the National Institute on Deafness and other Communication Disorders, National Institutes of Health.

References

Bohmer A (1993) Hydrostatic pressure in the inner ear fluid compartments and its effects on inner ear function. Acta Otolaryngol Suppl 507: 3–24.

Bohmer A (1996) Experimental endolymphatic hydrops and inner ear pressure. In Ernst A, Marchbanks R, and Samii M (eds) Intracranial and Intralabyrinthine Fluids. Berlin/Heidelberg: Springer Verlag.

Couloigner V, Loiseau A, Sterkers O, Amiel C, Ferrary E (1995) Electrochemical composition of the luminal fluid in the endolymphatic sac and genesis of the trans-epithelial potential. Proceedings of the 18th Midwinter Meeting of the Association for Research in Otolaryngology, 1995, 29.

Demott J E and Salt A N (1997) Undetectable pressure increase during induction of acute endolymphatic hydrops by microinjection. Proceedings of the 20th Midwinter Meeting of the Association for Research in Otolaryngology, 12.

Friberg U, Wackym PA, Bagger-Sjoback D, Rask-Andersen H (1986) Effect of labyrinthectomy on the endolymphatic sac. Acta Otolaryngol 101: 172–82.

Guild S R (1927) The circulation of the endolymph. Am J Anat 39: 57–81.

Kimura R S and Schuknecht H F (1965) Membranous hydrops in the inner ear of the guinea pig after obliteration of the endolymphatic sac. Pract Oto-Rhino-Laryngol 27: 343–54.

Konishi T, Hamrick PE and Walsh PJ (1978) Ion transport in guinea pig cochlea. Acta Otolaryngol 86: 22–34.

Lawrence M, Wolsk D, Litton WB (1961) Circulation of the inner ear fluids. Ann Otol 70: 753–76.

Lundquist PG, Kimura R, Wersall J (1964) Experiments in endolymph circulation. Acta Otolaryngol 188 :194–201.

Naftalin L, Harrison MS (1958) Circulation of the labyrinthine fluids. J Laryngol Otol 72: 118–36.

Salt AN and Thalmann R (1989) Rate of longitudinal flow of cochlear endolymph. In Nadol JB (ed.) Ménière's Disease. Amsterdam: Kugler, pp. 69–73.

Salt AN, DeMott JE (1995) Endolymph volume changes during osmotic dehydration measured by two marker techniques. Hear Res 1995; 90: 12-23.

Salt AN, DeMott JE (1997a) Longitudinal endolymph flow associated with acute volume increase in the guinea pig cochlea. Hear Res 107: 29–40.

Salt AN, Demott JE (1997b) Longitudinal endolymphatic movements induced by perilymphatic injections. Hear Res.

Takeuchi S, Takeda T, Saito H (1991) Pressure relationship between perilymph and endolymph associated with endolymphatic infusion. Ann Otol Rhinol Laryngol 100: 244–8.

Takumida M, Bagger-Sjoback D, Rask-Andersen H (1989) The endolymphatic sac and inner ear homeostasis I: effect of glycerol on the endolymphatic sac with or without colchicine pretreatment. Hear Res 40: 1–16.

Takumida M, Hirakawa K and Harada Y (1995) Effect of glycerol on the guinea pig inner ear after removal of the endolymphatic sac. ORL 57: 5–9.

Wangemann P, Schacht J (1996) Homeostatic mechanisms in the cochlea. In Dallos P, Popper AN, Fay RR (1996) The Cochlea. Springer-Verlag, pp. 130–85.

*Correspondence/reprint requests to Alec N Salt, Department of Otolaryngology, Washington University School of Medicine, 517 South Euclid Avenue, St Louis MO 63110, USA. Tel. (+1)314-362-7560. Fax (+1)314-362-7522.
E-mail salt_a@kids.wustl.edu.*

Chapter 15
Cellular mechanisms of volume control in the labyrinth

PHILINE WANGEMANN, Cell Physiology Laboratory, Boys Town National Reasearch Hospital, Omaha, USA

ABSTRACT. The transduction of sound into nerve impulses in the cochlea is dependent on a standing current that is generated by the stria vascularis. The stria vascularis is a multi-layered epithelium and is part of the epithelial barrier between endolymph and perilymph. The present model, designed to explain the generation of the endocochlear potential, assumes that the molecular mechanism for the generation of the endocochlear potential is a K^+ conductance localized in the inner membrane of basal cells or in intermediate cells. The endocochlear potential would thus be generated by this K^+ conductance in conjunction with a steep K^+ concentration gradient across the cell membranes harbouring the K^+ conductance. The concept that basal and/or intermediate cells are responsible for the generation of the endocochlear potential contrasts with previous models, which had focused solely on strial marginal cells. Strial marginal cells, however, play an indirect role in the generation of the endocochlear potential. This role is limited to the maintenance of a low K^+ concentration in the intrastrial space by absorbing K^+ from the intrastrial space and secreting it into endolymph. The molecular mechanisms for K^+ transport in strial marginal cells are now well established. Strial marginal cells absorb K^+ from the intrastrial space via the Na/K-ATPase and the $Na^+/2Cl^-/K^+$-cotransporter and secrete it across the apical membrane via a K^+ channel, the I_{sK} channel. The recently introduced concept that strial marginal cells transport K^+ but that the mechanism for the generation of the endocochlear potential is located in another cell type provided the basis to hypothesize that ion transport mechanisms and their regulation are similar in vestibular dark cells and strial marginal cells. Indeed, transport mechanisms in strial marginal cells and *vestibular dark cells are not only qualitatively similar but have far reaching similarities on the molecular level. K^+ secretion into endolymph by strial marginal cells and vestibular dark cells depends on the I_{sK} channel in their apical membrane. Strial marginal cells and vestibular dark cells of mice lacking this IsK channel have been shown to be unable to secrete K^+ and unable to maintain an appropriate volume of endolymph. Consequently, the endolymphatic compartment throughout the cochlear and vestibular labyrinth appeared collapsed in these mice due to ongoing reabsorptive processes during an uncompensated failure of K^+ secretion. Cellular mechanisms of volume control in the labyrinth thus consist of secretory and reabsorptive processes. Sites of K^+ secretion include strial marginal cells in the cochlea and vestibular dark cells in the vestibular labyrinth. Reabsorptive processes are currently under investigation. Sites include, but may not be limited to, the sensory hair cells, Reissner's membrane and outer sulcus cells in the cochlea and vestibular transitional cells in the vestibular labyrinth.*

Key words: cochlea, endolymph, perilymph, cellular mechanisms, volume control, labyrinth, stria vascularis.

Generation of the endocochlear potential

The transduction of sound into nerve impulses in the cochlea is dependent on a standing current that is generated by stria vascularis (Dallos 1973; Zidanic and Brownell 1990) (Figure 15.1). Stria vascularis is a multilayered epithelium that is part of the epithelial barrier between endolymph and perilymph. Endolymph and perilymph are the two most prominent extracellular fluids in the inner ear. Perilymph, like most lymphatic fluids, consists mainly of 150 mM NaCl whereas endolymph is an unconventional extracellular fluid consisting mainly of 150 mM KCl (Wangemann and Schacht 1996). Stria vascularis consists of two barriers that enclose the intrastrial space. The ionic composition of the intrastrial fluid is thought to be similar to perilymph (Salt et al. 1987). The barrier between endolymph and the intrastrial space is formed by an epithelial monolayer consisting of strial marginal cells and the barrier between the intrastrial space and perilymph is formed by basal cells, which, like strial marginal cells, are connected to each other by tight junctions (Jahnke 1975).

Indirect evidence supports the concept that stria vascularis generates the endocochlear potential of about 80 mV. The current model designed to explain the generation of the endocochlear potential assumes that the molecular mechanism for the generation of the endocochlear potential is a K^+ conductance localized in the inner membrane of basal cells or in intermediate cells (Salt et al. 1987; Schulte and Steel 1994; Wangemann 1995). Accordingly, the endocochlear potential would be a K^+ equilibrium

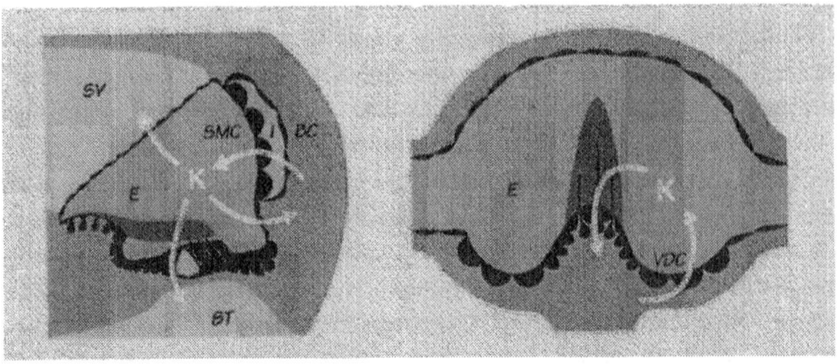

Figure 15.1: pathways of K^+ secretion and K^+ reabsorption in the cochlea and vestibular labyrinth. Schematic cross-section of the cochlear duct (left) and vestibular ampulla (right). Abbreviations: Scala vestibuli (SV), scala tympani (ST), endolymph (E), strial marginal cells (SMC), intrastrial compartment (I), basal cells (BC), vestibular dark cells (VDC).

potential, which is generated by a K^+ conductive cell membrane facing a high K^+ concentration on the cytosolic side and a low K^+ concentration on the extracellular side (Figure 15.2). The larger the concentration difference is between the cytosol and the intrastrial space, the larger would be the endocochlear potential. The hypothesis that intermediate cells may be involved in the generation of the endocochlear potential is based on the finding that mutants lacking intermediate cells lack the customary large positive endocochlear potential (Schulte and Steel 1994). Intermediate cells are connected to basal cells through a dense group of gap junctions suggesting that the membrane of intermediate cells may be a functional extension of the inner membrane of basal cells and thus part of the basal cell barrier across which the endocochlear potential is generated (Kikuchi et al. 1995). The concept that basal and/or intermediate cells are responsible for the generation of the endocochlear potential contrasts with previous models, which had assumed that strial marginal cells are directly responsible for the generation of the endocochlear potential (Offner et al. 1987). A direct role of strial marginal cells in the generation of the endocochlear potential, however, has now been ruled out (Wangemann et al. 1995a). Even though the concept that stria vascularis generates the endocochlear potential is well accepted, the cellular mechanisms involved are still unknown and the current model is speculative in this aspect (Wangemann and Schacht 1996; Wangemann 1995). Further investigations are warranted.

Strial marginal cells secrete K^+ into endolymph

Strial marginal cells, according to the current model, play an indirect role in the generation of the endocochlear potential. This role is limited to the maintenance of a low K^+ concentration in the intrastrial space by absorbing K^+ from the intrastrial space and secreting it into endolymph (Salt et al. 1987). The molecular mechanisms of K^+ transport in strial marginal cells are now well established (Wangemann et al. 1995a; Wangemann 1995) (Figure 15.3). Strial marginal cells take up K^+ from the intrastrial compartment via the Na/K-ATPase and the $Na^+/2Cl^-/K^+$-cotransporter. Na^+ and Cl^- recycle across the basolateral membrane via the Na/K-ATPase and via Cl^- channels (Takeuchi et al. 1995). K^+ taken up across the basolateral membrane is secreted across the apical membrane via a K^+ channel, the I_{sK} channel (Wangemann et al. 1995b; Shen et al. 1997). K^+ secretion into endolymph is driven by the voltage difference across the apical membrane of strial marginal cells, which *in vivo* is up to 10 mV positive with respect to scala media (Offner et al. 1987). This mechanism of transepithelial K^+ transport is theoretically very energy efficient. For each ATP hydrolysed by the Na/K-ATPase, 5 K^+ are transported from the intrastrial space into endolymph assuming that the stoichiometries of the Na/K-ATPase and the $Na^+/2Cl^-/K^+$-cotransporter are the usual 3 Na^+ to 2 K^+ and 1 Na^+ to 2 Cl^- to 1 K^+, respectively.

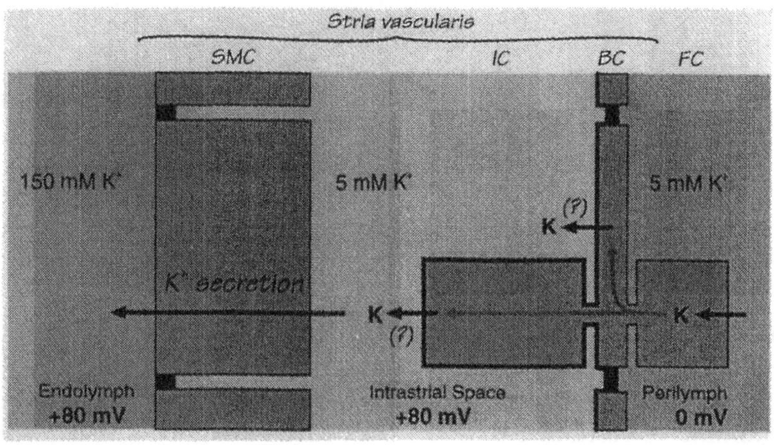

Figure 15.2: generation of endocochlear potential. Schematic diagram. The molecular mechanism for the generation of the endocochlear potential is assumed to be K^+ conductance in the inner membrane of basal cells (BC) or in intermediate cells (IC). The membrane barrier across generating the endocochlear potential is highlighted. Further abbreviations: SMC = strial marginal cells, FC = fibrocytes in spiral ligament.

Similarity between strial marginal cells and vestibular dark cells

Morphological similarities between strial marginal cells and vestibular dark cells have long been recognized (Kimura, 1969) and it has long been accepted that both of these cell types are involved in the secretion of K^+ into endolymph. Functional similarities of these two epithelia, however, were considered unlikely as long as strial marginal cells were assumed to generate the endocochlear potential which has no equivalent in the vestibular labyrinth. The concept that strial marginal cells transport K^+ but that the mechanism for the generation of the endocochlear potential is located in another cell type provided the basis to hypothesize that ion transport mechanisms and their regulation are similar in vestibular dark cells and strial marginal cells (Wangemann, 1995). In both cells, K^+ is taken up across the basolateral membrane by the $Na^+/2Cl^-/K^+$-cotransporter and the Na/K-ATPase and secreted across the apical membrane via the IsK channel (Figure 15.3). Na^+ and Cl^- taken up via the $Na^+/2Cl^-/K^+$-cotransporter are transported out via the basolateral Na/K-ATPase and released via basolateral Cl^- channels, respectively.

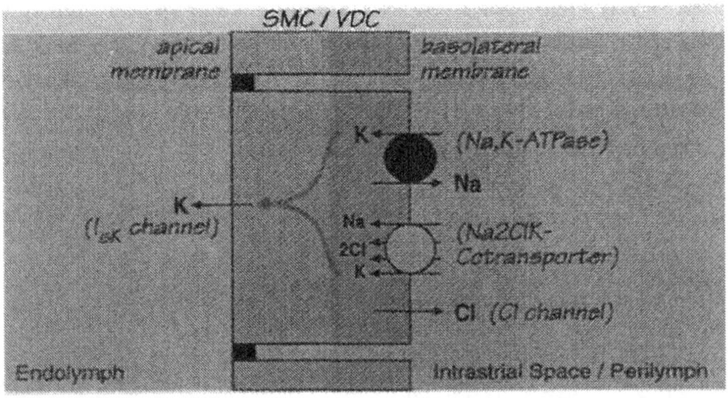

Figure. 15.3: K^+ secretion in strial marginal cells (SMC) and vestibular dark cells (VDC). Schematic diagram.

Transport mechanisms in strial marginal cells and vestibular dark cells are not only qualitatively similar but have far-reaching similarities on the molecular level (Wangemann, 1995). The Na/K-ATPase in the basolateral membrane of vestibular dark cells and strial marginal cells consists in both cell types of the same subtypes (Schulte and Steel, 1994; Ryan et al. 1994) and is in both cells responsible for uptake of K^+ and for establishing a Na^+ gradient across the basolateral membrane. This Na^+ gradient drives in both cell types additional K^+ uptake via the $Na^+/2Cl^-/K^+$-cotransporter (Wangemann and Marcus 1990; Wangemann et al. 1995a; Marcus et al. 1987). Further, the basolateral membrane of both cell types is primarily Cl^- conductive and the apical membrane of both cell types is primarily K^+ conductive (Marcus and Shen 1994; Wangemann and Marcus 1992; Wangemann et al. 1995a). The basolateral Cl^- conductance consists in both cell types primarily of a 80–95 pS Cl^- channel (Wangemann and Marcus 1992; Marcus et al. 1993; Takeuchi et al. 1995) and the apical K^+ conductance consists in both cell types of the I_{sK} channel (Marcus and Shen 1994; Sakagami et al. 1991; Sunose et al. 1994; Shen et al. 1997; Wangemann 1995). The apical I_{sK} channel is in both cell types essential for K^+ secretion and the formation of endolymph (Vetter et al. 1996).

The I_{sK} channel is essential for the formation of endolymph

The maintenance of the high K^+ concentration in endolymph is crucial for the function of the cochlear and vestibular sensory cells of the inner ear. Indeed, sensory hair cells are lost when K^+ secretion fails (Vetter et al. 1996). The loss of sensory cells is most likely due to a flooding of the cells with Na^+ via the apical transduction channel and a collapse of the Na^+ gradient across their basolateral membrane which disables Ca^{2+} extrusion mechanisms and may lead to Ca^{2+}-induced cell death. Thus, the sensory cells depend on the high K^+ concentration in endolymph. K^+ secretion into endolymph by strial marginal cells and vestibular dark cells has been shown to be dependent on the I_{sK} channel in their apical membrane (Shen et al. 1997; Marcus and Shipley 1994). Strial marginal cells and vestibular dark cells of mice lacking the I_{sK} channel have been shown to be unable to secrete K^+ and to maintain an appropriate volume of endolymph (Vetter et al. 1996). Consequently, the endolymphatic compartment throughout the cochlear and vestibular labyrinth appeared collapsed (Figure 15.4) since endolymph volume is governed by reabsorptive as well as secretory processes. Ongoing reabsorptive processes during an uncompensated failure of K^+ secretion led most likely to the observed collapse in endolymph volume.

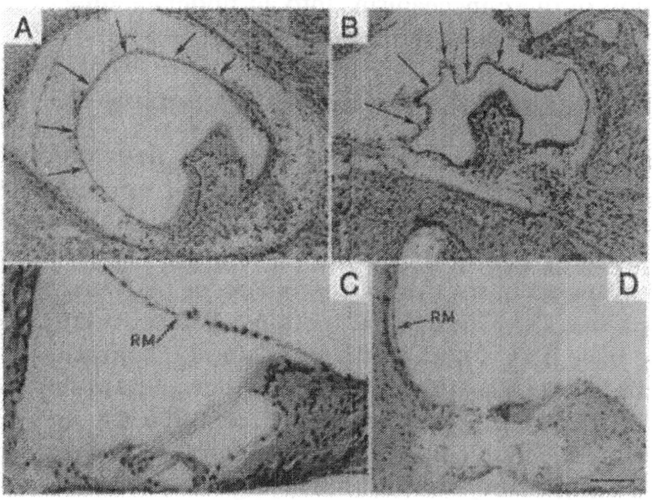

Figure 15.4: the endolymphatic compartment appeared collapsed in mice lacking the IsK channel. The ampulla (A) and cochlea (C) of a heterozygote control mouse appeared normal although this mouse had only one I_{sK} allele. In contrast, the ampulla (B) and cochlea (D) of a homozygote mouse lacking the I_{sK} gene appeared collapsed. Reprinted from Vetter et al. with permission from Cell Press.

The collapsed endolymphatic compartment of mice lacking the I_{sK} channel resembles Jervell and Lange-Nielsen cardio-auditory syndrome, a rare disease that combines congenital bilateral deafness with a pathologically prolonged QT interval of the heart action potential leading to life-threatening cardiac arrhythmia. A defect of the I_{sK} channel was a likely candidate for a cause of this disease since the I_{sK} channel is responsible for K^+ secretion in the inner ear and contributes to the repolarization of the heart. Indeed, recent studies have determined that the I_{sK} channel consists not only of the I_{sK} gene product but also of KVLQT1 (Barhanin et al. 1996; Roberts et al. 1994) and that 12–50% of patients with long-QT syndrome have mutations in KVLQT1 (Tanaka et al. 1997; Wang et al. 1996). It has been speculated that specific mutations in KVLQT1 will lead in the heterozygous form to long-QT syndrome and in the homozygous form to Jervell and Lange-Nielsen syndrome (Mannens and Wilde 1997). This speculation was based on the finding that heterogeneous carriers of the Jervell and Lange-Nielsen syndrome show slight to moderate prolongation of the QT interval (Fraser et al. 1964).

Regulation of K^+ secretion

The rate of K^+ secretion in strial marginal cells and vestibular dark cells is under the control of hormones and local mediators. Purinergic (Liu et al. 1995), cell volume-mediated (Wangemann et al. 1995c), β– adrenergic (Wangemann and Liu 1996) and K^+-mediated regulation of K^+ secretion (Wangemann et al. 1996) are most well known. A satisfactory integration of these mechanisms into an understanding of labyrinthine fluid regulation is not yet available.

Adrenergic regulation of K^+ secretion

Norepinephrine and epinephine are both neurotransmitters and hormones. Norepinephrine has been shown to be present in the cochlea and vestibular labyrinth (Gil-Loyzaga et al. 1997; Gil-Loyzaga and Pares-Herbute 1989). Receptors for norepinephrine and epinephrine are termed adrenoceptors. Adrenoceptors occur in a variety of types and subtypes that have different functions: β-adrenoceptors of the β1-subtype have been found in the basolateral membranes of vestibular dark cells and strial marginal cells where they mediate an increase in the rate of K^+ secretion (Wangemann and Liu 1996; Liu and Wangemann 1997). It is conceivable that β1-adrenoceptors in vestibular dark cells are part of a regulatory mechanism designed to maintain a constant endolymphatic K^+ concentration in the presence of varying levels of vestibular transduction.

Transduction in hair cells is known to be associated with an outflow of K^+ from endolymph into perilymph via the sensory hair cells

(Johnstone et al. 1989; Valli et al. 1990). This loss of K^+ from endolymph must be restored to maintain a constant endolymphatic K^+ concentration. An ideal regulatory mechanism for endolymphatic K^+ homeostasis would conceivably not only work as a feedback mechanism to restore K^+ after it has been lost from endolymph but would anticipate a loss and increase the capacity for K^+ secretion before the K^+ is actually lost from endolymph. The recently identified β1-adrenoeptors in vestibular dark cells may be part of such a mechanism. Support for this hypothesis comes from the finding that the norepinephrine concentration in plasma (0.7 nM) increases by a factor of two when a human subject rises from the supine position to the standing position (Lake et al. 1997). Clearly, a high frequency of head movements involving an elevated level of vestibular stimulation is more likely to occur when the subject is alert and standing than when a subject is resting in the supine position. The increase in the plasma norepinephrine concentration would most likely arrive at the β1-adrenoeptors in vestibular dark cells since the blood-perilymph barrier is not a strict barrier and has been shown to be permeable to organic compounds of similar molecular weight (Jung et al. 1989). This hypothesis is supported by the observation that norepinephrine at concentrations comparable to those present in plasma stimulates K^+ secretion in vestibular dark cells (Wangemann and Liu 1996; Liu and Wangemann 1997).

K^+-mediated regulation of K^+ secretion

Strial marginal cells are, according to the present model, indirectly involved in the generation of the endocochlear potential by maintaining the low K^+ concentration in the intrastrial space. By homology, vestibular dark cells maintain the low K^+ concentration of perilymph. A requirement of these functions is that the rate of K^+ secretion is adjusted to small variations in the K^+ concentration in the intrastrial space or in perilymph. Indeed, it has been shown that the rate of K^+ secretion by vestibular dark cells is most sensitively regulated by the basolateral K^+ concentration in the physiologically relevant range between 3.5 and 6.5 mM K^+ (Wangemann et al. 1996). Thus, vestibular dark cells play a critical role in the homeostasis of the perilymphatic K^+ concentration. It has been suggested that fibrocytes in the connective tissue beneath vestibular dark cells form a network which aids transport of K^+ from the basolateral membrane of hair cells toward the basolateral membrane of vestibular dark cells. This hypothesis was based on the finding that fibrocytes are connected to each other by gap junctions (Kikuchi et al. 1994) and that they contain Na/K-ATPase as a potential uptake mechanism of K^+ from perilymph (Schulte and Steel 1994). If such a network mechanism existed, it would be conceivable that the amplitude (0.2 mM) of the K^+ concentration changes measured in perilymph during mechanical stimulation of hair cells (Valli et al. 1990) is a significant underestimation of the K^+ concentration changes, which occur at the basolateral membrane of vestibular dark cells. Similar sensitive mechanisms for K^+-mediated regulation of K^+ secretion may be present in the cochlea. Indeed, a gap junction network has been identified in the cochlea (Kikuchi et al. 1995) and K^+ secretion by strial marginal cells has been shown to be stimulated by an increase in the basolateral K^+ concentration (Wangemann et al. 1995a).

Cellular mechanisms of volume control in the labyrinth

Volume control in the labyrinth is necessary on different levels. On the organ level the control of labyrinthine fluid volumes is important to ensure the optimal mechanical properties for the transduction process and on the cellular level volume control is crucial to ensure optimal cell function. Cell volume regulation is especially critical in cells that transport ions at a high rate such as vestibular dark cells and strial marginal cells. High rates of ion transport pose a threat to the maintenance of a constant cell volume as water follows the transport of ions. Small imbalances between uptake and release of ions can cause major changes in cell volume.

In response to cell shrinking, vestibular dark cells have been shown to perform a regulatory volume increase that involves uptake of ions via the $Na^+/2Cl^-/K^+$-cotransporter. This uptake of ions is followed by water to restore cell volume (Wangemann and Shiga 1994). Similarly, in response to swelling, the cells perform a regulatory volume decrease that involves the release of K^+ and Cl^- with water following (Wangemann and Shiga 1994; Shiga and Wangemann 1995). Cl^- is released via Cl^- channels in the basolateral membrane (Wangemann et al. 1995c; Shiga and Wangemann 1995) and K^+ is released in part via the IsK channel in the apical membrane and in part via a yet undefined mechanism most likely located in the basolateral membrane.

Interestingly, the I_{sK} channel, which is essential for K^+ secretion (see above), is not essential for regulatory volume decrease. Indeed, this regulatory response occurs in vestibular dark cells of mice lacking the I_{sK} channel (Vetter et al. 1996). Thus, the apical I_{sK} channel and the putative basolateral K^+ channel provide alternative K^+ release mechanisms that may allow vestibular dark cells to meet the challenge to regulate their own cell volume, to maintain a constant K^+ concentration in endolymph and to regulate endolymph volume during a hypoosmotic challenge.

In the event that perilymph becomes hypotonic, water will enter cells including vestibular dark cells. Water will also enter the endolymphatic compartment causing swelling which, if not prevented, will lead to an endolymphatic hydrops. Regulatory mechanisms are thus necessary not only to maintain a constant cell volume but also to maintain a constant endolymph volume. Vestibular dark cells will reduce their volume by releasing K^+ and Cl^-. Swelling-induced stimulation of the I_{sK} channel will allow an improved equilibration between the K^+ concentration in endolymph and in cytosol and will increase the K^+ permeability of the barrier enclosing endolymph. An increase in this K^+ permeability will, according to Starling's principle, reduce the osmotic pressure difference that drives water influx into endolymph. In addition, opening of the putative basolateral K^+ release mechanism may be able to drain K^+ from endolymph and thereby reduce endolymph volume. Further studies are necessary to verify these hypothetical mechanisms of endolymph volume regulation.

Endolymph volume regulation

Mechanisms of endolymph volume control in the labyrinth consist of secretory and reabsorptive processes (Figure 15.1). Strial marginal cells and vestibular dark cells secrete K^+. K^+ reabsorption occurs via the cochlear and vestibular sensory hair cells since the transduction current is carried mainly by K^+ (Valli et al. 1990; Johnstone et al. 1989). In addition, Reissner's membrane has been shown to be another site of K^+ reabsorption in the cochlea (Salt and Ohyama 1993; Zidanic and Brownell 1990). Preliminary evidence suggests that outer sulcus cells in the cochlea and transitional cells in the vestibular labyrinth are additional sites which reabsorb K^+ from endolymph (Marcus et al. 1997). Reabsorptive processes in the cochlea and vestibular labyrinth are currently under investigation.

References

Barhanin J, Lesage F, Guillemare E, Fink M, Lazdunski M, Romey G (1996) KVLQT1 and IsK (minK) proteins associate to form the I_{sK} cardiac potassium current. Nature 384: 78-80.

Dallos P (1973) Cochlear potentials. In Dallos P (ed.) The Auditory Periphery: Biophysics and Physiology. New York: Academic Press, pp. 218-390.

Fraser GR, Froggatt P, Murphy T (1964) Genetical aspects of the cardio-auditory syndrome of Jervell and Lange-Nielsen (congenital deafness and electrocardiographic abnormalities). Ann Hum Genet 28:133–57.

Gil-Loyzaga P, Vicente-Torres MA, Garcia-Bonacho M, Esquifino A (1997) Presence of catecholamines and serotonin in the rat vestibule. Brain Res 746: 265–8.

Gil-Loyzaga P, Pares-Herbute N (1989) HPLC detection of dopamine and noradrenaline in the cochlea of adult and developing rats. Brain Res Dev Brain Res 48:157–60.

Jahnke K (1975) The fine structure of freeze-fractured intercellular junctions in the guinea pig inner ear. Acta Otolaryngol Suppl (Stockh) 336: 1–40.

Johnstone BM, Patuzzi R, Syka J, Sykova E (1989) Stimulus-related potassium changes in the organ of Corti of guinea-pig. J Physiol (Lond) 408: 77–92.

Jung WK, Gattaz G, Schön FJ (1989) Kinetic experiments with radionucleotides concerning the perilymph-blood barrier in a guinea pig model. Arch Otolaryngol 246: 11–19.

Kikuchi T, Adams JC, Paul DL, Kimura RS (1994) Gap junction systems in the rat vestibular labyrinth: immunohistochemical and ultrastructural analysis. Acta Otolaryngol (Stockh) 114: 520–8.

Kikuchi T, Kimura RS, Paul DL, Adams JC (1995) Gap junctions in the rat cochlea: immunohistochemical and ultrastructural analysis. Anat Embryol (Berl) 191: 101–18.

Kimura RS (1969) Distribution, structure, and function of dark cells in the vestibular labyrinth. Ann Otol Rhinol Laryngol 78: 542–61.

Lake CR, Chenow B, Feuerstein G, Goldstein DS, Ziegler MG (1997) The sympathetic nervous system in man: its evaluation and the measurement of plasma NE. In Ziegler MG and Lake CR (eds) Norepinephrine. Baltimore/London: Williams & Wilkins, pp. 1–26.

Liu J, Kozakura K, Marcus DC (1995) Evidence for purinergic receptors in vestibular dark cell and strial marginal cell epithelia of the gerbil. Audit Neurosci 1: 331–40.

Liu J, Wangemann P (1997) K^+ secretion in vestibular dark cells and strial mariginal cells from the gerbil is stimulated via beta-1 adrenergic but not via vasopressin receptors. Assoc Res Otolaryngol 20: (Abstract).

Mannens M, Wilde A (1997) KVLQT1, the rhythm of imprinting [news; comment]. Nat Genet 1997; 15: 113–15.

Marcus DC, Marcus NY, Greger R (1987) Sidedness of action of loop diuretics and ouabain on nonsensory cells of utricle: a micro-Ussing chamber for inner ear tissues. Hear Res 30: 55–64.

Marcus DC, Takeuchi S, Wangemann P (1993) Two types of chloride channel in the basolateral membrane of vestibular dark cell epithelium. Hear Res 69: 124–32.

Marcus DC, Chiba T, Wangemann P (1997, in press) Vibrating probe measurements demonstrate active cation reabsorption in inner ear epithelia. J Gen Physiol.

Marcus DC, Shen Z (1994) Slowly activating, voltage-dependent K^+ conductance is apical pathway for K^+ secretion in vestibular dark cells. Am J Physiol 267: C857–C864.

Marcus DC, Shipley A (1994) Potassium secretion by vestibular dark cell epithelium demonstrated by vibrating probe. Biophys J 66: 1939–42.

Offner FF, Dallos P, Cheatham MA (1987) Positive endocochlear potential: mechanism of production by marginal cells of stria vascularis. Hear Res 29: 117–24.

Roberts SK, Yano M, Ueno Y, et al. (1994) Cholangiocytes express the aquaporin CHIP and transport water via a channel-mediated mechanism. Proc Natl Acad Sci USA 91: 13009–13.

Ryan AF, Luo L, Fina M, Hepler KM (1994) Gene expression in the tissues of the endolymphatic compartment. Proc Sendai Symp 4: 47–52.

Sakagami M, Fukazawa K, Matsunaga T, et al. (1991) Cellular localization of rat Isk protein in the stria vascularis by immunohistochemical observation. Hear Res 56: 168–72.

Salt AN, Melichar I, Thalmann R (1987) Mechanisms of endocochlear potential generation by stria vascularis. Laryngoscope 97: 984–91.

Salt AN, Ohyama K (1993) Accumulation of potassium in scala vestibuli perilymph of the mammalian cochlea. Ann Otol Rhinol Laryngol 102: 64–70.

Schulte BA, Steel KP (1994) Expression of alpha and beta subunit isoforms of Na, K-ATPase in the mouse inner ear and changes with mutations at the Wv or Sld loci. Hear Res 78: 65-76.

Shen Z, Marcus DC, Sunose H, Chiba T, Wangemann P (1997) IsK channel in strial marginal cell: voltage-dependence, ion selectivity, inhibition by 293B and sensitivity to clofilium. Audit Neurosci 3: 215–30.

Shiga N, Wangemann P (1995) Ion selectivity of volume regulatory mechanisms present during a hypoosmotic challenge in vestibular dark cells. Biochim Biophys Acta 1240: 48–54.

Sunose H, Ikeda K, Suzuki M, Takasaka T (1994) Voltage-activated K channel in luminal membrane of marginal cells of stria vascularis dissected from guinea pig. Hear Res 80: 86–92.

Takeuchi S, Ando M, Kozakura K, Saito H, Irimajiri A (1995) Ion channels in basolateral membrane of marginal cells dissociated from gerbil stria vascularis. Hear Res 83: 89–100.

Tanaka T, Nagai R, Tomoike H, et al. (1997) Four novel KVLQT1 and four novel HERG mutations in familial long-QT syndrome. Circulation 95: 565–7.

Valli P, Zucca G, Botta L (1990) Perilymphatic potassium changes and potassium homeostasis in isolated semicircular canals of the frog. J Physiol (Lond) 430: 585–94.

Vetter DE, Mann JR, Wangemann P, et al. (1996) Inner ear defects induced by null mutation of isk gene. Neuron 17: 1251–64.

Wang Q, Curran ME, Splawski I, et al. (1996) Positional cloning of a novel potassium channel gene: KVLQT1 mutations cause cardiac arrhythmias. Nat Genet 12: 17–23.

Wangemann P (1995) Comparison of ion transport mechanisms between vestibular dark cells and strial marginal cells. Hear Res 90: 149–57.

Wangemann P, Liu J, Marcus DC (1995a) Ion transport mechanisms responsible for K^+ secretion and the transepithelial voltage across marginal cells of stria vascularis in vitro. Hear Res 84: 19–29.

Wangemann P, Liu J, Shen Z, Marcus DC (1995b) Similarity of ion transport properties between vestibular dark cells and strial marginal cells. Proc Sendai Symp 5: 145–50.

Wangemann P, Liu J, Shen Z, Shipley A, Marcus DC (1995c) Hypo-osmotic challenge stimulates transepithelial K^+ secretion and activates apical IsK channel in vestibular dark cells. J Membr Biol 147: 263–73.

Wangemann P, Shen Z, Liu J (1996) K^+-induced stimulation of K^+ secretion involves activation of the IsK channel in vestibular dark cells. Hear Res 100: 201–10.

Wangemann P, Liu J (1996) Beta-adrenergic receptors but not vasopressin-receptors stimulate the equivalent short circuit current in K^+ secreting inner ear epithelial cells. J Gen Physiol 108:13a (Abstract).

Wangemann P, Marcus DC (1990) K^+-induced swelling of vestibular dark cells is dependent on Na^+ and Cl^- and inhibited by piretanide. Pflügers Arch 416: 262–9.

Wangemann P, Marcus DC (1992) The membrane potential of vestibular dark cells is controlled by a large Cl^- conductance. Hear Res 62: 149–56.

Wangemann P, Schacht J (1996) Cochlear Homeostasis. In Dallos P, Popper AN, Fay R (ed.) Springer Handbook of Auditory Research: The Cochlea. New York: Springer, pp. 130–85.

Wangemann P, Shiga N (1994) Cell volume control in vestibular dark cells during and after a hyposmotic challenge. Am J Physiol Cell Physiol 266: C1046–C1060.

Zidanic M, Brownell WE (1990) Fine structure of the intracochlear potential field. I. The silent current. Biophys J 57: 1253–68.

Acknowledgments

Supported by research grant R01-DC01098 to P.W. from the National Institutes on Deafness and Other Communication Disorders, National Institutes of Health.

Correspondence to Philine Wangemann, Associate Professor, Cell Physiology Laboratory at Boys Town National Research Hospital, 555 North 30th Street, Omaha NE 68131, USA. Tel. (+1)402 498-6721. Fax (+1)402 498-6351. Email: wange@boystown.org.

Chapter 16
Non-invasive perilymphatic pressure measurements using the MMS-10 Tympanic Displacement Analyser in patients with Ménière's disease

HJ ROSINGH, HP WIT, FWJ ALBERS, Department of Otorhinolaryngology, University Hospital Groningen, Netherlands

ABSTRACT. Non-invasive perilymphatic pressure was performed in patients with Ménière's disease and in normal hearing subjects using the MMS-10 Tympanic Displacement Analyser. No significant differences were found between the perilymphatic pressure measurement results with patients and normal hearing controls. Following a rapid change of body position, changes in intracranial cerebrospinal fluid pressure were found to influence perilymphatic pressure within one minute. With respect to the time course of perilymphatic pressure changes, as measured non-invasively, no significant differences were found between affected ears, non-affected ears and control ears.

Key words: inner ear pressure, non-invasive perilymphatic pressure measurement, Ménière's disease.

Introduction

The hydrostatic pressure of the endolymph and perilymph forms a complex facet in the physiology and pathophysiology of the inner ear. Inner ear hydrostatic pressure alternations can affect audiovestibular symptoms (Allen et al. 1971; Böhmer 1993; Densert et al. 1975; Densert et al. 1995; Sismanis 1987; Walsted 1991). The regulation of inner-ear pressure is complex. The hydrostatic pressure is regulated by the production and resorption of the fluids, osmotic and metabolic processes and external influences. Anatomically, the labyrinth is a rigid cavity with several communications to the environment. The oval and round windows form the border with the middle ear. The cochlear aqueduct forms together with the endolymphatic duct and sac the connection with the cranial cavity (Palva 1969). The cochlear aqueduct transverses the petrous bone from the perilymphatic compartment of the scala tympani to the cerebrospinal fluid compartment of the fossa posterior. Through the cochlear aqueduct perilymphatic pressure is directly influenced by the intracranial cerebrospinal fluid pressure (Marchbanks 1990; Beentjes 1972). The hydrostatic pressure has been extensively measured in animal experiments using invasive measurement methods. However, these invasive methods are not suitable for human inner ears. Recently the MMS-10 Tympanic Displacement Analyser (Marchbanks Measurement Systems Ltd, United Kingdom) has been developed, which is a commercially available device for non-invasive perilymphatic pressure measurements in human inner ears (Reid et al. 1989; Reid et al. 1990; Marchbanks 1991). The measurement method is based on the principle that movements of the tympanic membrane, induced by the stapedial reflex, produce small volume displacements in the external auditory canal.

The displacement of the tympanic membrane after contraction of the stapedius muscle depends on the resting position of the stapes footplate within the oval window and this alters with different perilymphatic pressures. The displacement of the tympanic membrane after contraction of the stapedius muscle can be quantified by two measurements parameters, Vi and Vm, both expressed in nanolitres. Vi is defined as the maximum inward displacement of the tympanic membrane while the acoustic stimulus is present and Vm is the mean displacement from the time at which Vi is reached until the acoustic stimulus offset. Large negative values of Vi and Vm are thought to correlate with high perilymphatic pressure, whereas Vi values near zero and positive Vm values correspond with normal or low perilymphatic pressures (Marchbanks 1991; Reid et al. 1989; Reid et al. 1990; Rosingh et al. 1996).

Ménière's disease is characterized by a hydrops of the endolymphatic compartment that leads to distension and bulging of Reissner's membrane. Distension and bulging of Reissner's membrane towards the perilymphatic space, as seen in endolymphatic hydrops, can occur only when a mechanical force acts from the endolymphatic to the perilymphatic compartment. In fluid compartments the acting force on a surface is identical with hydrostatic pressure (Böhmer 1993). In normal ears the endolymphatic and perilymphatic pressure are found to be equal, because the highly compliant Reissner's membrane equalizes every pressure difference by shifts towards the compartment with the lower pressure. However, in long-standing endolymphatic hydrops, as seen in Ménière's disease, endolymphatic perilymphatic pressure gradients may occur (Böhmer 1993).

In this study we performed extensive measurements in patients with Ménière's disease and in normal hearing subjects in a stable sitting position and following change of posture in order to detect possible differences in the dynamics of perilymphatic pressure between both groups.

Materials and methods

In this study two groups of patients with Ménière's disease were compared with normal hearing subjects. All patients suffered from a chronic Ménière's disease for at least one year. The diagnosis of Ménière's disease was defined by simultaneous fulfilment of three symptoms: cochlear hearing loss, tinnitus and periodic attacks of vertigo. During a four-day stay in hospital, the diagnosis of Ménière's disease was confirmed according a fixed diagnostic protocol. The diagnostic protocol consisted of routine ENT examination, audiovestibular tests, laboratory investigations and magnetic resonance imaging (MRI) of the temporal bones and the cerebellopontine angle.

Group 1 consisted of 25 patients (13 male, 12 female, mean age: 47 years, range: 27–64 years) from which 22 patients suffered from unilateral and three from bilateral Ménière's disease. The non-invasive perilymphatic pressure measurement results of group 1 were compared with a group of 50 normal hearing subjects (30 male, 20 female, mean age: 25 years, range 20–34). The perilymphatic pressure of all 25 patients and 50 normal hearing subjects (100 ears) were measured non-invasively by means of the MMS-10 tympanic displacement analyser. Test stimuli of 1000 Hz with intensities of 10, 20 and 25 dB above stapedial reflex threshold were used for the measurements. Stimulus duration was 500 milliseconds, measurement duration 1.5 seconds and the inter-stimulus interval was 7 seconds. Responses to 10 acoustic stimuli were averaged and Vi and Vm were determined.

In Group 2 the dynamics of perilymphatic pressure changes following posture changes were examined. Group 2 consisted of 27 patients with Ménière's disease (15 male, 12 female, mean age 47 years, range 28–68 years). Twenty patients suffered from unilateral Ménière's disease and seven from bilateral disease. The results of this group were compared with the results of 12 normal hearing subjects (eight male, four female, mean age: 24, range 18-32 years). Only ears with a clearly measurable stapedial reflex were included. Therefore 28 affected ears and 16 non-affected ears of patients and 18 ears of normal hearing subjects were included.

In all ears (patients and normal hearing subjects), the change in perilymphatic pressure after posture change was assessed non-invasively using the MMS-10 Tympanic Displacement Analyser. The perilymphatic pressure in each ear was measured seven times consecutively without removing the headset: (1) in upright-sitting position; (2) in supine position immediately after lying down; (3) 1 minute after lying down;

(4) 3 minutes after lying down; (5) 5 minutes after lying down; (6) 10 minutes after lying down, and finally (7) in an upright sitting position again. For each of the seven measurements, 10 test stimuli of 20 dB above stapedial reflex threshold were presented contralaterally. The acoustic stimulus presentation lasted 0.5 seconds and the inter-stimulus duration was 5 seconds. For each measurement the average tympanic membrane displacement of 10 test stimuli was quantified by two measurement parameters, Vi and Vm, both expressed in nanoliters.

The results of both groups of patients were compared with the normal hearing subjects and statistically evaluated with analysis of variance (ANOVA). Results with p values < 0.01 were considered to be statistically significant.

Results

The Vi and Vm parameters of the non-invasive perilymphatic measurement results in both groups of patients with Ménière's disease showed large inter-individual differences. Nevertheless in this study, the correlation of the test results within one subject and one ear was high (about 0.8).

In group 1 no significant differences were found in perilymphatic pressure measurements results between affected, non-affected ears and the results of 50 normal hearing subjects. The averaged values for Vi and Vm with standard deviation for affected, for non-affected and for normal hearing ears are shown in Table 16.1. In patients with Ménière's disease, no relationship was found between non-invasive perilymphatic pressure measurement results and hearing thresholds, blood pressure, gender or age.

Table 16.1: averaged values with standard deviation (SD) for Vi and Vm in the affected, non-affected and normal hearing ears

	Vi		Vm	
	Mean	SD	Mean	SD
Affected ears (n = 28)	–174	182	–19	214
Non-affected ears (n = 22)	–197	194	–39	215
Normal hearing ears (n = 100)	–167	149	33	212

Values in nl.

In group 2 the dynamics of perilymphatic pressure changes were examined. After a rapid change of body position, the perilymphatic pressure changed in affected, in non-affected as well in normal hearing ears within one minute. In order to study the changes in perilymphatic pressure despite large inter-individual differences of the Vi and Vm parameters, all values were related to the values of the first measurement of the same ear in upright sitting position. The averaged changes of Vi related to the first measurement are shown in Figure 16.1 (affected ears), in Figure 16.2 (non-affected ears) and in Figure 16.3 (normal hearing ears). The time course of Vm change following change of body position was comparable to the Vi change. After an initial decrease of Vi and Vm, related to an increase of perilymphatic pressure, in the first minute after lying down, the values did not change significantly until the sitting position was restored.

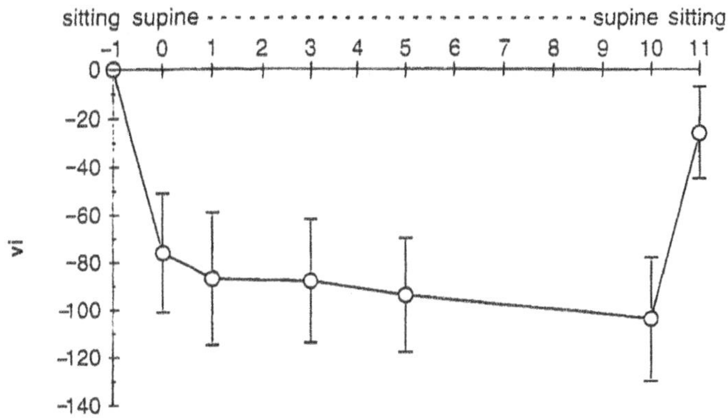

Figure 16.1: change of Vi in the affected ears (n = 28) after change of body position. The X-axis represents the time in minutes after change of body position. The Y-axis represents the Vi change averaged over all affected ears with standard deviation of the mean in nanolitres with respect to the first measurement in upright sitting position.

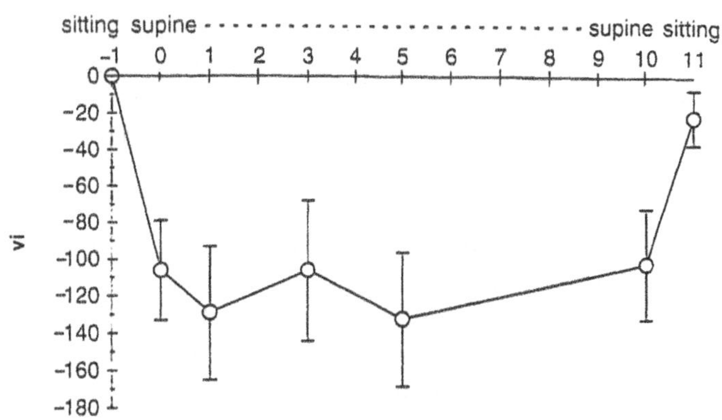

Figure 16.2: change of Vi in the non-affected ears (n = 16) after change of body position. The X-axis represents the time in minutes after change of body position. The Y-axis represents the Vi change averaged over all non-affected ears with standard deviation of the mean in nanolitres with respect to the first measurement in upright sitting position.

After change of body position no significant differences were found in perilymphatic pressure measurement results between affected, non-affected and normal hearing ears ($p > 0.01$).

Differences between the first and the last measurement, both in sitting position, were not significant ($p > 0.01$). No conclusions can be made for perilymphatic pressure changes in the first minute after change of body position because our measurement technique takes about one minute for each single measurement.

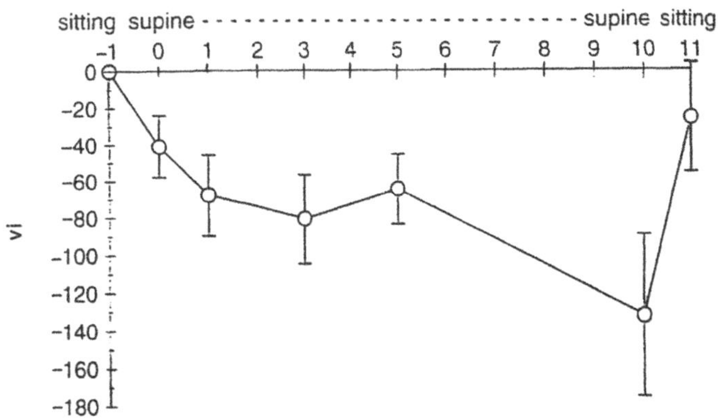

Figure 16.3: change of Vi in the normal hearing ears (n = 18) after change of body position. The X-axis represents the time in minutes after change from upright sitting position to supine position. The Y-axis represents the Vi change averaged over all normal hearing ears with standard deviation of the mean in nanolitres with respect to the first measurement in upright sitting position.

Discussion

Perilymphatic pressure alterations in the human inner ear can be assessed in a non-invasive way using the tympanic membrane displacement technique. The method has been used in several studies in intracranial hypertension and peri-operative monitoring of ventricular shunt surgery and it also has been used in patients with otological disorders, such as Ménière's disease, Gusher syndrome and round window rupture (Philips and Marchbanks 1989; Reid et al. 1989; Ernst 1993; Ernst 1995; Rosingh et al. 1996; Rosingh et al. 1997).

We found in our series no significant differences in non-invasive perilymphatic pressure measurement results between patients with Ménière's disease and normal hearing subjects. In normal ears endolymphatic pressure and perilymphatic pressure are equal (Böhmer 1993). However, in ears with a long-standing hydrops, as would be expected in our patients with a chronic Ménière's disease for at least one year, the endolymphatic pressure can be higher than the perilymphatic pressure. Potential endolymphatic pressure variations in affected and non-affected ears among patients with Ménière's disease could not be investigated using our non-invasive measurement technique, which provides the assessment of perilymphatic pressure.

The perilymphatic compartment of the inner ear is connected with the cerebrospinal fluid compartment of the fossa posterior by the cochlear aqueduct (Palva 1969). The hydrostatic pressure of the perilymphatic compartment is closely related to the cerebrospinal fluid pressure. The time course of inner ear pressure change following rapid change of intracranial CSF pressure is influenced by the patency of the cochlear aqueduct. In group 2 the patency of the cochlear aqueduct in patients with Ménière's disease and in normal hearing subjects was assessed non-invasively by means of the tympanic membrane displacement technique.

No significant differences between affected, non-affected and normal hearing subjects were found. However, in normal hearing ears an interesting trend can be detected of a slightly more delayed change of perilymphatic pressure after lying down. In contrast with the affected and non-affected ears, the perilymphatic pressure in controls, as measured non-invasively, still increase after 5 minutes. The mean age of the controls was more than 20 years younger than the investigated group of patients with Ménière's disease. Since the patency of cochlear aqueduct is suggested to decrease at an advanced age, this delayed change in the controls was not expected (Wlodyka 1978). When the cochlear aqueduct in the affected as well as the non-affected ears may be more patent than in normal hearing ears, the so-called low pass filter capacity of the cochlear aqueduct may have deteriorated. The low pass filter capacity of the cochlear aqueduct is suggested to protect the inner ear against large and rapid pressure changes. In case of a deteriorated function of the aqueduct, large and rapid variations of cerebrospinal fluid pressure can affect the inner ear pressure more easily, which can contribute to inner ear damage as in Ménière's disease. The influence of the cochlear aqueduct patency and the effects of pressure changes on audiovestibular symptoms might be subjects for further investigation.

References

Allen GW, Dallos P, Sakamoto S, Homma T (1971) Cochlear microphonic potential in cats. Arch Otolaryng 93: 388–96.

Beentjes BIJ (1972) The cochlear aqueduct and the pressure of the cerebrospinal and endolabyrinthine fluids. Acta Otolaryngol (Stockh) 73: 112–20.

Böhmer A (1993) Hydrostatic pressure in the inner ear fluid compartments and its effect on the inner ear function. Acta Otolaryngol (Stockh) 507 (Suppl): 3–24.

Densert O, Ingelstedt S, Ivarsson A, Pedersen K (1975) Immediate restoration of basal sensorineural hearing (Mb Ménière) using a pressure chamber. Acta Otolaryngol (Stockh) 80: 93–100.

Densert B, Sass K, Arlinger S (1995) Short term effects of induced middle ear pressure changes on the electrocochleogram in Ménière's disease. Acta Otolaryngol (Stockh) 115: 732–7.

Ernst A, Lenarz T (1993) Non-invasive intracochlear and intracranial pressure measurement: an additional tool in intraoperative monitoring. In: Höhman D (ed.) ECoG, OAE and Intra-operative Monitoring. Amsterdam/New York: Kugler Publications, pp. 185–7.

Ernst A, Issing PR, Bohndorf M (1995) Die nicht-invasive Beurteilung des intracochleären Druckes: II. Untersuchungsbefunde von Patienten mit M. Ménière, fluktuierendem Tieftongehör sowie periphervestibulären Schwindelattacken. Laryngo-Rhino-Otol 74: 13–20.

Marchbanks RJ, Reid A (1990) Cochlear and cerebrospinal fluid pressure: their interrelationship and control mechanisms. Br J Audiol 24: 179–87.

Marchbanks RJ (1991) Implications of measuring intracranial pressure. Hearing Instrument 42: 30–1.

Palva T, Dammert K (1969) Human cochlear aqueduct. Acta Otolaryngol (Stockh) 246 (Suppl): 1–57.

Philips AJ, Marchbanks RJ (1989) Effects of posture and age on tympanic membrane displacement. Br J Audiol 23: 279–84.

Reid A, Marchbanks RJ, Bateman DE, Martin AM, Brightwell AP, Pickard JD (1989) Mean intracranial pressure monitoring by a non-invasive audiological technique; a pilot study. J Neurol Neurosurg Psychiatry 52: 610–12.

Reid A, Cottingham CA, Marchbanks RJ (1993) The prevalence of perilymphatic hypertension in subjects with tinnitus: a pilot study. Scand Audiol 22: 61–3.

Rosingh HJ, Wit HP, Albers FWJ (1996) Non-invasive perilymphatic measurement in normal hearing subjects using the MMS-10 Tympanic Displacement Analyser. Acta Otolaryngol (Stockh) 116: 382–7.

Rosingh HJ, Wit HP, Sulter AM, Albers FWJ (1997) Longitudinal non-invasive perilymphatic pressure measurement in patient with Ménière's disease. ORL 59: 135–40.

Sismanis A (1987) Otologic manifestations of benign intracranial hypertension syndrome: diagnosis and management. Laryngoscope 97 (Suppl):1–17.

Walsted A, Salomon G, Thomson J, Tos M (1991) Hearing decrease after loss of cerebro-spinal fluid. A new hydrops model? Acta Otolaryngol (Stockh) 111: 468–75.

Wlodyka J (1978) Studies on cochlear aqueduct patency. Ann Otol 87: 22–8.

Correspondence to HJ Rosingh, Department of Otorhinolaryngology, University Hospital Groningen, PO Box 30.001, NL-9700 RB Groningen, Netherlands. Tel (+31) 503612540. Fax (+31) 503696726.

Chapter 17
Modelling labyrinthine fluid dynamics

JAMES C ANDREWS, DAVID STRELIOFF, MARVIN BERGSNEIDER, DONALD BECKER, Division of Head and Neck Surgery (Drs Andrews and Strelioff) and the Division of Neurosurgery (Drs Bergsneider and Becker) UCLA School of Medicine, Los Angeles, California, USA

ABSTRACT. The transmission of pressures through the labyrinthine fluids of the inner ear has been considered anatomically based on the structure and interconnecting channels of the temporal bone. Further knowledge of labyrinthine fluid dynamics has been gained by measuring the actual pressures of these fluids, the physiologic factors that influence these pressures and the ability of the various anatomic channels to transmit pressure.The development of a mathematical model that can account for the dynamics of the labyrinthine fluid system becomes the next step in further understanding the physiology of inner ear fluids. The model simplifies each fluid into formation, absorption and storage and considers the compliance of each compartment. The dependence and connection of the endolymph, perilymph and cerebro-spinal fluid are provided by the model, Other factors such as atmospheric, middle ear, vascular and osmotic pressures provide only an indirect influence within the current model and further refinements in the model may need to directly account for these influences. The future of the development of an effective treatment for inner-ear fluid balance dynamics. This application of modelling and labyrinthine fluid dynamnics measurements provides a mechanism for determining whether a therapy has a beneficial effect or any effect on the inner ear.

Key words: labyrinth, fluid dynamics, inner ear, mathematical model, endolymph, perilymph, cerebrospinal fluid.

Introduction

The transmission of pressures through the labyrinthine fluids of the inner ear has been considered to be anatomically based on the structure and interconnecting channels of the temporal bone (Goodhill and Harris 1979). Further knowledge of labyrinthine fluid dynamics has been gained by measuring the actual pressures of these fluids, the physiologic factors that influence these pressures and the ability of the various anatomic channels to transmit pressure (Andrews et al. 1991; Carlborg and Farmer 1983). The development of a mathematical model that can account for the dynamics of the labyrinthine fluid system is the next step in further understanding the physiology of the inner ear fluids.

Mathematical model

This model is a hydrodynamic consideration of the inner ear. Some of the basic features include a description of CSF dynamics, fluid pressure transmission between the cerebrospinal fluid (CSF) and the labyrinth via the cochlear aqueduct, and displacement of membranes and tissues separating the various compartments. All pressures are considered relative to atmospheric pressure at the level of the cisterna magna.

Each fluid (CSF, perilymph and endolymph) is housed in a compartment and has a constant storage rate which is equal to a normal or baseline steady state rate of formation minus a normal or baseline steady state rate of absorption and is modelled as a leakage from a higher to a lower pressure via a resistance (Figure 17.1). In this model we have considered the cochlear aqueduct patent and a resistive connection between the CSF and the perilymph (Salt and Stopp 1979). Each compartment considered (intracranial, perilymph and endolymph) has a compliance that is defined by the rate of compartment volume change per unit pressure change.

For each compartment, a perturbation is considered in which there is infusion, at a constant rate, of an artificial fluid of a similar osmotic and electrolytic content into a compartment (artificial endolymph is infused into the endolymph space). Eventually a new steady state is achieved in which the compartmental storage rate remains constant and the rate of fluid formation (including normal production and artificial infusion) equals the rate of absorption.

CSF compartment

In the normal steady-state condition, CSF is continually formed and absorbed. The rate of CSF storage (I_{SC}) is equal to the rate of CSF formation (I_{FC}) less the rate of absorption (I_{AC}):

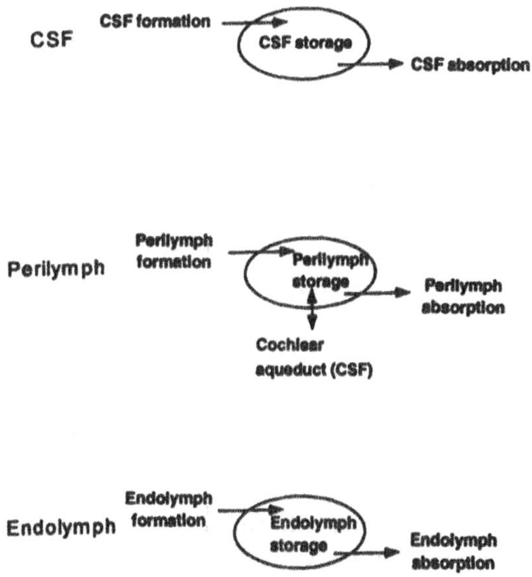

Figure 17.1: compartmental consideration for the CSF, perilymph and endolymph and the fluid formation and absorption.

$$I_{SC} = I_{FC} - I_{AC}$$

I_{FC} is assumed to be constant and I_{AC} is modelled as a leakage from a higher pressure, the CSF pressure (P_C), to a lower or absorption pressure (P_{AC}), via a resistive connection R_{AC}:

$$I_{AC} = \frac{P_C - P_{AC}}{R_{AC}}$$

where R_{AC} is the resistance to CSF absorption.

CSF volume is constant in the steady-state condition therefore $I_{SC} = 0$ or $I_{FC} = I_{AC}$.

$$I_{FC} = I_{AC} = \frac{P_{OC} - P_{AC}}{R_{AC}}$$

where P_{OC} is defined as the normal or steady-state CSF pressure.

Substituting equations and algebraic reduction results in:

$I_{SC} = I_{OC} = \dfrac{P_C}{R_{AC}}$ where by definition

$$I_{OC} = \frac{P_{OC}}{R_{AC}}$$

In descriptive terms, I_{oc} represents the normal or steady-state CSF formation.

Intracranial compliance is C_C defined as the change in CSF volume per unit change in CSF pressure:

$$C_C = \frac{dV_C}{dP_C} \quad \text{or } dP_C = \left[\frac{1}{C_C}\right] dV_C .$$

$$\text{Since } \frac{dV_C}{dt} = I_{SC}, \text{ then } \frac{dP_C}{dt} = \left[\frac{1}{C_C}\right] I_{SC} .$$

Consideration of an infusion into the CSF at a constant rate, I_I, results in changes in the CSF volume (V_C) and pressure (P_C). These are reflected as:

$$\frac{dP_C}{dt} = \left[\frac{1}{C_C}\right] \frac{dV_C}{dt} = \left[\frac{1}{C_C}\right]\left[I_I + I_{OC} - \frac{P_C}{R_{AC}}\right]$$

By the experimental findings of Marmarou et al. (1975), intracranial compliance is inversely proportional to CSF pressure:

$$C_C = \frac{1}{K_C P_C} .$$

The resulting CSF pressure from an artificial infusion into the subarachnoid space as a function of time $P_C(t)$ can then be described as:

$$P_C(t) = \frac{P_{OC} + I_I R_{AC}}{1 + \frac{I_I}{I_{OC}} e^{-(I_I + I_{OC}) K_C t}}$$

The expected responses are graphically represented in Figure 17.2. It is evident that if the infusion time is long enough, the above relationship reduces to a linear equation describing the steady state:

$$P_{IC}(I_I) = P_{OC} + I_I R_{AC} .$$

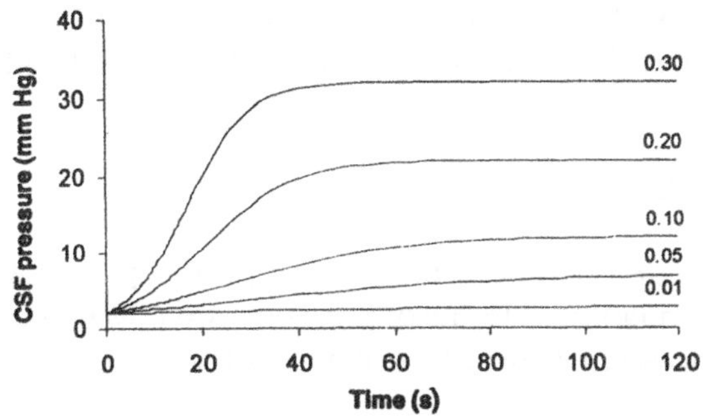

Figure 17.2: modelled infusion of artificial CSF into the subarachnoid space and the resulting change in pressure over time, $P_C(t)$. Infusion rates of 0.01, 0.05, 0.1, 0.2, and 0.3 are considered.

This relationship is shown in the Figure 17.3. P_{OP}, the normal or baseline steady state pressure of the CSF is represented by the intersection of the curve with the pressure axis. The intersect of the curve with the infusion axis indicates the normal or baseline steady state CSF formation and absorption rates. The slope of the curve represents the absorption resistance, R_{AC}.

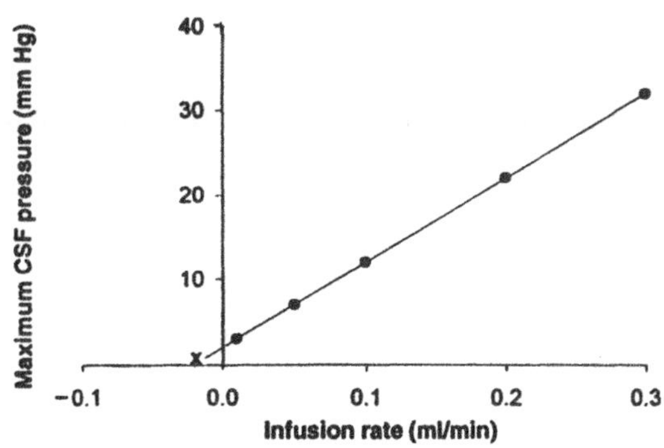

Figure 17.3: modelled plot of maximum CSF pressure vs. infusion rate. In this plot, the intersect of the curve with the pressure axis intersect indicates the normal or baseline pressure, the infusion rate axis intersect indicates the baseline steady state rate of CSF formation (and absorption) and the slope of the curve is the CSF resistance to absorption.

Perilymph

Similar to the CSF, the perilymph is continuously being formed and absorbed at rates I_{FP} and I_{AP} respectively and there is a baseline or normal perilymph pressure, P_P. The resistance to perilymph absorption (R_{AP}) is assumed to be constant. Pressure transmission from the CSF to the perilymph via the cochlear aqueduct (I_{CA}) occurs with a resistance (R_{CA}).

Compliance is twofold and must consider the perilymph compartment (C_P) and the partition between the CSF and the perilymph (C_{CP}). The rate of perilymph storage, I_{SP}, is described by:

$$I_{SP} = I_{FP} + I_{CA} - I_{AP}$$

In the steady-state condition, there is no perilymph storage or $I_{SP} = 0$. Rewriting the equation in terms of pressure and resistance to flow under normal or steady-state CSF and perilymph pressures (P_{OC} and P_{OP} respectively):

$$I_{FP} = \frac{P_{OP}}{R_{AP}} - \frac{P_{OC} - P_{OP}}{R_{CA}}$$

$$IFP = \frac{P_{OP}}{R_{PP}} - \frac{P_{OC}}{R_{CA}}$$

where R_{PP} is R_{AP} and R_{CA} acting in parallel,

$$\frac{1}{R_{PP}} = \frac{1}{R_{AP}} + \frac{1}{R_{CA}}$$

The normal or steady-state perilymph exchange rate can be described as:

$$I_{OP} = \frac{P_{OP}}{R_{PP}} = I_{FP} + \frac{P_{OC}}{R_{CA}}$$

and substituting into the above equation results in:

$$I_{SP} = I_{FP} + \frac{P_C}{R_{CA}} - \frac{P_P}{R_{PP}}$$

Since the compliance, C_P, of the perilymph compartment is defined as:

$$C_P = \frac{dV_P,}{dV_P,}$$

and the rate of perilymph storage by definition is:

$$\frac{dVP}{dt} = I_{SP}\,.$$

Substituting equations results in:

$$\frac{dP_P}{dt} = \frac{1}{C_P}\frac{dV_P}{dt}, \text{ or}$$

$$\frac{dP_P}{dt} = \frac{1}{C_P}\left[I_{FP} + \frac{P_C}{R_{CA}} - \frac{P_P}{R_{PP}}\right].$$

If a constant infusion of fluid (I_I) into the perilymphatic space is considered then formation and absorption rates, resistance to absorption and even the compliance of the perilymph compartment can be determined. By blocking the cochlear aqueduct, there is no fluid leakage through this structure which simplifies the relationship:

$$I_{SP} = I_{FP} + I_I - \frac{P_{IP}}{R_{AP}} = 0 \text{ or}$$

$$P_{IP}(I_I) = P_P + I_I R_{AP} \text{ where } P_P = I_{AP}R_{AP} = I_{FP}R_{AP}$$

In this equation $P_{IP}(I_I)$ is the perilymph pressure produced by an infusion at rate I_I. A modelled plot of this relationship is shown in Figure 17.4. I_{AP} and I_{FP} are represented by the curve intersect of the infusion axis and the normal or steady-state perilymph pressure, P_{OP}, represents the curve intercept of the pressure axis while the slope of the line is the resistance to absorption.

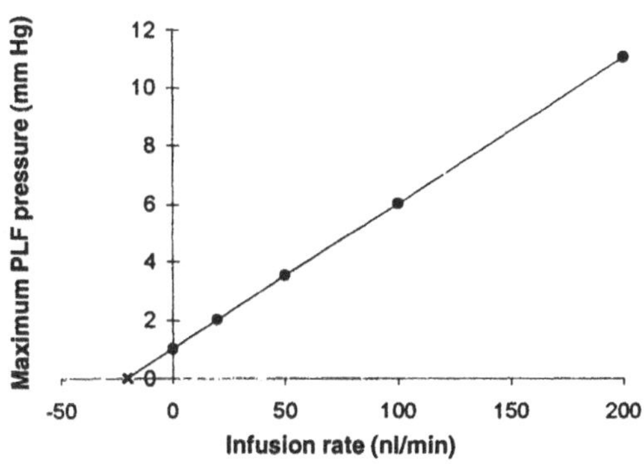

Figure 17.4: modelled plot of maximum perilymph pressure vs. infusion rate. In this plot, the intersect of the curve with the pressure axis intersect indicates the normal or baseline pressure, the infusion rate axis intersect indicates the baseline steady-state rate of perilymph formation (and absorption) and the slope of the curve is the perilymph resistance to absorption. A similar curve would appear for endolymph.

To understand the influence of the cochlear aqueduct, it is necessary to determine the resistance to fluid flow through this structure. This can be determined by studying the effect of constant CSF infusion at several rates while measuring the resulting CSF pressure (P_{IC}) and perilymph pressure (P_{IP}). This is provided by the following relationship:

$$P_{IP} = I_{FP} R_{PP} + \left[\frac{R_{PP}}{R_{CA}} \right] P_{IC}$$

The slope of a curve of P_{IP} vs. P_{IC} (Figure 17.5) provides the ratio of R_{AP} to R_{CA}. If R_{AP} is known, then R_{CA} can be determined.

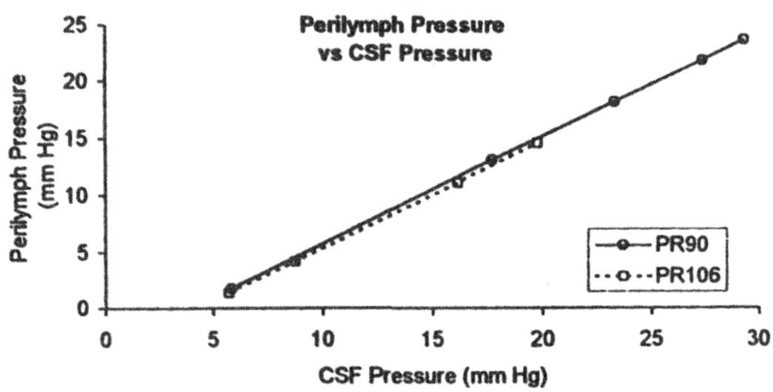

Figure 17.5: perilymph pressure vs. intracranial pressure during various infusions of artificial CSF up to 0.4 ml/min in two animals.

Endolymph

The parameters that describe the endolymph system include the rate of endolymph formation (I_{FE}), the rate of endolymph absorption (I_{AE}), the fluid pressure of the endolymph (P_E), the normal or steady state endolymph pressure (P_{OE}), the resistance of endolymph absorption (R_{AE}). The rate at which endolymph is stored (I_{SE}) is described by the following:

$$I_{SE} = I_{FE} - I_{AE}$$

This can be rewritten in terms of pressure and resistance as:

$$I_{SE} = \frac{P_{OE}}{R_{AE}} - \frac{P_E}{R_{AE}}$$

The compliance, C_E, of the endolymph compartment is defined as

$$C_E = \frac{dV_E}{dP_E}$$

The change of endolymph volume is

$$\frac{dV_E}{dt} = I_{SE}$$

Combining these equations, the rate of endolymph pressure change due to endolymph formation and absorption is:

$$\frac{dP_E}{dt} = \frac{1}{C_E}\frac{dV_E}{dt} = \frac{1}{C_E}\left[I_{OE} - \frac{P_E}{R_{AE}}\right]$$

If a constant infusion of fluid (I_I) into the endolymphatic space produces a constant pressure (P_{IE}). The above equation can be reduced to the following relationship:

$$P_{IE}\,(I_I) = P_{OE} + I_I R_{AE}, \text{ where } P_{OE} = I_{AE}R_{AE} = I_E R_{AE}.$$

$P_{IE}(I_I)$ is the maximum endolymph pressure as a function of the infusion rate. A modelled plot of the relationship of P_{IE} vs. infusion would appear as for Figure 17.4 as described for the perilymph. The parameters of the endolymph can be analogously derived from such a curve with P_{OE} representing the intersection of the curve with the ordinate, $I_{AE} = I_{FE}$ representing the intersection of the curve with the abscissa and R_{AE} as the slope of the curve.

Methods

Normal multicoloured guinea pigs with normal ears were studied. General anaesthesia was achieved with intraperitoneal pentobarbital (Nembutal®) 20 mg/kg requiring mechanical ventilator support with room air through a tracheostomy. Simultaneous monitoring of the electrocardiograph (Tektronix® Electrocardiograph monitor), arterial oxygen saturation, capnograghy (Criticare® pulse oximeter), and intracranial pressure (Abbott® microchip transducer) were performed. Endolymph and perilymph pressures were measured utilizing a WPI® model 900A virtual no-flow microcompartment pressure measuring device. Simultaneous measurement of the endocochlear potential allowed confirmation of the compartment being measured. All measurements were calibrated, digitized and continuously monitored through a computer with a 12-bit sampling rate at 200 Hz running a customized version of LabWindows® CVI software.

For CSF dynamic modelling, four animals were studied. Two 20-gauge flexible catheters were introduced into the fourth ventricle; one for artificial CSF infusion via a 10 ml syringe connected to a Harvard® pump and the other for pressure measurement. Artificial CSF infusion ranged from 0.05 to 0.3 ml/min. A steady-state intracranial pressure was achieved before increasing to the next step of infusion rate.

Measurement of the transmission of fluid flow from the CSF to the perilymph was performed in two animals. Two catheters were introduced into the fourth ventricle; one for artificial CSF infusion and the other for pressure measurement. Perilymph pressure was measured through a pipette introduced into the scala tympani. Artificial CSF infusion rates ranged up to 0.4 ml/min.

Endolymph pressure dynamics was studied in five animals. Two pipettes were introduced through the lateral wall of the cochlea for infusion of artificial endolymph into the second turn and for measurement of pressure in the basal turn. Endocochlear potential measurement in each pipette anatomic localization of the pipette tip and stabilization of the preparation. The round window membrane was perforated to remove any potential interactions with the perilymph system and to effectively isolate the endolymph system. Artificial endolymph infusion was performed at step rates of 5 to 300 nl/min via a 10 ml syringe utilizing a Harvard® pump. Control of the infusion rate was performed with a data acquisition feedback loop through the computer by measuring pressure across the infusion pipette tip. At each infusion rate, a steady-state pressure was achieved prior to advancing to the next infusion step rate.

Results

For the measurement of the intracranial fluid dynamics, a steady-state CSF pressure was achieved at each of the various infusion rates of artificial CSF. A linear plot of the CSF pressure achieved for the various infusion rates was made (Figure 17.6). With this plot the following were calculated: a CSF absorption resistance of 79.5 mmHg/ml/min (slope of the curve), a CSF formation/absorption rate of 0.1 ml/min (abscissa intercept) and a baseline CSF pressure of 8.0 mmHg (ordinate intercept).

The influence of the cochlear aqueduct was studied by measuring the pressure transmission through this structure. CSF and perilymph pressures remained linear during all artificial CSF infusion rates considered in this study (Figure 17.5). The resistance of the cochlear aqueduct could have been calculated had the perilymph resistance to absorption been determined.

Endolymph fluid dynamics measurements demonstrated that a steady-state endolymph pressure was achieved at each of the artificial endolymph infusion rates up to 75 nl/min. Above this rate, a sudden drop in endolymph pressure along with alteration in the endocochlear potential occurred. Using only the data achieved for artificial endolymph infusion rates up to 75 nl/min, a linear regression of endolymph pressure achieved for the various infusion rates was performed (Figure 17.7). With this regression an endolymph absorption resistance of 0.015

mmHg/nl/min (slope of the curve), an endolymph formation/absorption rate of 13 nl/min (abscissa intercept) and a baseline endolymph pressure of 0.19 mmHg (ordinate intercept) were determined.

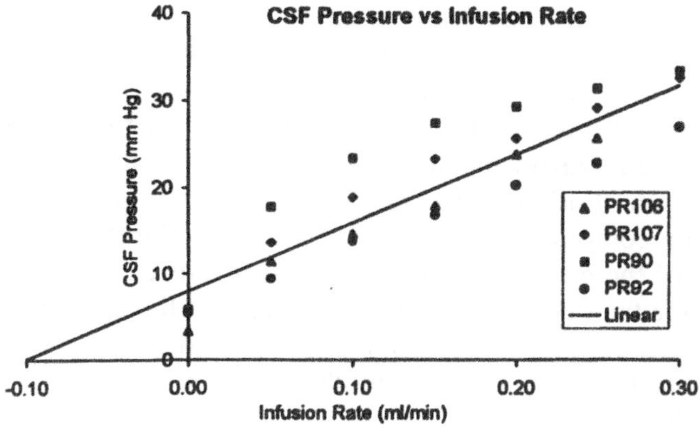

Figure 17.6: measured maximum steady-state CSF pressure achieved vs. infusion rate. A linear fit of the data reveals a CSF absorption resistance of 79.5 mmHg/ml/min, a normal or baseline CSF formation/absorption rate of 0.1 ml/min and a normal or baseline CSF pressure of 8.0 mmHg.

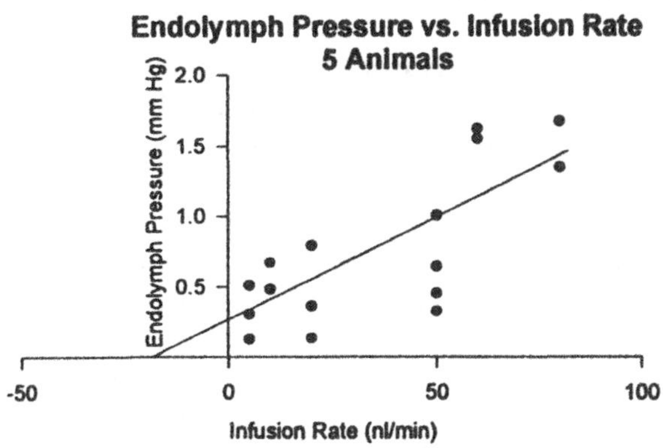

Figure 17.7: measured maximum steady-state endolymph pressure achieved vs. infusion rates up to 75 ml/min. A linear fit of the data reveals a endolymph absorption resistance of 0.015 mmHg/ml/min, a normal or baseline endolymph formation/absorption rate of 13 nl/min and a normal or baseline endolymph pressure of 0.19 mmHg.

Discussion

Marmarou et al. (1975, 1978) applied the principles of hydrodynamics to intracranial pressure and provided a completely different insight into the physiology of the CSF and intracranial pressure. Such modelling serves as a foundation for understanding the pathophysiology and has advanced the diagnosis and management of disorders of intracranial pressure. If a parallel analogy can be made for the inner ear, such modelling may provide the basis to advance our understanding of the labyrinthine fluids and lead to rational and effective therapy of imbalances within this system, especially Ménière's disease.

The work presented in this paper represents a first step at applying the principles of hydrodynamics to the labyrinthine fluids. The model simplifies each fluid into formation, absorption and storage and considers the compliance of each compartment. The dependence and connection of the endolymph, perilymph and CSF are provided by the model. Other factors such as atmospheric, middle-ear, vascular and osmotic pressures provide only an indirect influence within the current model. Future refinements in the model may need to directly account for these influences.

The CSF dynamics, predicted by the model, demonstrated good correlation with actual measurements made. The values calculated for CSF production and absorption of 0.1 ml/min, steady state pressure of 8.0 mmHg and resistance to absorption of 79.5 mmHg/ml/min are, to our knowledge, the first data of this kind reported for the guinea pig. The values for CSF production and steady-state pressure correlate with known data for other species including cats, dogs and rats (Davson et al. 1987).

Perilymph dynamics measurements are a current project of our laboratory and in this report are only modelled. Measurements comparing perilymph and intracranial pressure during infusion of artificial CSF showed a linear response in the range studied indicating that the cochlear aqueduct could be considered a resistive connection. Controversy remains as to the variability in the patency of the cochlear aqueduct, however, in the two animals studied here the aqueduct appeared patent.

The endolymph dynamics, as measured, correlated well with those predicted by the model. A linear response was seen with artificial endolymph infusions up to 75 nl/min. Infusion rates above this rate showed evidence of damage to the membranous labyrinth. The value of endolymph production (and absorption) of 13 nl/min correlates with other techniques for determining endolymph production (Salt and Thalmann 1988). A steady-state endolymph pressure of 0.2 mmHg is similar to measurements made previously (Andrews et al. 1991). The endolymph resistance to absorption was determined to be 0.015 mmHg/nl/min.

It would be beneficial to consider endolymphatic hydrops in terms of fluid dynamics, as this may provide a better understanding of the pathophysiology of this condition. Basic questions as to whether hydrops is produced by abnormal production or absorption of endolymph may be answered by comparing the normal ear with the hydrops ear. The manner in which hydrops is produced, either surgically or medically, may influence altered dynamics. The future of developing an effective treatment for inner ear fluid balance disorders lies in understanding the therapeutic alteration of the labyrinthine fluid dynamics. This application of modelling and labyrinthine fluid dynamics measurements provides a mechanism for determining whether a therapy has a beneficial effect or any effect on the inner ear.

Acknowledgement

Supported by NIH NIDCD grant DC000945.

References

Andrews JC, Bohmer A, Hoffman L (1991) The measurement and manipulation of endolymphatic pressure in experimental endolymphatic hydrops. Laryngoscope 101: 661–8.

Carlborg BI, Farmer JC (1983) Transmission of cerebrospinal fluid pressure via the cochlear aqueduct and endolymphatic sac. Am J Otolaryngol 4: 273–82.

Davson H, Welch K, Segal MB (1987) The Physiology and Pathophysiology of the Cerebrospinal Fluid. New York: Churchill Livingstone.

Goodhill V, Harris I (1979) Sudden hearing loss syndrome. In Goodhill V (ed.) Ear Diseases, Deafness and Dizziness. Hagerstown: Harper & Row, pp. 664–81.

Marmarou A, Shulman K, La Morgese J (1975) Compartmental analysis of compliance and outflow resistance of the cerebrospinal fluid system. J Neurosurg 43: 523–34.

Marmarou A, Shulman K, Rosende RM (1978) A nonlinear analysis of the cerebrospinal fluid system and intracranial pressure dynamics. J Neurosurg 48: 332–44.

Salt AN, Stopp PE (1979) The effect of cerebrospinal fluid pressure on perilymphatic flow in the opened cochlea. Acta Otolaryngol 88: 198–202.

Salt AN, Thalmann R (1988) Cochlear fluid dynamics. In Jahn AF, Santos-Sacchi J (eds.) Physiology of the Ear, New York: Raven Press.

Correspondence to Dr. James C. Andrews, M.D., Division of Head and Neck Surgery, 62-132 CHS, UCLA Medical Center, 10833 Le Conte Avenue, Los Angeles, CA 90095-1624, USA. Tel (1) 310-794-7289. Fax (1) 310-206-1393.
E-mail: JANDREWS@SURGERY.MEDSCH.UCLA.EDU.

Chapter 18
The effect of changes in cochlear fluid pressure on audiological measurements

AJ PHILLIPS, Audiology Department, Woolmanhill Hospital, Aberdeen UK; GB FARRELL, NHS Management Executive, Winchester, UK; ARD THORNTON, Medical Research Council Institute of Hearing Research Outstation, Royal South Hants Hospital, Southampton, UK

ABSTRACT. A series of experiments were conducted over several years to examine the effect of changes in cochlear fluid pressure on audiological measurements. Increases in cochlear fluid pressure were induced by changing posture, and decreases in fluid pressure obtained by administration of glycerol. This paper reports the effect of changing cochlear fluid pressure on tympanic membrane displacement (TMD) and auditory brainstem response travelling wave velocity (ABR TWV) measurements and otoacoustic emission results. The experimental subjects were a mix of patients undergoing glycerol dehydration as a prognostic indicator for saccus surgery and normal healthy young adults. The ABR TWV measurements demonstrated an increase in travelling wave velocity with increase in cochlear fluid pressure and a decrease with decreasing cochlear fluid pressure. The otoacoustic emissions showed a decrease in latency of individual peaks when the cochlear fluid pressure increased, and an increase in latency as cochlear fluid pressure decreased. TMD measurements showed the changes that have consistently been reported with postural change. When TMD measurements were performed on patients undergoing glycerol dehydration, these also showed changes in the opposite direction to those induced by increasing cochlear fluid pressure by posture. Since each type of measurement was performed on individual patients, it was possible to compare the sensitivity of the techniques. It is suggested that the three different measurement techniques measure changes in cochlear fluid pressure by three separate, but related mechanisms.

Key words: cochlear fluid pressure, tympanic membrane displacement, otoacoustic emissions, auditory brainstem responses.

Introduction

This work arose from a study at the Institute of Hearing Research in Southampton into the development of a test of endolymphatic hydrops (Farrell 1992). Our work there showed an increase in basilar membrane travelling wave velocity of high frequency sounds for Ménière's subjects compared with a group of matched normals (Thornton and Farrell 1991). We wanted to see how sensitive the test was to changes in cochlear fluid pressure and to compare this sensitivity with other non-invasive audiological techniques such as evoked otoacoustic emissions and tympanic membrane displacement measurements (TMD).

To do this, we have taken measurements from subjects where we have increased cochlear fluid pressure by changing posture (Phillips and Farrell 1992) and decreased cochlear fluid pressure by glycerol dehydration (Dauman and Cazals 1993). When we began this work we were aware of posture studies on evoked otoacoustic emissions (Antonelli and Grandori 1991) and had done a similar study using TMD (Phillips and Marchbanks 1989). More recently, other investigators have done similar work on glycerol dehydration. We knew about the effect of presumed cochlear aqueduct patency, middle-ear pressure effects, time lags and myriad other pitfalls (Phillips and Farrell 1988) and took pains to optimise the experimental protocol and procedure to minimise experimental error.

Method

Other papers will describe the TMD and otoacoustic emission techniques. The ABR travelling wave velocity technique involves recording Wave V of the ABR in response to a high intensity click with added ipsilateral high-pass masking noises. Pair-wise subtraction of these high-pass masked responses gives derived bands corresponding to discrete regions of the cochlea from which travelling wave velocities can be obtained (Parker and Thornton 1978).

In our posture experiment we tested both ears of six healthy adults between the ages of 23 and 40. All the following results refer to a change in posture from sitting to lying head down 10 degrees below horizontal, an assumed increase in CSF pressure of 150 mm saline.

The evoked otoacoustic emission results showed that increasing CSF pressure led to an average decrease in peak latency of 300 microseconds (Table 18.1). The largest changes in the emission occurred at around 11 ms after click delivery, corresponding to the 1 kHz region of the cochlea.

Table 18.1: maximum, minimum and average changes in EOAE peaks (ms) with change in posture

Subject	Ear	Max ▲E	Min ▲E	Av ▲E
AP	L	0·234	0·078	0·18
	R	0·078	0	0·039
IG	L	0·156	0·078	0·137
	R	0·312	0·156	0·253
GF	L	0·12	0·04	0·08
	R	0·48	0·04	0·21
AG	L	0·234	0·156	0·208
	R	0	0	0
GS	L	0·108	0	0·027
	R	0·078	0	0·020
AB	L	0·078	0·078	0·078
	R	0·234	0·078	0·156

TMD measurements showed the well-documented changes in 'inward-goingness' and reflex growth function, with increase in CSF pressure, the biggest occurring at the highest stimulus intensities (Table 18.2).

Table 18.2: maximum, minimum and average changes in TMD parameters (nl), maximum inward displacement (v_I and mean displacement (v_m) with change in posture

Sub.	Ear	Max ▲V_I	Min ▲V_I	Av ▲V_I	Max ▲V_m	Min ▲V_m	Av ▲V_m
AP	L	192	63	140·33	501	95	295
	R	A	A	A	78	78	78
IG	L	156	127	141·5	644	337	490·5
	R	82	69	78·5	87	128	107·5
GF	L	95	4	36	121	–92	–9
	R	*	*	*	359	195	277
AG	L	10	6	8	92	–78	7
	R	9·5	7·5	8·5	5	31	18
GS	L	33	33	33	–26	–26	–26
	R	5	5	5	–5	–5	–5
AB	L	127	127	127	86	86	86
	R	268	268	268	–57	–57	–57

A These waveforms were too noisy to analyse.
* This subject had outward reflex configurations in both ears in both postures so no V_I measurement was possible.

Of the ABR travelling wave velocity measures, only those at the highest centre frequency used – around 6 kHz – showed significant changes with posture, of around 12 metres per second (Table 18.3). This is consistent with our previous work that showed TWV differences between Ménière's patients and normals only at this centre frequency.

Table 18.3: changes in TWV value (ms^{-1}) at centre frequencies of 5.68 (a), 2.84 (b), 1.42 (c) and 0.771 (d) kHz with change in posture

Sub.	Ear	▲TWV(a)	▲TWV(b)	▲TWV(c)	▲TWV(d)
AP	L	4·19	2·77	–1·15	–1·29
	R	–3·99	0·66	0·19	–0·58
IG	L	7·18	–2	–2·51	3·43
	R	7·18	3·47	–0·28	–0·79
GF	L	5·38	2·64	–1·4	0·05
	R	17·46	–3·08	–1·02	0·33
AG	L	50·26	–1·03	–0·74	0·58
	R	0	–3·62	–0·61	–0·34
GS	L	15·56	1·7	–2·71	0·47
	R	*	–4·43	–2	0·3
AB	L	0	4·9	0·41	0
	R	31·42	5·68	–1·33	–0·44

*This waveform was too noisy to analyse.

Statistical analysis of the data showed change in posture to produce change in each of the three different types of measurement that was significant at the 1% level (Table 18.4). There was no statistically significant correlation between measurements taken for the three techniques, although the same ears generally showed large posturally induced changes for all three techniques.

Table 18.4: Wilcoxon signed-ranks test statistics for maximum v_i values, maximum EOAE peak latency shifts and changes in TWV values obtained at the highest centre frequency, with change in posture

Test	z statistic	Two-tailed probability of no difference between sitting and lying values
Evoked otoacoustic emissions	–2·9341	0·0033
TMD V_i parameter	–2·8031	0·0051
TWV obtained at highest centre frequency	–2·5471	0·0109

In our second experiment we decreased cochlear fluid pressure by glycerol dehydration. Because of time pressures, we gained TMD results on a limited number of subjects and we applied our clinical test protocol to gaining ABR TWV results rather that the full travelling wave velocity measurement that would have been too time consuming.

The subjects for this experiment were 15 patients with clinically confirmed endolymphatic hydrops undergoing glycerol dehydration as a prognostic indicator for saccus decompression surgery. Most of these patients had moderate to severe cochlear hearing losses and it would not normally be expected to record an otoacoustic emission from them. However, we reasoned that for patients with early cochlear hydrops, we might be able to record emissions since the cochlear sensory structures would be essentially intact. We further reasoned that when the hydropic loading of the basilar membrane was removed by glycerol dehydration, we should see changes in the emission, maybe due to increased hair-cell mobility. Since we were dealing with patients with relatively large hearing losses, we expected to record poor-quality emissions. We therefore optimised recording conditions to achieve the best possible results.

Having collected the data we needed an objective description of the quality of the otoacoustic emissions. From knowledge of signal processing and statistical properties of the emission, we determined that a cross-correlation of 0.42 between sample points stored within two separately averaged buffers, gives a 95% chance of the presence of an emission signal. The upper trace in Figure 18.1 shows the cross-correlation for the normally hearing person's emission (shown in the lower trace). The parameters that we felt best described the emission were: the power in the emission, number of replicable peaks and the proportion of the waveform that was well cross correlated. We also believed that we might see changes in the amplitude and latency of emission peaks and in the frequency content of the emission during glycerol dehydration. To investigate this, software was written to analyse the data appropriately.

Results

From a total of 15 patients we obtained good, repeatable otoacoustic emissions from 17 ears of 10 patients. Thirteen of these ears had abnormal travelling wave velocities, which have been shown by our previous work to be a diagnostic indicator for endolymphatic hydrops. Of these 13 ears, five had a best audiometric threshold of greater than or equal to 35 dB, the remaining eight had best thresholds of less than or equal to 20 dB.

The first finding of note was the presence of otoacoustic emissions in patients with hearing loss. The lower trace in Figure 18.2 shows the emission taken from a patient with a best audiometric threshold of 40 dB, where there is good cross-correlation up to 12 ms post-stimulus latency. In agreement with our hypothesis we found that in some patients with poor emissions prior to dehydration, good emissions appeared during the dehydration. Figure 18.3 shows the change in an emission during dehydration. The thick lines represent an emission signal, whereas the thin lines represent noise. Another set of patients had good emission prior to dehydration and following glycerol ingestion, the waveform changed markedly. In these cases, one could not relate emission peaks observed pre-dehydration to those recorded during the dehydration procedure. A third pattern of response was found for peaks to increase in latency during dehydration. An example of this is shown in Figure 18.4.

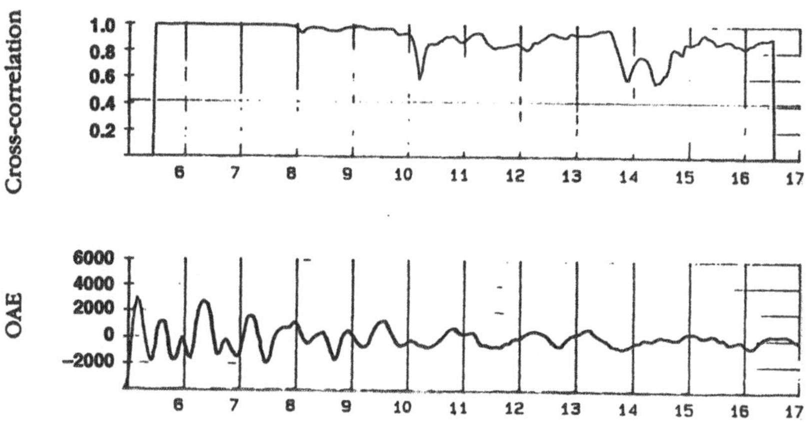

Figure 18.1: moving window cross-correlation analysis for normally-hearing person's otoacoustic emission.

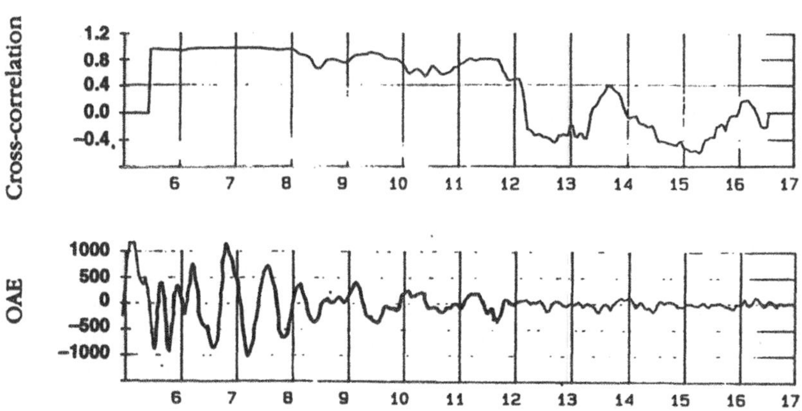

Figure 18.2: moving window cross-correlation analysis for sensorineural hearing loss patients' otoacoustic emission.

On further examining changes in peak latencies, we observed that the biggest changes occurred for peaks at about 11 ms post-stimulus latency consistent with our posture study. Peaks in this area have been shown to represent the part of the cochlea responsible for mid-frequency audiometric thresholds. The largest change in emission peak latencies were observed during the first hour following dehydration, as shown in the top left of Figure 18.5.

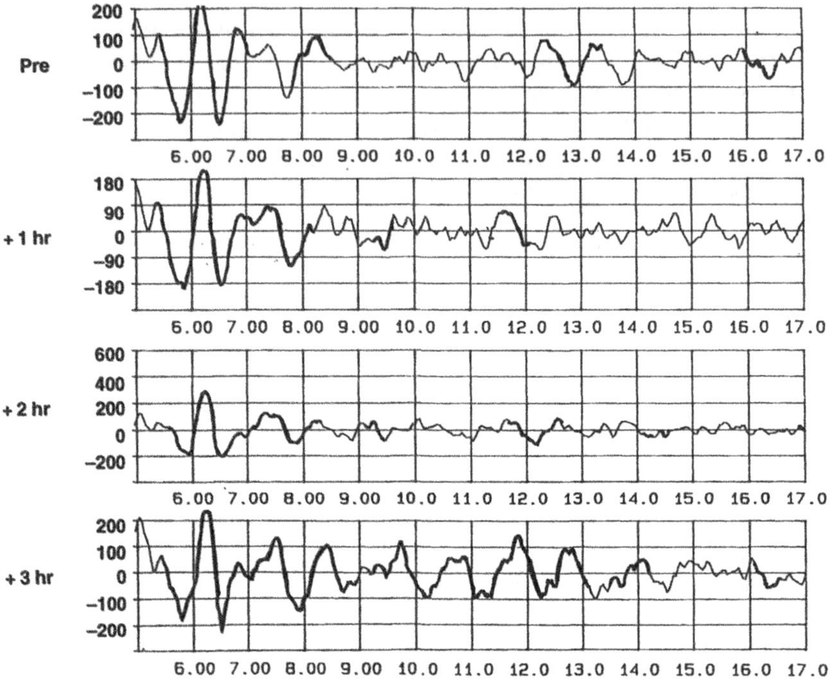

Figure 18.3: patient RC (right ear): changes in cochlear emission during glycerol dehydration.

On average, the quality of the emission, as determined by the proportion of well cross-correlated waveform, increased following glycerol ingestion. The largest improvements were seen in the emission recorded an hour into the dehydration (as shown in the top right of Figure 18.5). This increase in proportion of well cross-correlated emissions was accompanied by an increase in the number of emission peaks. This is shown in the bottom left of Figure 18.5.

Figure 18.5 also shows a summary of our 'TWV' estimate results, demonstrating a decrease in travelling wave velocity but shown as an increase in latency in the figure. The TMD measurements obtained on a limited number of patients all showed results that were more 'outward-going' post-dehydration.

We next performed statistical analysis to investigate a possible relationship between changes in any of the otoacoustic emission parameters during dehydration. Statistically significant correlations were found between changes in emission spectral bands and peak latency and amplitude changes. This supports our supposition that peaks around 7 ms post-stimulus latency equate to the high frequency parts of the emission, 11 ms for the mid-frequency emission and 15 ms for the low frequency part of the emission spectrum. The best description of the quality of the emission was the percentage of well cross-correlated waveform, changes of which were significantly correlated with changes of all other parameters.

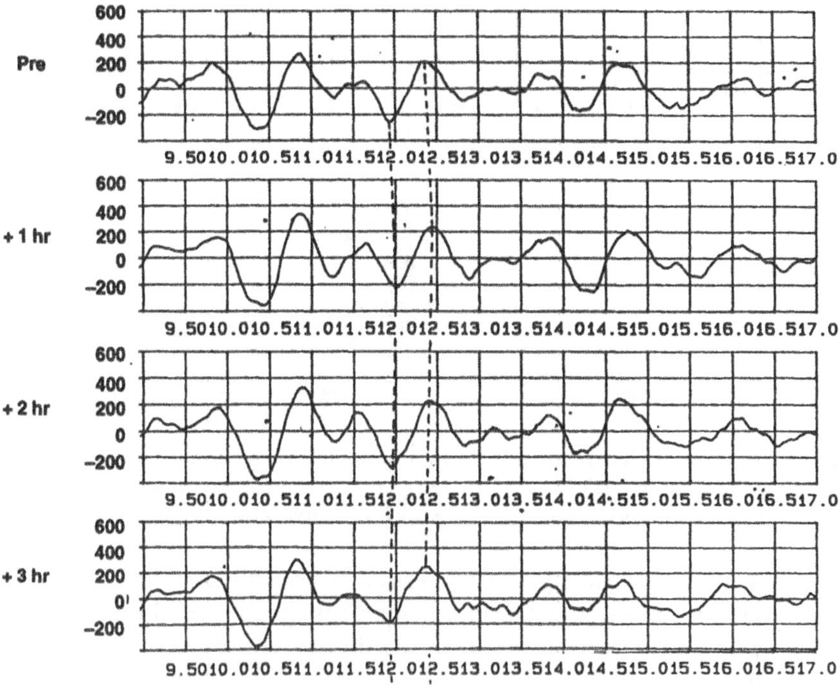

Figure 18.4: patient JB (left ear): changes in emission peak latency during dehydration.

The relationship between changes in the energy in emission spectral bands and changes in audiometry threshold was complex and requires further analysis.

Conclusions

The conclusions we have reached from the two experiments are that each technique – TMD, otoacoustic emissions and ABR travelling wave velocity – is sensitive to changes in cochlear fluid pressure. The TMD technique shows an increase in 'inward-goingness' with increase in cochlear fluid pressure and movement becomes more outward with decrease in pressure.

Evoked otoacoustic emissions show a decrease in peak latency with increasing cochlear fluid pressure and an increase in latency with decreasing pressure. The ABR travelling wave velocities are increased with increasing cochlear fluid pressure and are decreased with lessening pressure.

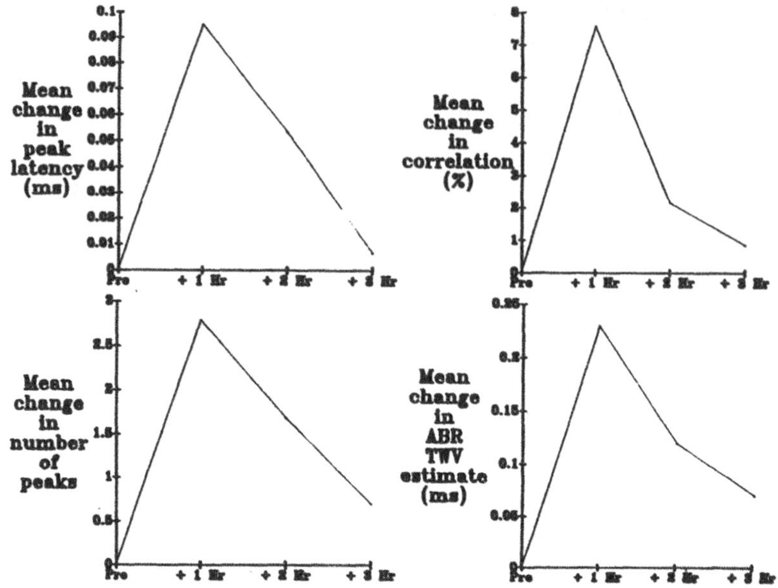

Figure 18.5: mean change after 1, 2 and 3 hours following glycerol dehydration in otoacoustic emission peak latency (top left), cross-correlation (top right), number of peaks (bottom left) and in ABR travelling wave velocity (bottom right).

In this study we found good agreement between the techniques for group data but poor correlation between results for individual subjects. We conclude from this that anatomical differences between subjects lead to the function of different structures being affected differently by changes in cochlear fluid pressure. That is, the TMD measures changes in the dynamics of the stapedius muscle reflex and tympanic membrane. We believe that the TWV test measures changes in the stiffness of the high frequency portion of the basilar membrane, and the evoked otoacoustic emission changes might well be due to changes in the generation of OAEs at low to mid frequencies.

The time course of the changes in emission parameters during glycerol dehydration is similar to the travelling wave velocity changes observed in the same patients (as shown in the bottom right of Figure 18.5). The same function has been demonstrated by other investigators who directly measured changes in CSF and cochlear fluid pressures in animals during glycerol dehydration. That is, an initial steep decrease in cochlear fluid pressure followed by a gradual increase to the pre-dehydration level.

We observed several different types of changes in the emission after glycerol ingestion and found that these were not simply related to audiometric changes. This suggests that we were not only observing decreases in cochlear fluid pressure during dehydration. Glycerol ingestion has been shown by other workers to alleviate the deterioration of cochlear blood flow, long suspected to be an important factor in Ménière's disease. Moreover, in dehydrating the cochlea we will be changing the ionic concentrations of endolymph and perilymph and may thus affect the electrical transduction system responsible for generating otoacoustic emissions.

References

Antonelli A, Grandori F (1991) Long term stability, influence of the head position and modelling considerations for evoked otoacoustic emissions. Scand Audiol (Suppl 25) 97–108.

Dauman R, Cazals Y (1993) Evoked otoacoustic emissions and psychoacoustic measurements in Ménière's Disease: the influence of spontaneous or glycerol-induced changes in hearing sensitivity. Presented at International Congress on Otoacoustic Emissions, Lyons.

Farrell G (1992) Objective estimation of endolymphatic hydrops using auditory brainstem response measures. University of Southampton: PhD thesis.

Parker DJ, Thornton ARD (1978) Cochlear travelling wave velocities calculated from the derived components of the cochlear nerve and brainstem evoked responses of the human auditory system. Scand Audiol 7: 67–70.

Phillips AJ, Farrell G (1988) A review of the effect of changes in csf pressure on audiological measurement. IHR Internal Report Series A No. 5.

Phillips AJ, Farrell G (1992) The effect of posture on three objective audiological measures. Brit J Audiol 26: 339–45.

Phillips AJ, Marchbanks RJ (1989) Effects of posture and age on tympanic membrane displacement measurements. Brit J Audiol 23: 279–84.

Thornton ARD, Farrell G (1991) Apparent travelling wave velocity changes in cases of endolymphatic hydrops. Scand Audiol 20: 13–18.

Correspondence to AJ Phillips, Audiology Department, Aberdeen Royal Hospitals NHS Trust, Woolmanhill Hospital, Aberdeen AB9 1GS, UK. Tel (+44) (0)1224 840764. Fax (+44)(0) 1224 840751.

Chapter 19
Ionic transports in the endolymphatic sac of the guinea pig

V COULOIGNER, M TEIXEIRA, P HULIN, O STERKERS, G PLANELLES and E FERRARY, INSERM U426, Faculté X, Bichat, Université Paris, France and Inserm U.467, Faculté Necker Enfants-Malades, Université Paris, France

ABSTRACT. The endolymphatic sac is a part of the inner ear involved in the regulation of the volume and composition of endolymph. Its dysfunction may be responsible for some cases of Ménière's disease. Conversely, the pharmacological stimulation of the epithelial transports in a functional endolymphatic sac may allow the reduction of the endolymphatic hydrops. The aim of the present study was to examine the effects of locally applied transport inhibitors on the endolymphatic sac transepithelial potential (ESP) in order to determine the membranous transporters involved in ionic transports through the endolymphatic sac epithelium. The results demonstrate that the blockade of acid-base transporters including carbonic anhydrase, a vacuolar-type H^+-ATPase, and an anionic transport system blocked by DIDS induced a decrease of ESP. The inhibition of Na^+,K^+-ATPase by ouabain has only minor effects on the ESP, although this transporter is known to be present on the basolateral membrane of the epithelium.

Key words: ionic transports, endolymphatic sac, transepithelial potential, guinea pig, Ménière's disease.

Introduction

The endolymphatic sac is a monolayer epithelium composed of polarized cells linked by tight junctions and surrounded by a loose connective tissue rich in fenestrated capillaries (Lundquist 1965; Rask-Andersen

1979). Partly located within the petrous bone and partly within the posterior fossa dura, it is linked to the vestibule by the endolymphatic duct. One of its main functions is the regulation of the volume and composition of endolymph. According to the theory of the longitudinal flow of endolymph, the endolymph secreted in the cochlea and the vestibule reaches the endolymphatic sac where it is reabsorbed. Indeed, macromolecules (colloidal silver, horseradish peroxydase, lanthanum) injected in the cochlea were observed after several hours in the lumen of the endolymphatic sac (for review, see Manni 1987), and the destruction of the endolymphatic sac induces an increase of the volume of endolymph called endolymphatic hydrops (Kimura and Schucknecht 1965). Experimental results have already been reported concerning the hydroelectrolytic transports across the endolymphatic sac. The trans-epithelial potential of the endolymphatic sac (ESP) is 20 mV lumen positive (Vosteen and Morgenstern 1986). The chemical composition of the luminal fluid of the endolymphatic sac (Na = 100–130 mM; K = 15 mM; pH = 6.7) (Vosteen and Morgenstern 1986; Tsujikawa et al. 1993; Couloigner et al. unpublished data) differs from that of the basolateral compartment which is similar to any extracellular fluid (Na = 140 mM; K = 5 mM; pH = 7.4). The intravenous administration of acetazolamide, inhibitor of carbonic anhydrase, decreases the ESP and the luminal pH of the endolymphatic sac (Tsujikawa et al. 1993).

Immunohistochemical studies have localized a Na^+, K^+-ATPase (α1- and β2- subunits) basolaterally (Mizukoshi et al. 1988; ten Cate et al. 1994). In a murine study, Takumida and Bagger-Sjöbäck (1989) found that amiloride, an inhibitor of Na channels or Na/H antiporters, can induce a collapse of the endolymphatic sac. Whole cell voltage clamp studies upon isolated endolymphatic sac epithelial cells have found an amiloride-sensitive Na^+ conductance (Wu and Mori 1996a), and a delayed rectifier outward K^+ current (Wu and Mori 1996b). In this work, we study the effects of ion transport inhibitors, especially diuretics, applied on the basolateral side of the endolymphatic sac, upon ESP.

Material and methods

The methods have been previously described (Couloigner et al. 1998). Briefly, adult male pigmented guinea pigs (305 ± 10 g body weight) were anaesthetized by intramuscular injection of 2 ml/kg body weight of a mixture of ketamine (50 mg/ml, Ketalar©, Panpharma, Fougères, France) and xylazine (2%, Rompun©, Bayer, Leverkusen, Germany) (2/1, vol/vol). The same dose was reinjected after 30 min and then a half dose every hour. The breathing was spontaneous. The endolymphatic sac was approached through the posterior cerebral fossa. The ESP was measured with a glass microelectrode with an outside tip diameter ranging from 3 to 5 μm. The microelectrode was inserted in a holder containing an Ag-AgCl pellet (World Precision Instrument, Sarasota, USA), and was moved with a De Fonbrune micromanipulator (CIT Alcatel, Paris, France) under stereomicroscopic observation. The whole system was filled with 150 mM NaCl and connected to an electrometer amplifier (Intra 767, World Precision Instrument). An Ag/AgCl reference electrode was placed on the neck muscles. The zero potential was determined in the fluid surrounding the endolymphatic sac. The ESP was discontinuously measured before and up to 150 min after the perisaccular injection of 5 ml of solutions containing various drugs known to inhibit ionic transport systems. The control solution injected was saline solution (150 mM, n = 12; Biopharma, Paris, France). The tested drugs were: ouabain (10^{-3}M, n = 3; Sigma), bumetanide (10^{-6} M, n = 6; 10^{-4} M, n = 6; Sigma), chlorothiazide (10^{-4} M, n = 3; Sigma), amiloride (10^{-4} M, n = 7; Sigma), 4,4' - diisothocyano- 2,2'- disulfonic acid stilbene (DIDS, 10^{-4} M, n = 4; Sigma), acetazolamide (Diamox®, 10^{-3} M, n = 4; Théraplix, Paris, France), bafilomycin A1 (10^{-5} M, n = 4; 10^{-6} M, n = 3; 10^{-8} M, n = 4; gift of K. Altendorf, Osnabrück, Germany). The drugs were dissolved in 150 mM NaCl except for bafilomycin which was dissolved in dimethylsulfoxid (DMSO) (final DMSO dilution: 1/1000 to 1/1000000 in 150 mM NaCl, vol/vol).

Results

The mean ESP value before the injection was +8.4±0.3 mV (n = 78). No difference was observed between the different groups of animals (Newman–Keuls analysis).

The control basolateral injection of a 5 µl solution containing 150 mM NaCl (n = 12) decreased the ESP of 64 %±6 at 5 min ($p < 0.001$). Then the ESP returned to its initial value at 60 min and overpassed it at 120 min ($p < 0.01$). The same biphasic evolution was observed in all tested groups, with an initial ESP decrease induced by the basolateral injection, followed after 5 min by a progressive total or partial ESP recovery. This surprising pattern suggests a pressure effect. This interesting point remains to be further elucidated by measuring the local pressure after the injection.

The ESP recovery was partly inhibited by ouabain (10^{-3} M; n = 3), inhibitor of Na^+, K^+-ATPase ($p < 0.01$), acetazolamide (10^{-3} M, n = 4), inhibitor of carbonic anhydrase ($p < 0.001$), amiloride (10^{-4} M, n = 7), inhibitor of Na channel or Na/H exchanger ($p < 0.05$), DIDS (10^{-4} M, n = 4), inhibitor of transporters involving HCO_3– ($p<0.001$), and bafilomycin A1, inhibitor of the vacuolar type H^+-ATPase. A log-dose effect (10^{-5} M, n = 4; 10^{-6} M, n = 3; 10^{-8} M, n = 4) upon the ESP was observed with bafilomycin A1 (r = 0.968; $p < 0.001$). At variance, bumetanide (10^{-6} M, n = 6; 10^{-4} M, N = 6), inhibitor of Na-K-Cl cotransport, and chlorothiazide (10^{-4} M, n = 3), inhibitor of Na-Cl cotransport, did not alter the ESP recovery as compared to the control injection of 150 mM NaCl.

Discussion

The endolymphatic sac is involved in the resorption of endolymph (Kimura and Schucknecht 1965) and its dysfunction is probably one of the mechanisms of Ménière's disease (review by Merchant et al. 1995). The aim of this study was to examine the effects of locally applied drugs upon the ESP in guinea pigs and thus to determine the membranous transporters eventually involved in ionic transports across the endolymphatic sac. The method of local drugs administration was preferred to the intravenous injection because it avoided systemic effects of the tested drugs, such as an alteration of the body fluids homeostasis or a modification of the CSF secretion, which could have indirect effects on inner ear fluids. Another advantage of this method is that it allowed the testing of drugs like ouabain or bafilomycin whose systemic administration are potentially lethal. Limitations of this experimental method are that: (i) the exact location of the drugs targets (apical or basolateral) remain unsure; (ii) what is compared between the control (150 mM NaCl) and the tested groups (drugs dissolved in 150 mM NaCl) is not the effect upon the ESP at the steady-state but the differences in the ESP recovery after the initial ESP, which follows the basolateral injection.

The genesis of the ESP appears to be highly dependent on acid-base transport systems including carbonic anhydrases, a vacuolar-type H^+-ATPase, and an anionic transport system blocked by DIDS. The acidity of the endolymphatic sac luminal pH (Tsujikawa et al. 1993) may result from an active luminal proton secretion by the H^+-ATPase. The slight inhibition of the ESP recovery by ouabain, inhibitor of Na^+-K^+-ATPase, suggests that this pump, which is known to be present in the basolateral membrane of the endolymphatic sac (Mizukoshi et al. 1988) may be less directly involved in the genesis of the ESP than acid-base transporters.

Further studies are in progress in order to: (i) inject drugs in the luminal compartment of the endolymphatic sac; (ii) determine the effects of these drugs upon the luminal pH of the endolymphatic sac.

Acknowledgments

This work was supported by grants from Fondation pour la Recherche Médicale (V. Couloigner) INSERM, Université Paris 7, and Faculté Xavier Bichat.

References

Couloigner V, Loiseau A, Sterkers O, Amiel C, Ferrary E (in press) Effect of locally applied drugs upon the endolymphatic sac potential. Laryngoscope.

Kimura RS, Schucknecht HF (1965) Membranous hydrops of the inner ear of the guinea pig after obliteration of the endolymphatic sac. Practical in Otorhinolaryngol 27: 343–54.

Lundquist PG (1965) The endolymphatic duct and sac in the guinea pig. An electron microscopic and experimental investigation. Acta Otolaryngol Suppl 201.

Manni JJ (1987) The endolymphatic duct and sac in the rat. Thesis, Nijmegen, Netherlands.

Merchant SN, Rauch SD, Nadol JB (1995) Ménière's disease. Eur Arch Otorhinolaryngol 252: 63–75.

Mizukoshi F, Bagger-Sjöbäck D, Rask-Andersen H, Wersall J (1988) Cytochemical localization of Na^+, K^+-ATPase in the guinea pig endolymphatic sac. Acta Otolaryngol 105: 202–8.

Rask-Andersen H (1979) The vascular supply of the endolymphatic sac. Acta Otolaryngol 88: 315–27.

Takumida M, Bagger-Sjöbäck D (1989) Effects of amiloride on the endolymphatic sac. J Laryngol Otol 103: 466–70.

ten Cate WJF, Curtis LM, Rarey (1994) KEb Na^+, K^+-ATPase subunit isoform expression in the guinea pig endolymphatic sac. ORL 56: 257–62.

Tsujikawa S, Yamashita T, Tomoda K, Iwai H, Kumazawa H, Cho H, Kumazawa T (1993) Effects of acetazolamide on acid–base balance in the endolymphatic sac of the guinea pig. Acta Otolaryngol Suppl 500: 50–3.

Vosteen KH, Morgenstern C (1986) Biochemical aspects of inner ear pathophysiology. In: Pfaltz CR, ed. Controversial aspects of Ménière's disease. Stuttgart: Thieme-Verlag 16–28.

Wu D, Mori N (1996a) Amiloride-sensitive Na^+ currents in epithelial cells isolated from the endolymphatic sac of guinea pigs. Proc. Sendai Ear Symp6: 37–40.

Wu D, Mori N (1996b) Outward current in epithelial cells isolated from intermediate portion of endolymphatic sac of guinea pigs. Am J Physiol 271: C1765–73.

Correspondence to Evelyne Ferrary, INSERM U.426; Faculté X. Bichat BP 416, 16 rue Henri Huchard 75870 Paris Cedex 18, France. Tel. (+33) 1 44 85 62 70. Fax (+33) 1 42 28 15 64. E-mail ferrary@bichat.inserm.fr.

Chapter 20
Beta-trace protein in human perilymph

G BACHMANN, SE BALDUS, M NEKIC, O MICHEL, Universität Zu Köln, Köln, Germany

ABSTRACT. The beta-trace protein, a lipocalin, which has recently been identified as glutathion-independent prostaglandin D-synthase [PGS-S, prostaglandin-H2 D-isomerase, (5Z, 13E)-(15S)-9a, 11-epidioxyd-15-hydroxyprosta-5, 13-dienoate D-isomerase, EC 5,3,99,2], was found in human cerebrospinal fluid, aqueous humour, sperm and urine. Beta-trace protein is a very specific and practical marker for cerebrospinal fluid, which is important for the clinical diagnosis of cerebrospinal fluid fistulae because only very small samples are necessary (5 µl) to perform the test.

To answer the question of whether beta-trace protein is contained in human perilymph, 20 samples were collected post mortem. Between 17 and 107 hours after death the footplates of stapes ossicles were removed and samples were taken with capillary tubes under microscopic view. In order to prevent contamination with CSF or endolymph, the samples were drawn by the power of adhesion without aspiration. The sample volumes were limited to 5 µl per ear. With a one-dimensional immunelectrophoresis all specimens precipitated with a monospecific, polyclonal antibody. Beta-trace protein is contained in human perilymph. This might be a further sign for the cerebrospinal-fluid genesis of human perilymph and it is possible to diagnose a perilymph fistula with beta-trace protein.

Introduction

In clinical routine the beta-trace test confirms the diagnosis of a cerebrospinal fluid fistula. Hoffmann et al. (1993) found that beta-trace protein is identical to glutathion-independent prostaglandin D-synthase. Beta-trace protein with a molecular mass of 23500 Dalton belongs to a lipocalin superfamily with suggested lipophilic ligand-carrier and enzymatic functions. It is the second most abundant protein in human cerebrospinal fluid and is found in aqueous humour, sperm and urine, too. Usually, the protein is not detectable in serum except in patients with acute glomerulonephritis or renal insufficiency. The choroid plexus is the major site of mRNA expression for beta-trace protein (Blödorn et al. 1996). In animal studies, Tachibana et al. (1987) showed that beta-trace protein occurs in the rat cochlea.

Material and method

To answer the question of whether beta-trace protein is contained in human inner-ear fluid, 20 perilymph samples were collected post mortem. After opening the footplate of the stapes, the samples were taken with capillary tubes (DRUMMOND® Wiretrol) under microscopic view. In order to prevent contamination with cerebrospinal fluid, the samples were drawn only by the power of adhesion without aspiration and the volumes were limited to 5 µl per ear. The capillary tubes were placed just to the edges of the vestibular window and the membranous labyrinths were not touched. The samples were investigated by an one-dimensional immunelectrophoresis with a monospecific, polyclonal antibody.

Result

By immunelectrophoresis all specimens precipitated with a mono-specific, polyclonal antibody. Beta-trace protein is contained in human perilymph. First quantitative measurements (Figure 20.1) revealed that beta-trace protein was very concentrated in perilymph (Mean 148.5 mg/l SD 65.5), with a high inter-individual variation. In correlation with the time of death, there was no clear evidence for a circadian fluctuation. The highest concentration was found in a sample that was taken two days after death (Figure 20.2).

Discussion

Apparently the beta-trace protein is more concentrated in perilymph than in cerebrospinal fluid and varies much inter-individually. Hayaishi et al. (1993) found that prostaglandin D synthase activity in the rat brain exhibits a circadian fluctuation in parallel with the sleep/wake cycle. In relation to the time of death, there was no clear evidence for a circadian fluctuation but there are not enough data in order to answer this question. Furthermore, the influence of post mortem changes are not clarified.

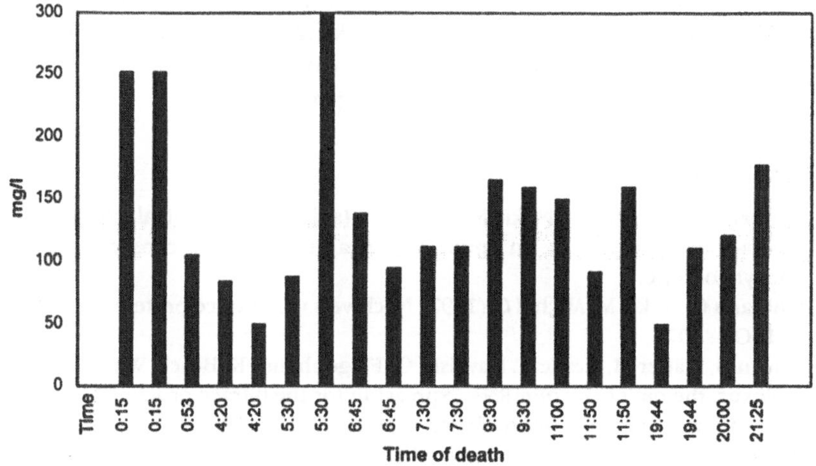

Figure 20.1: beta-trace protein in perilymph (n = 20) (mg/l versus hours time of death).

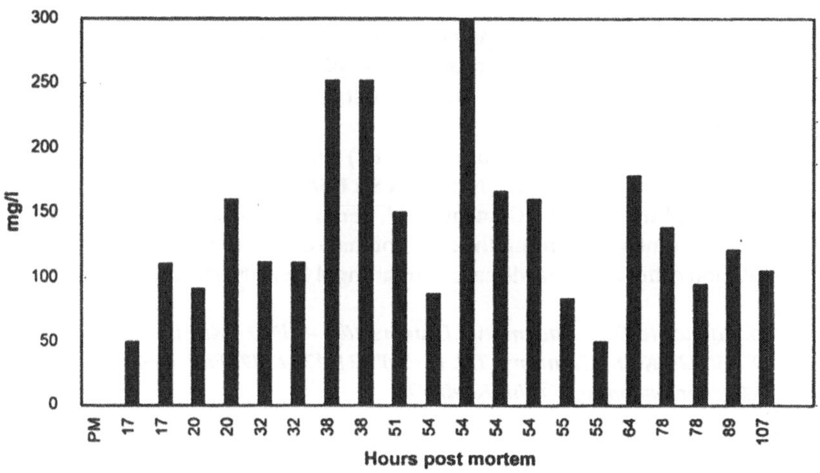

Figure 20.2: stability of beta-trace protein post mortem (n = 20) (mg/l versus hours post mortem).

Felgenhauer et al. (1987) described the normal value for beta-trace protein in cerebrospinal fluid (33 mg/l SD 11; n = 162) but there are no data available about the stability of the protein and normal values post mortem.

There can be no doubt that beta-trace protein is contained in human perilymph. This might be a further sign for the cerebrospinal fluid genesis of human perilymph. In the search for markers for the diagnosis of perilymph fistula, Thalmann et al. (1994) detected β2-transferrin in perilymph at a concentration roughly equal to that in cerebrospinal fluid. To diagnose a perilymph fistula, small sample volumes could be collected with little pieces of nasal packing. In room temperature, the beta-trace protein remains unchanged over several days. A perilymph fistula can be diagnosed with beta-trace protein.

References

Bachmann G, Achtelik R, Nekic M, Michel O (submitted) Clinical experience with beta-trace protein as an immunological marker for cerebrospinal fluid. Laryngoscope.

Bachmann G, Nekic M, Michel O (1997) Nachweis von β-trace protein in Perilymphe. HNO 4: 307.

Blödorn B, Mäder M, Urade Y, Hayaishi O, Felgenhauer K, Brück W (1996) Choroid plexus: the major site of mRNA expression for the β-trace protein (prostaglandin D synthase) in human brain. Neurosci Lett 209: 117–20.

Felgenhauer K, Schädlich H J, Nekic M (1987) β-trace protein as marker for cerebrospinal fluid fistula. Klin Wochenschr 65: 764–8.

Hayaishi O, Matsumura H, Urade Y (1993) Prostaglandin D synthase is the key enzyme in the promotion of physiological sleep. J. Lipid Mediat Cell Signal 6:429–31.

Hoffmann A, Conradt H S, Gross G, Nimtz M, Lottspeich F, Wurster U (1993) Purification and chemical characterization of β-trace protein from human cerebrospinal fluid: Its identification as prostaglandin D synthase. J Neurochem 61(2): 451–6.

Tachibana M, Fex J, Urade Y, Hayaishi O (1987) Brain-type prostaglandin D synthase occurs in the rat cochlea. Proc Natl Acad Sci USA 84: 7677–80.

Thalmann I, Kohut R I, Ryu J, Comegys T H, Senarita M, Thalmann R (1994) Protein profile of human perilymph: In search of markers for the diagnosis of perilymph fistula and other inner ear disease. Otolaryngol Head Neck Surg 111(3):273–80.

Correspondence to Dr G Bachmann, Universitäts – HNO – Klinik, Joseph Stelzmann Str 9, D – 50924 Köln, Germany. Tel. (+49) 221 4784787. Fax (+49) 221 4784793. E-mail: Gregor.Bachmann@Uni-Koeln.DE.

Chapter 21
Characteristics of infrasonic otoacoustic emissions generated by the cardiovascular system

MARK P HUGHES, Institute of Laryngology and Otology, University College London and ROBERT J MARCHBANKS, University Child Health, University Hospital, Southampton, UK

ABSTRACT. This study examines the characteristics of cardiovascular-generated infrasonic otoacoustic emissions (CVSEs), which are measured in terms of low frequency movements of the tympanic membrane (TM). The amplitude of these emissions was studied in 10 normal subjects (23–50 years). The measurements were made using the TMD Cerebral and Cochlear Fluid Pressure Analyser (TMD analyser).

The study of these CVSEs is important for a number of reasons: they are the main source of noise with TMD cerebral and cochlear pressure measurements; a component of this activity increases in amplitude with intracranial pressure (ICP) and is considered to be the transmission of cerebral CVS to the labyrinth via the cochlear aqueduct; they may be a generator of vertigo and tinnitus found in patients with intracranial hypertension.

The CVSE amplitude was observed in relation to body posture, intracochlear fluid pressure, middle-ear compliance and pressure, cerebral-to-labyrinth fluid pressure transfer. The main positive findings were definite changes in amplitude with body posture and a highly significant relationship with peak compliance of the middle ear ($p = 0.001$). Results also indicate that the CVS emissions change if the TMD measurements are undertaken with the external ear canal pressure set to that of the middle ear. Further investigations should be undertaken to see if the intersubject variability in TMD cerebral and cochlear pressure measurements might be reduced by controlling for middle-ear compliance and pressure.

Key words: tympanic membrane displacement, infrasonic, vertigo, pulsatile tinnitus, cardiovascular, intracranial pressure.

Introduction

This chapter reports some exploratory work on cardiovascular infrasonic otoacoustic emissions (CVSEs), which are generated by the beating heart (usually in the range 1–2 Hz, which corresponds to heart rate of 60–120 beats per minute). The pressure wave is transmitted from the heart to the ear via the arteries. The understanding of these emissions is considered to be important to such conditions as pulsatile tinnitus (Andreasson et al. 1978) and the Tullio phenomenon in which sound entering the ear causes vertigo (Erlich and Lawson 1980). The chapter also aims to publicise the tympanic membrane displacement analyser (TMD analyser) as a tool for this work, and for related work such as intracranial pressure monitoring.

These CVSEs are not related to the 'cochlear echo' type of otoacoustic emissions publicised by Kemp's work (Kemp 1978). They are otoacoustic emissions in that they are sound (albeit infrasound) emitted from the ear, and they were reported earlier by Marchbanks (1984).

Interest in CVSEs arises from the fact that externally applied low-frequency sound and infrasound produce vertical nystagmus and sensations of falling forwards (Evans 1976). It is hypothesised that internally generated infrasound from the heart beat may be linked to some forms of vertigo, and low frequency or pulsatile tinnitus that exists with intracranial hypertension. (Marchbanks 1996, 1997; Sismanis 1987). Studies of a patient with positional-related vertigo (Marchbanks and Davies 1994) using the tympanic membrane displacement (TMD) analyser, indicated a more than tenfold increase in the amplitude of these infrasonic emissions from the ear after the Hallpike manoeuvre (a manoeuvre from sitting to supine head-hanging right or left position – Dix and Hallpike 1952). In this study we investigate these findings of changes in CVSE amplitude with posture change further.

Materials and methods

The work was conducted at the Institute of Laryngology and Otology, and the Royal National Throat Nose and Ear Hospital, 330/332 Gray's Inn Road, London WC1X 8EE, England.

Equipment

The apparatus used for recording CVSEs was the tympanic membrane displacement analyser MMS-10 (Marchbanks Measurements Systems UK Ltd). This apparatus measures the eardrum displacement in nanolitres using a null pressure technique – that is, eardrum displacement is inversely matched by a reference diaphragm that maintains constant pressure by means of servo control. The eardrum displacement is quantified in terms of the voltage to the reference diaphragm, which is required to maintain a constant pressure within the sealed external auditory meatus (Tweed et al. 1986; Marchbanks 1980, 1984).

Subject selection

The ten subjects comprised four males and six females, with an age range of 23 to 50 years. A balanced experimental design was adopted according to gender and the downmost ear in the Hallpike manoeuvre.

Procedure – measurements made

The TMD analyser provided an indirect measurement of intracochlear fluid pressure as well as amplitude of the CVSE.

Intracochlear fluid pressure

This measurement uses the acoustic stapedial reflex. Acoustic stimulation above the threshold results in a displacement of the tympanic membrane, the direction and magnitude of which is dependent on the intracochlear fluid pressure (Tweed et al. 1986; Marchbanks 1980; Casselbrant et al. 1978).

Further, if the cochlear aqueduct, which passes between the perilymph of the labyrinth and the intracranial cerebrospinal fluid, is patent, then the TM displacement will also reflect the intracranial fluid pressure (Phillips and Marchbanks 1989; Beentjes 1972).

The ipsilateral reflex is elicited at 15 dB, then at 20 dB, above the acoustic reflex threshold at 1000 Hz; the stimulus duration is 0.5 seconds. The TMD is recorded for a total time of 1.5 seconds. In order to improve the signal-to-noise ratio, 10 records of each measurement are averaged. The intracochlear pressure was quantified in terms of the configuration and magnitude of the TM displacements. Vm is an indirect measure of intracochlear pressure, and is the mean displacement of the TM from the point of maximum inward displacement whilst the acoustic stimulus is present to acoustic stimulus offset.

Cardiovascular-induced TMD

The variation of the TM displacement over a 6 s period was measured in terms of the peak-to-trough amplitude. The periodicity was checked against the radial pulse to ensure the record is truly CVS TMD.

Ipsilateral acoustic reflex threshold and tympanometry

These were elucidated using the GSI 33 middle ear analyser. The acoustic reflex threshold at 1000 Hz was measured.

Tympanometry yields a middle ear pressure (MEP) measurement and peak compliance measurement. The GSI 33 'zeros' the compliance at the initial starting pressure; therefore the peak compliance is equivalent to static compliance (that is, without the external ear component).

Procedure – testing a subject

A brief history from each subject enabled them to be considered under the 'benign paroxysmal positional vertigo (BPPV) group' or the 'non-BPPV group'. Subjects were questioned on past middle-ear problems.

Otoscopy was performed in all subjects. The ipsilateral acoustic reflex was elicited in each ear, both at 15 and 20 dB above threshold in the sitting posture. These measurements were repeated in the supine (15 degrees head-down) posture.

The CVSE was recorded twice in each ear in the sitting posture. The Hallpike manoeuvre was then performed. The CVSE was again recorded twice in each ear in the supine posture.

Statistical methods

Regression was performed to look for correlations:

- between CVSE amplitude and peak compliance;
- between CVSE amplitude and both intracochlear pressure and intracochlear pressure change with Hallpike manoeuvre;
- between CVSE amplitude change and intracochlear pressure change (that is, change with Hallpike manoeuvre).

The MINITAB computer statistics program was used. Analysis of variance was carried out to provide a multivariate analysis of any correlation between CVSE amplitude, peak compliance and middle-ear pressure.

Results

Results are shown in Tables 21.1, 21.2 and 21.3, and Figs 21.1–21.4.

The CVSE amplitude is about 195 nl (median) but has a wide range. The emission measurements from right and left ears of a subject are independent of each other. Change in CVSE amplitude with posture, from sitting to supine, by the Hallpike manoeuvre has a median value of −40 nl with a right-skewed distribution as shown in Figure 21.2. The relationship between CVSE amplitude and peak compliance is highly significant ($p = 0.001$); however, the correlation coefficient ($r = 0.61$) is not exceptionally high. This indicates that factors such as MEP should be taken into account, hence the multivariate analysis of the three variables (CVSE amplitude, peak compliance and MEP).

Table 21.1: results

CVSE amplitude Characteristic	Result Med	IQR (nl)	Range	Med	IQR (%)	Range
Intra-subject variability *	28	12→65	0→318	13·5	9→29	0→82
Inter-subject variability **	195·5	133·5→ 349	51→ 1136	–	–	–
Inter-ear difference	133·5	59·5→ 250·5	27·5→ 745·5	2·01	1·41→ 2·63†	1·21→ 6·32†
Change after Hallpike manoeuvre	–40	88→ +133·5	–352→ +857·5	–14	–35→ +133	–73→ +718

Med = Median
IQR = Interquartile range
* Difference (in nl) and percentage difference between two measurements under the same conditions
** Distribution of all the measurements made (n = 100)
† Ratio larger : smaller CVSE

Table 21.2: results

Correlation of CVSE amplitude with	Result
Intracochlear pressure	None (p values) 0·48 (sitting) 0·41 (supine) 0·69 (patient)
Peak compliance	Highly significant p = 0·001, r = 0·61
Pulse rate	No correlation; large inter-ear difference in a subject yet same pulse

Table 21.3: results

Correlation of change in CVSE amplitude with Hallpike manoeuvre with:	Result
Intracochlear pressure change	None

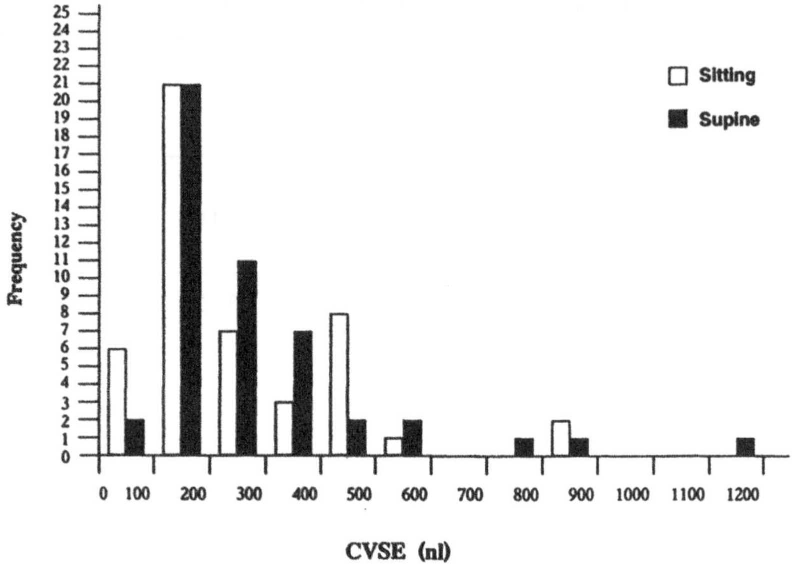

Figure 21.1: distribution of CVSE amplitude measurements.

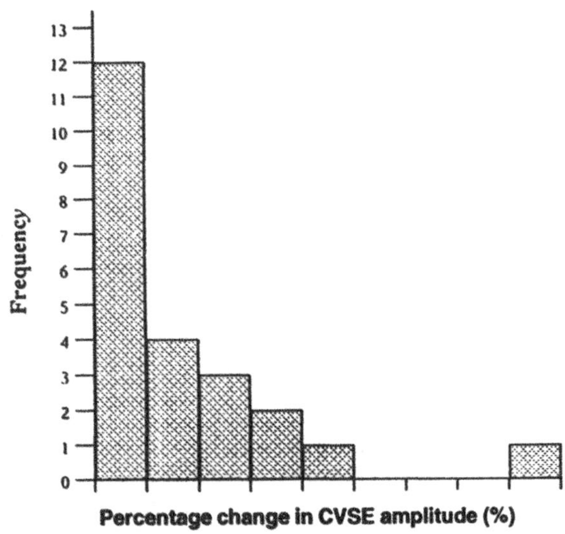

Figure 21.2: distribution of % change in CVSE amplitude due to the Hallpike manoeuvre.

Discussion

Measurement repeatability

The median intra-subject variability of the CVSE measurements was 28 nl (13.5%). The CVSE is a dynamic physiological parameter, which is

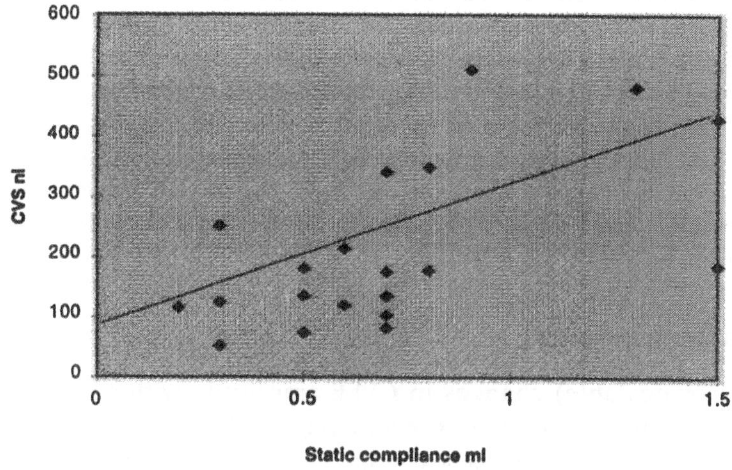

Figure 21.3: correlation between CVSE amplitude in the sitting posture and peak compliance. Regression equation is CVSsit = 72.6 + 207 peak comp.

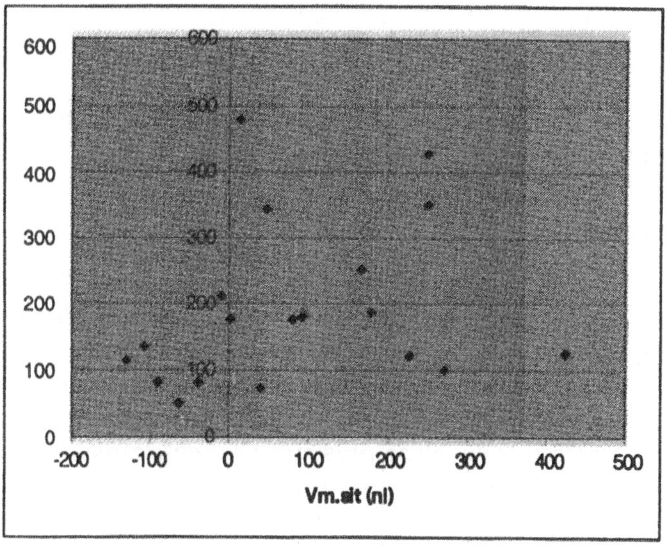

Figure 21.4: scatterplot of CVSE amplitude in the sitting posture and Intracochlear pressure (indirect).

expected to be continuously variable over time due to changes, such as in the force of contraction of heart muscle and in cranial arterial dilation. The intra-subject variation is much smaller than inter-subject variation, suggesting that a true phenomenon is being observed.

The observed CVSE is likely to be the superimposition of CVS activity from a number of sources, including vasculature within the external and middle ear cavities, as well as a component transmitted from the intracranial fluid to the perilymph by means of the cochlear aqueduct (Marchbanks 1996). There are also a number of procedural factors that should be considered. It is observed that the CVSE is usually greater in magnitude if testing is performed immediately after physical exertion, or when the patient is anxious. It is normal to allow at least 10 minutes for the patient to relax before attempting to undertake TMD measurements. The exact position of the seal within the external auditory meatus also appears to be a factor. This may relate to pulse synchronous probe movement translating from the blood vessels of the meatus walls, which in turn is expected to vary with the lateral probe position.

Effect of body posture

Definite (repeatable) changes in CVSE amplitude with change from sitting to supine posture are observed (Figure 21.2). The inter-subject variability is large (percentage change ranges from –73% to +718%. Table 21.1). In addition to the various factors mentioned above, a body manoeuvre from sitting to supine increases the intracranial pressure and causes cardiovascular changes.

Intracochlear pressure

There is not a significant correlation between CVSE amplitude and intracochlear pressure (Figure 21.4, $p = 0.48$) (Table 21.2).

Peak compliance

There is a striking correlation ($r = 0.61$) between peak TM compliance and CVSE amplitude in the same ear (sitting posture). The higher the peak compliance, the higher the aural emission amplitude. This suggests that the CVSE is likely to be principally due to TM displacements and not of external auditory meatus origin. The TM displacements are a combination of intracranial-cochlear pressure changes transmitted from the oval window by means of the ossicles and will also reflect CVS activity within the middle ear cleft, which will correspondingly alter middle ear pressure. For both sources a more compliant middle ear and TM will be expected to increase the magnitude of the CVSE. Peak compliance is measured when pressures either side of the TM are equal; the effect of MEP on TM compliance is abolished in the peak compliance measurement. The CVSE measurements were made at atmospheric pressure. In those cases with high or low MEP the measured peak compliance will be higher than the actual compliance when the CVSE amplitude measurement was made because the pressure versus compliance curve (tympanometry) has a peak, falling away to lower compliance at lower and higher pressures.

Middle-ear pressure

Plotting CVSE amplitude (sitting) against MEP does not display an obvious relation between them. The effect of MEP on CVSE amplitude appears to be mediated by its effect on compliance as previously mentioned.

Cerebral-to-labyrinth fluid pressure transfer

The parameter [Vm supine – Vm sitting] indicates intracochlear pressure change with posture change. An increase in intracranial pressure by moving from sitting to supine may be transmitted via the cochlear and vestibular aqueducts to cochlear fluids. Cardiovascular pulsations producing CVSEs may also be transmitted by these aqueducts. It was thought that the CVSE amplitude might rise with rising [Vm supine – Vm sitting].

Plots of CVSE (sitting) against [Vm supine – Vm sitting] and of [CVSE supine – CVSE sitting] against [Vm supine – Vm sitting] did not display a recognisable relationship.

It may be that changes in cochlear window compliance and the cochlear aqueduct frequency filter bandwidth also affect transmission of intracranial cardiovascular pulsations to the cochlea (Marchbanks 1996).

Conclusion

Data characterising cardiovascular-generated infrasonic otoacoustic emissions have been collected. The results indicate the emission amplitudes from the right and left ears of a subject to be independent of one another, suggesting factors local to each ear influence the amplitude (for example, middle ear compliance and pressure).

Further work may look at the relationship between CVSE amplitude and middle ear/TM compliance at atmospheric pressure (rather than peak compliance). The CVSE amplitude may also be studied at various degrees of external ear canal pressure. The results may enable improvements in measurement technique with the TMD analyser by reducing CVSE 'noise'.

Acknowledgements

My thanks go to Professor L. Luxon for her help as second supervisor and to Mrs J Bird who helped with the acquisition of data. The work was done as the project for the MSc in Audiological Medicine at the Institute of Laryngology and Otology, using the facilities of the hospital with R Marchbanks supervising.

References

Andreasson L, Harris S, Ivarsson A (1978) Pulse volume recordings in outer ear canal in pulse synchronous tinnitus. Acta Otolaryngologica 86: 241–7.

Beentjes B I J (1972) The cochlear aqueduct and the pressure of cerebrospinal and endolabyrinthine fluids. Acta Otolaryngologica 73: 112–20.

Casselbrant M, Ingelstedt S, Ivarsson A (1978) Volume displacement of the tympanic membrane at stapedius reflex activity in different postures. Acta Otolaryngol 85: 1–9.

Dix R, Hallpike CS (1952) The pathology, symptomatology, and diagnosis of certain common disorders of the vestibular system. Proceedings of the Royal Society of Medicine 45: 341–54.

Erlich MA, Lawson W (1980) The incidence and significance of the Tullio phenomenon in man. Otolaryngol Head Neck Surg 88: 630–5.

Evans MJ (1976) Physiological and Psychological Effects of Infrasound at Moderate Intensities. In: Tempest W (ed.) Infrasound and Low Frequency Vibration, London: Academic Press pp. 97-113.

Kemp DT (1978) Stimulated acoustic emissions from within the human auditory system. J Acoust Soc Am 64(5): 1386–91.

Marchbanks RJ (1980) A study of tympanic membrane displacement. PhD Thesis, Brunel University, England.

Marchbanks RJ (1984) Measurement of tympanic membrane displacement arising from aural cardiovascular activity, swallowing, and intra-aural muscle reflex. Acta Otolaryngol 98: 119–29.

Marchbanks RJ (1996) Hydromechanical Interactions of the Intracranial and Intralabyrinthine Fluids. In Ernst A, Marchbanks R, Samii M (eds) Intracranial and Intralabyrinthine Fluids: Basic Aspects and Clinical Applications. Berlin: Springer Verlag.

Marchbanks RJ (1997) Why monitor perilymphatic pressure in Ménière's disease? Acta Otolaryngol (Stockh) Suppl 526: 27–9.

Phillips AJ, Marchbanks RJ (1989) Effects of posture and age on tympanic membrane displacement measurements. British Journal of Audiology 23: 279–84.

Sismanis A (1987) Otologic manifestations of benign intracranial hypertension syndrome : diagnosis and management. Laryngoscope 97: 1–17.

Tweed JJ, Marchbanks RJ, Martin AM (1986) The effect of changes in cerebrospinal fluid pressure on the labyrinth in terms of tympanic membrane displacement. ISVR (Southampton University) Technical Report 135.

Correspondence to Mark P Hughes, 78 Belmont Road, Wealdstone, Harrow, Middlesex HA3 7PN, England. Tel (+44)(0)181 863 3349.

Chapter 22
Morphological effects of low extracellular calcium on stereociliary bundles in the guinea pig cochlea

BN KUMAR, DN FURNESS, CM HACKNEY, Department of Communication and Neuroscience, Keele University, UK

ABSTRACT. Mechano-sensitive hair cells of the mammalian cochlea have an apical bundle of stereocilia, deflections of which control the opening of mechano-electrical transduction channels and thus generate receptor potentials in the cell below. These channels may be gated either by an extracellular filament (tip link) or by a mechanism associated with a junction-like contact region near the stereociliary tips. In lower vertebrates (turtles) low calcium levels have been shown to produce loss of tip links that run between stereocilia, supporting the suggestion that these are the gating springs. However, the tip junction region has also been seen to be disrupted by low calcium conditions. The present study aims to investigate whether morphological changes take place in low-calcium conditions in mammalian hair bundles, which may provide some evidence about the role of these structures in mammalian hair cell transduction. The preliminary results are similar to those in lower vertebrates, implying that the process leading to tip-link loss is likely to be the same in mammals and that the role of calcium in the transduction mechanism may be similar.

Key words: calcium, mechano-transduction, stereocilia.

Introduction

Mechano-sensitive hair cells of the mammalian cochlea have an apical bundle of stereocilia, deflections of which control the opening of mechano-electrical transduction channels and thus generate receptor potentials in the cells below. These channels may be gated either by an extracellular filament (tip link) that runs between the tips of the shorter stereocilia and the sides of the stereocilia in the row behind, or, as shown by Hackney and Furness (1995), by a mechanism associated with a junction-like region just below their tips. In lower vertebrates, the transduction mechanism is sensitive to alteration of Ca^{2+} levels around the hair bundles. Assad et al. (1991) reported that in frog saccular hair cells exposed to very low calcium conditions, transduction currents were abolished and tip links were absent. Crawford et al. (1991) showed that in isolated turtle hair cells, transduction currents obtained in artificial perilymph solution containing 2.8 mM Ca^{2+} are increased following exposure to 0.05 mM Ca^{2+} and irreversibly abolished in 1μM Ca^{2+}. Furness et al. (1997) showed, in a scanning electron microscopic study examining the morphology of turtle hair bundles, that after exposure to these conditions the highest proportion of tip links was present after 50 μM Ca^{2+}, intermediate after exposure to 2.8 μM Ca^{2+} (artificial perilymph) and lowest after 1 μM Ca^{2+}. Changes also occurred in the tip junctions, with fusion of membranes being observed in the 1 μM Ca^{2+}.

Although experiments on transduction have yet to be performed on mammalian hair cells in different Ca^{2+} levels, these cells have similar structural features, including both tip links and tip junctions, and are presumed to have a similar transduction mechanism to that of lower vertebrates. The present study has therefore been performed to investigate whether morphological changes take place in these conditions in mammalian hair bundles that may provide some evidence about the role of these structures in mammalian hair cell transduction.

Materials and methods

Guinea pig cochleae were excised in mammalian artificial perilymph (AP) containing (in μM) NaCl (142), KCl (4), $MgCl_2$ (1.5), glucose (4), Na-HEPES (5) and $CaCl_2$ (1). They were incubated in either AP (N = 6), 0.05 mM Ca^{2+} solution (mM: NaCl (136), KCl (4), Na-HEPES (5), glucose (4) and $CaCl_2$ (0.05); n = 6), or 1 μM Ca^{2+} solution (mM: NaCl (136), KCL (4), Na-HEPES (5), glucose (4), Na-HEDTA (5) and $CaCl_2$ (2.5); n = 6) for 20 minutes before fixation for electron microscopy. Two cochleae in each group were prepared for scanning electron microscopy (SEM) and the remaining ones were prepared for transmission electron microscopy (TEM).

Scanning electron microscopy: the cochlear spiral was impregnated with osmium using the OTOTO technique and dissected. Segments were dehydrated, critical point dried, and viewed using a Hitachi S-4500 cold field emission SEM operated at 2–20 kV. Tip links were counted in 10 randomly selected hair bundles from each condition and the percentage present was calculated as a function of the expected number.

Transmission electron microscopy: the cochlear spiral was dissected and segments were dehydrated and embedded in Spurr resin. Ultrathin (70 nm) radial sections were cut, mounted on copper grids, stained in 2% uranyl acetate followed by aqueous lead citrate and examined in a Jeol 100CX EM at 100 kV. Micrographs of the tip region of stereocilia were taken at × 66 000 magnification and measurements were made of the length of the tip link and the minimum spacing in the contact region in order to provide normative data against which comparisons could subsequently be made between the different calcium conditions.

Results

Scanning electron microscopy: one cochlea in each of the three conditions have so far been examined. Tip links were clearly present in the 1mM and 50 μM Ca^{2+} but were difficult to observe in the 1 μM Ca^{2+}. Counts confirmed that the lowest percentage of tip links was observed in the 1 μM Ca^{2+} treatment where 4.6% were present compared with 51.8% and 41.4% in 1 μM and 50 μM Ca^{2+} treatments respectively. Statistical analysis showed that the difference in tip-link numbers was significant when compared between 50 μM and 1 mM ($p < 0.01$) and between 1 μM and 1 μM Ca^{2+} ($p < 0.001$) but not between 1 mM and 50 μM Ca^{2+} treatments.

Transmission electron microscopy: preliminary data have so far been obtained from three cochleae from different animals for the artificial perilymph condition only. Tip links and junctions were readily observed in all samples. The mean tip-link length was 130 nm ± SD = 47 nm, n = 10, and the minimum spacing measured in the junction region was 19 nm ± 20 nm, n = 10.

Conclusions

These preliminary results suggest that relatively few tip links are found in guinea pig cochleae exposed to low Ca^{2+} (1 μM) solution compared with 50 μM and 1 mM Ca^{2+} solutions. This has to be further confirmed in further specimens in order to rule out the possibility that these values represent inter-animal variability. However, these results are similar to those obtained in bull frog saccular hair cells (Assad et al. 1991) and turtle hair cells (Furness et al. 1997), implying that the process leading to tip-link loss is likely to be the same in mammals as in these lower vertebrates, and that the role of Ca^{2+} in the transduction mechanism may also be similar.

Acknowledgements

This work is supported by the Wellcome Trust and the Midland Institute of Otology.

References

Assad JA, Shepherd GMG, Corey DP (1991) Tip-link integrity and mechanical transduction in vertebrate hair cells. Neuron 7: 985–94.

Crawford AC, Evans MG, Fettiplace R (1991) The actions of calcium on the mechano-electrical transducer current of turtle hair cells. J Physiol 434: 369–98.

Furness DN, Hackney CM, Macnamara M (1997) Morphological effects of low extracellular calcium on stereociliary bundles in the terrapin cochlea. BRA fourteenth annual meeting abstracts.

Hackney CM, Furness DN (1995) Mechanotransduction in vertebrate hair cells: structure and function of the stereociliary bundle. Am J Physiol 268: C1–C13.

Correspondence to BN Kumar, 22 Greenside, Harborne, Birmingham B17 0BT, UK. Tel/fax (+44)(0)1782-416197.

Chapter 23
Dynamics of inner ear pressure release measured with a double-barrelled micropipette in the guinea pig

HP WIT, EO THALEN, JM SEGENHOUT, FWJ ALBERS, Department of Experimental Otolaryngology, University Hospital, Groningen, Netherlands

ABSTRACT. Inner-ear fluid pressure was measured in guinea pigs with a double-barrelled micropipette after sudden volume increase or decrease. Fitting the obtained results with a simple physical model shows that pressure release routes for the inner ear behave as a pressure-dependent flow resistance. It might even be that these routes close for very low pressure differences.

Key words: cochlea, Ménière's disease, aqueduct, patency.

Introduction

An insight into the relationship between inner-ear fluid volume and pressure is important in understanding the hydromechanical consequences of endolymphatic hydrops, which is thought to be the cause of Ménière's disease. The relation between endolymph flow and acute volume increase was recently investigated by Salt and DeMott (1997) with an elegant double-barrelled micropipette technique. We decided to use a similar technique for the measurement of inner-ear fluid pressure during acute volume increase or decrease. The way in which sudden volume change affects inner-ear pressure depends on the compliance of the bordering membranes of the inner-ear compartment and the flow resistance of pressure release routes. As a first approximation these factors have been incorporated in a very simple model.

Theoretical model

We consider the inner ear to be a rigid walled cavity, bordered by a compliant membrane at one side (Figure 23.1). The cavity is filled with fluid with volume V and pressure p. If p is larger than the outside pressure, fluid will escape through a flow resistance. (For simplicity reasons outside pressure is taken to be equal to zero. This will not affect the generality of the solution: a constant pressure may be added, both to p and the outside pressure.)

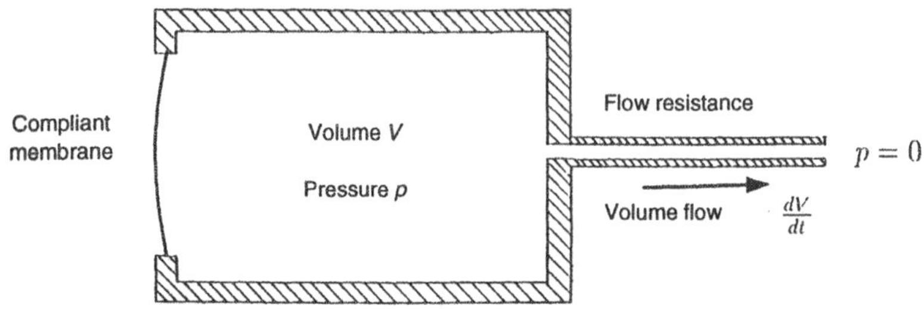

Figure 23.1: model for inner ear pressure release after sudden volume increase. Immediately after volume increase inner ear fluid volume is V and pressure is p. This pressure returns to zero by fluid flow through the tube.

We now assume volume flow $\frac{dV}{dt}$ to be a non-linear function of pressure:

$$-\frac{dV}{dt} = \alpha p + \beta p^2,$$

while changes of p and V with time are assumed to be proportional:

$$\frac{dp}{dt} = c\,\frac{dV}{dt}.$$

Combining these two assumptions yields the differential equation:

$$-\frac{dp}{dt} = \alpha p + \beta p2$$

with solution:

$$p = p_0 \frac{e^{-\alpha t}}{1 + \gamma\,(1 - e^{-\alpha t})},$$

in which $\gamma = \frac{\beta}{\alpha}\, p_0$.

This solution will be referred to as 'double exponential function'.

If $\beta = 0$ in the differential equation (which means that dV/dt is a linear function of p), its solution reduces to the well-known single exponential function:

$$p = p_0 e^{-\alpha t}.$$

For $\alpha = 0$ the differential equation turns into:

$$-\frac{dp}{dt} = \beta p_2, \text{ with solution } p = \frac{p_0}{1 + \beta p_0 t}.$$

This solution is no longer an exponential function, but it has the shape of a hyperbola.

Materials and methods

In this preliminary study experiments were performed on five albino guinea pigs, under general anaesthesia induced by intramuscular administration of ketamine/xylazine (60/3.5 mg/kg). Muscle relaxation was obtained with succinylcholine (2.5 mg/kg). The animals were artificially ventilated through a tracheostoma (Columbus Instruments, model 7950) and body temperature was maintained at 38°C.

The animal's head was kept in a stationary position by means of a steel bolt fixed to the skull with dental cement. The bulla was opened by a retrocochlear approach and the round window exposed. Through the round window membrane the tip of a double-barrelled micropipette was inserted into the scala tympani or into the scala media after subsequent perforation of the basilar membrane. Double-barrelled micropipettes were drawn from borosilicate glass (1.5/0.84 mm diameter) and the tips were bevelled (Narishige EG-40). Tip diameters were around 5 μm.

One barrel of the pipette was used to measure inner-ear pressure and DC potential (WPI 900A micropressure system). Fluid could be manually injected into the inner ear through the other barrel with a small steel plunger syringe. The connection between syringe and micropipette barrel was made of hard-walled teflon tubing.

During an experiment inner-ear pressure and DC potential values were stored digitally with a rate of 10 Hz. Fits to relevant portions of the obtained recordings were made off-line with an appropriate software package.

Results

A typical result of the first part of a recording is shown in Figure 23.2. After penetration of the round window at t=40 s a hydrostatic pressure of 3.3 cm water is measured inside scala tympani. At t=100 s fluid is injected to raise the pressure to 8 cm. Upon termination of fluid injection pressure returns to its initial value within a few seconds. One minute later fluid is withdrawn from scala tympani, making pressure fall to about 1 cm, and again pressure returns to its initial value along an exponential curve. It was quite consistently found that the time constant for return from overpressure was somewhat shorter than for return from underpressure. Similar results were obtained with fluid injection in scala media.

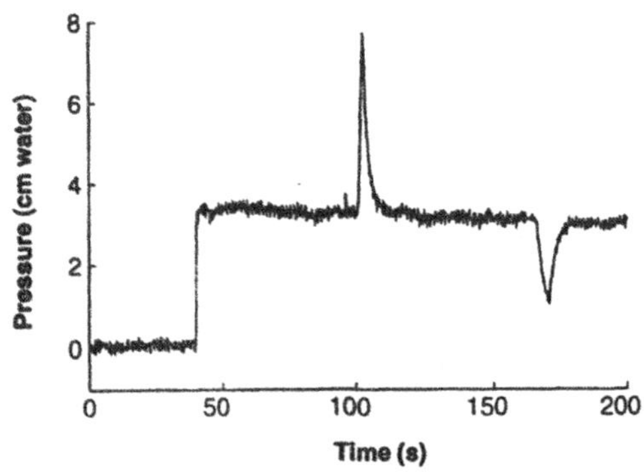

Figure 23.2: inner ear pressure as a function of time, as measured with one of the pipette barrels. At t = 40 s the pipette enters scala tympani. Around t = 100 s fluid is injected and around t = 160 s fluid is withdrawn.

The curves, along which pressure returns to its initial value after short over- or underpressure, have an exponential shape. Fits with a single exponential curve, however, are not good. This is shown in Figure 23.3a. An almost perfect fit is obtained with the double exponential function, as given in the theory section (Figure 23.3b). A fit with the hyperbolic function, also introduced in the theory section, is not much worse (Figure 23.3c). The results as shown in Figure 23.3 were consistently found, both for rising and falling curves.

Discussion

In order to compare our results in guinea pigs with results obtained in an other animal we digitised inner ear pressure recordings reported by Carlborg et al. (1982, their Figure 3). These authors studied inner-ear fluid pressure change after a sudden increase or decrease of middle-ear pressure in the cat. This resulted in an immediate increase or decrease of inner-ear pressure, followed by a pressure equalisation within a few seconds. Return of pressure to its initial value was most probably caused by fluid displacement through the cochlear aqueduct because no pressure release was seen if the aqueduct was blocked.

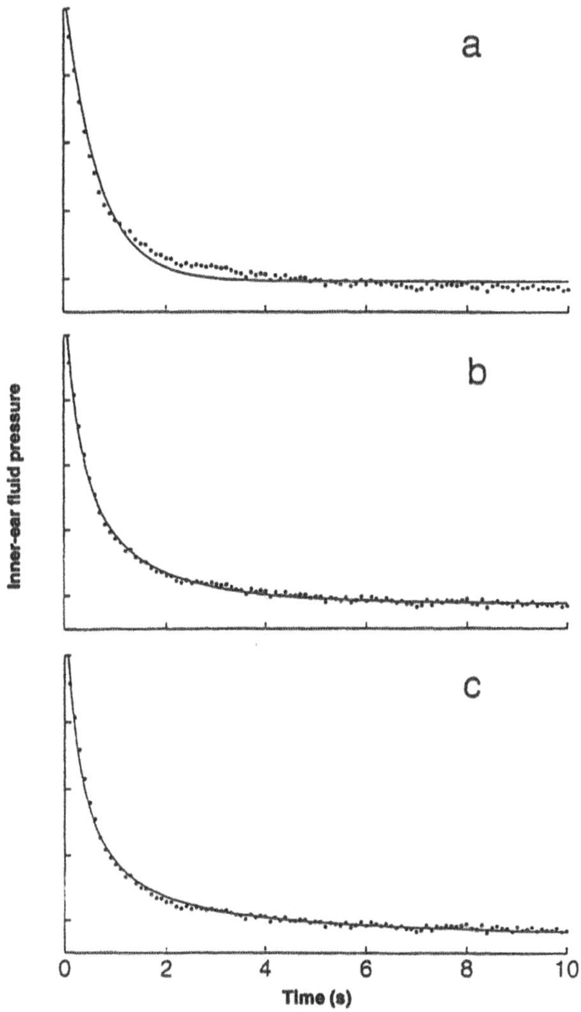

Figure 23.3: dots: measured pressure release curve after fluid injection (the curve is the same in all three panels). Solid lines: fit with single exponential curve (a), fit with double exponential curve (b), fit with hyperbolic curve (c). In this case the double exponential gave the best fit.

Figure 23.4 shows the time course of pressure release, as taken from Carlborg et al.'s results. In the upper panel these results are fitted with a single exponential curve. This fit shows systematic differences between measured and fitted curve, just like in our own results. The best fit was obtained with the hyperbolic curve (see theory section), as shown in the lower panel of Figure 23.4.

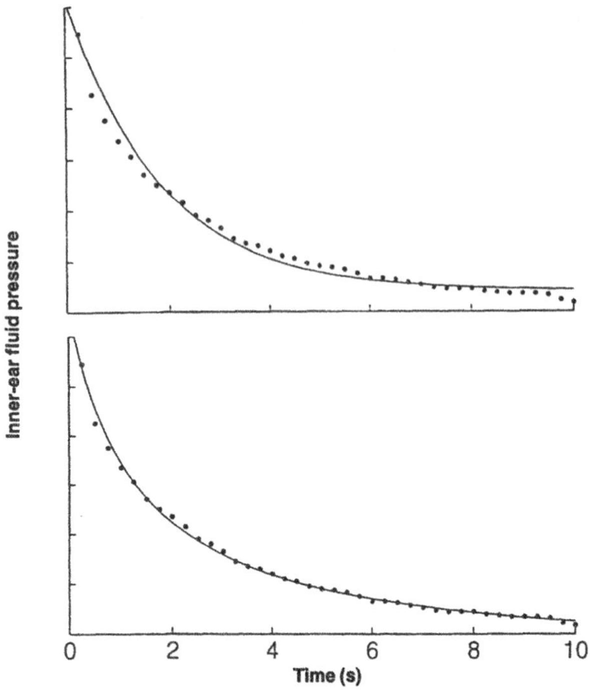

Figure 23.4: dots: measured pressure release curve after sudden middle ear pressure increase (data taken from Carlborg et al. 1982). Solid lines: fit with single exponential curve (upper panel), fit with hyperbolic curve (lower panel).

If volume flow *dV/dt* through the tube in Figure 23.1 is plotted against the pressure difference across the tube three situations can be distinguished (see theory section), as shown in Figure 23.5.

For $\beta = 0$ volume flow is a linear function of pressure, for $\alpha = 0$ the relation is purely quadratic and for α and β both different from zero the relation has a linear and a quadratic term. If we now define flow conductance as the slope of the *dV/dt* versus *p* curves in Figure 23.5, these three situations have different consequences, as shown in Figure 23.6: for $\beta = 0$ flow conductance (and its reciprocal: flow resistance) would be independent of pressure. This seems not to be the case in the inner ear, for then best fits in Figures 23.3 and 23.4 would have been obtained with a single exponential function. From the fact that best fits are obtained with a double exponential function (Figure 23.3) or with a hyperbolic function (Figure 23.4), we must conclude that conductance depends on pressure difference. In the limiting case of the hyperbolic fit conductance even goes to zero with pressure. Or, in other words, the tube in Figure 23.1 closes if the pressure difference across it becomes very small.

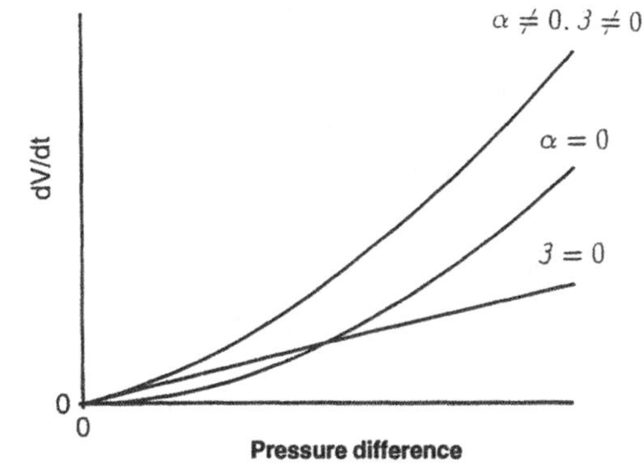

Figure 23.5: volume flow through the tube in Figure 23.1 as a function of pressure difference across it, for the different parameter conditions given in the theory section.

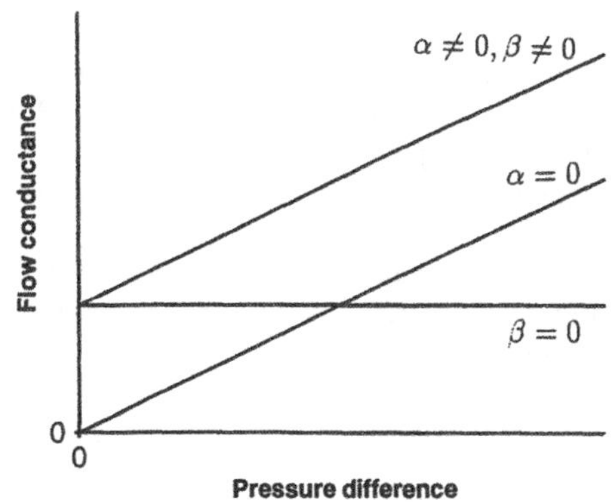

Figure 23.6: flow conductance (the reciprocal of flow resistance) of the tube in Figure 23.1 as a function of pressure difference across it, for the different parameter conditions given in the theory section.

From the results obtained by Carlborg et al. (1982) it is reasonable to conclude that the cochlear aqueduct is the main route for equalisation of inner-ear fluid pressure. If the assumptions that we made in the theory section are right, both our results and those from Carlborg et al. as shown in Figures 23.3 and 23.4 indicate that the cochlear aqueduct is not a simple flow resistance, like a narrow hard-walled tube would be.

Furthermore, this work shows that pressure differences between the inner ear and cerebrospinal fluid spaces are equalised within a few seconds, as was also found by Thalen et al. (see earlier, Chapter 3) with a different experimental approach.

The importance of the cochlear aqueduct for pressure release has been questioned because Wlodyka (1978) found a decreasing patency of the aqueduct in man with increasing age. This finding, however, could not be confirmed in a recent extensive study of the anatomy of the human cochlear aqueduct (Gopen et al. 1997).

Acknowledgement

This work was supported by the Heinsius Houbolt Foundation.

References

Carlborg B, Densert B, Densert O (1982) Functional patency of the cochlear aqueduct. Ann Otol 91:209–15.

Gopen Q, Rosowski JJ, Merchant SN (1997) Anatomy of the normal human cochlear aqueduct with functional implications. Hearing Res 107: 9–22.

Salt AN, DeMott J (1997) Longitudinal endolymph flow associated with acute volume increase in the guinea pig cochlea. Hearing Res 107: 29–49.

Thalen EO, Wit HP, Segenhout JM, Albers FWJ (1998) Dynamics of inner ear pressure change caused by CSF-pressure manipulation in guinea pigs. In Reid A, Marchbanks RJ, Ernst A. Intracranial and Inner Ear Physiology and Pathophysiology. London: Whurr. pp 29–34.

Wlodyka J (1978) Studies on cochlear aqueduct patency. Ann Otol 87: 22–8.

Correspondence to HP Wit, Department of Experimental Otolaryngology, University Hospital, PO Box 30.001, 9700 RB Groningen, Netherlands. Tel (+31) 50-3612550. Fax (+31) 50-3611698. E-mail h.p.wit@med.rug.nl.

Part 4
Inner ear Pathophysiology

Chapter 24
Treatment of otolith disease concepts and new development of laser microsurgery

M WESTHOFEN, Department of ENT and Head and Neck Surgery, University Hospital Aachen, Germany; U MANGOLD, Institute of Neuroanatomy, University Hospital Hamburg Eppendorf, Hamburg, Germany; W WÖLLMER, Department of ENT, University Hospital Hamburg Eppendorf, Hamburg, Germany

ABSTRACT. The examination of the otolith/ocular reflex is part of clinical vestibular testing. Macular function and central otolith dependent reflexes can be quantitatively examined in man. Diseases with recurrent perilymphatic and/or endolymphatic hypertension with vestibular dysfunction commonly show macular involvement. As a result of long duration and moderate outcome in older patients who are treated with labyrinth ablation or vestibular neurectomy a selective labyrinth ablation surgery was evaluated in an animal model. For this purpose transplatinal argon laser irradiation was performed in guinea pigs. A laser dosis window was determined so that macular function was ablated, cochlear and cristar function was preserved and only a three-day time interval for vestibular compensation was needed. Therapeutic effects in the otolith vesicles were evaluated by use of electron and light microscopy.

Key words: labyrinth, otolith, labyrinthectomy, argon laser, Nd:YAG laser, footplate, vestibular compensation, ear surgery.

Introduction

The number of vestibular tests increased during the last decade (Westhofen 1994a, b). New labyrinthine tests, which can determine the side of otolithic dysfunction, are of interest to otoneurosurgeons. While performing eccentric rotations for side differentiating macula tests, in addition to conventional bithermal bilateral caloric tests of canal function, we learned that nearly 50% of patients with clinical vestibular impairment have macular involvement (Figure 24.1). To determine whether surgical treatment is indicated in vestibular disorders we are using a simple classification, which is organized according to the vestibular diagnostic function tests (Table 24.1) and the diverse micro-otosurgical procedures (Table 24.2) used in our centre. The quantitative examination of the otolithic vestibulo-ocular reflex is performed with a newly constructed computer-driven rotation chair. Eccentric rotation with the axis in the non-stimulated labyrinth (Figure 24.2) and off-vertical axis rotation are performed (Figure 24.3), 3D-videooculography is used for registration of VOG (Figure 24.4). Diseases with abnormal inner-ear pressure are seen with high incidence. Observation of patients with chronic or recurrent labyrinthine disease before and after surgery showed that surgical destruction of labyrinthine or neural function was highly efficient in diminishing symptoms but a long time interval was needed for complete vestibular compensation, especially in older patients. Patients with unilateral macular disease showed quicker recovery of stand and gait than those who were suffering from unilateral canal paresis with or without macular involvement. The concept of selective macula destruction for surgical treatment of recurrent labyrinthine disease was therefore developed. Our first animal results were reported elsewhere (Westhofen 1996). The understanding of postoperative effects in animals was supported by studies about the vestibulo-ocular reflex and the macular contribution to that reflex in animals and men (Bronstein and Gresty 1988) (Fernandez and Goldberg 1976a, b). The macula controls not only the perception of static and dynamic linear acceleration but also the intensity of the crista response going to the brain stem.

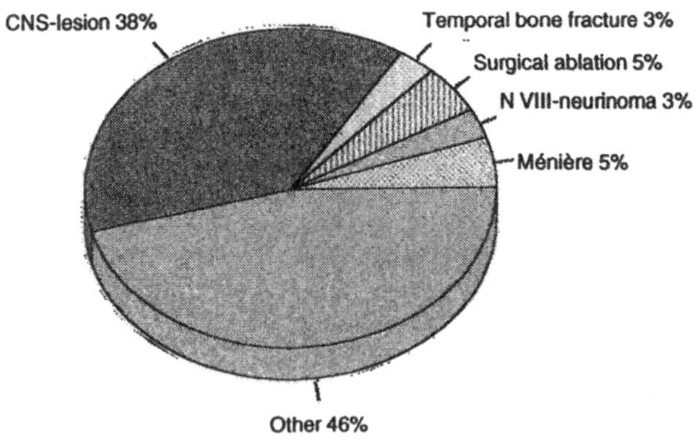

Figure 24.1: relative incidence of ear diseases in patients with pathological findings of dynamic macular function as seen in off-vertical-axis rotation (n = 526 patients).

Table 24.1:

Localization	**Diagnostic procedure**
Labyrinthine dysfunction	
• cristar	bilateral bithermal response (horizontal VOR)
• macular	eccentric rotation
	bilateral bithermal testing (torsional VOR)
Ocular dysfunction	smooth pursuit
	saccades
	optokinetic VOR
Proprioceptive dysfunction	clinical manual diagnostic of the neck
CNS deficits	Statomotor response
	harmonic pendular acceleration testing
	off vertical–axis rotation

Table 24.2: microotosurgical treatment of labyrinthine disorders

- tympanotomy and grafting
- endolymphatic shunt
- canal plugging
- cochleosacculotomy
- vestibular neurectomy

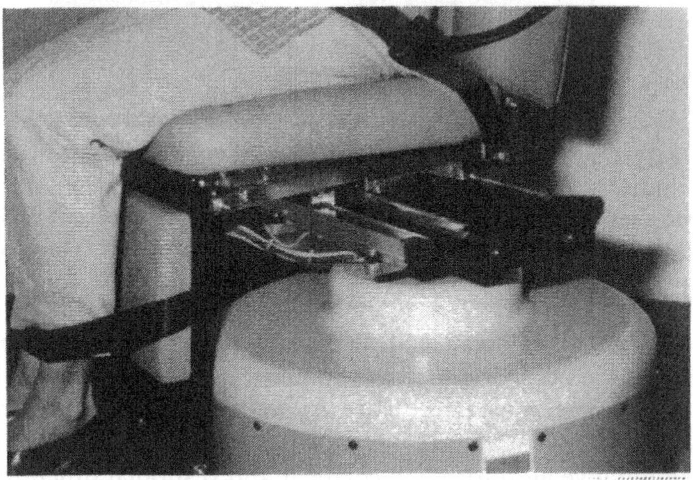

Figure 24.2: detail of the rotation chair for eccentric rotational testing. The seat is laterally moved during constant velocity rotation by a computer-driven step motor device to bring the labyrinth of the non-stimulated ear in the centre of rotation.

In developing surgical treatment of labyrinthine disease in man one must consider that most of the patients have only mild hearing impairment. We therefore avoided opening the vestibule in our animal studies.

Application of laser irradiation on the otolith organs in one patient was reported by Nomura (1993). The surgical procedure was performed by temporary removal of the whole footplate, laser irradiation of the opened vestibule and replacing the footplate after the irradiation. The author reported no postoperative hearing impairment. Our goal is development of non-invasive laser irradiation of the otolith organs. For this purpose we initiated our study in an animal model (Westhofen et al. 1995). The study had the following steps:

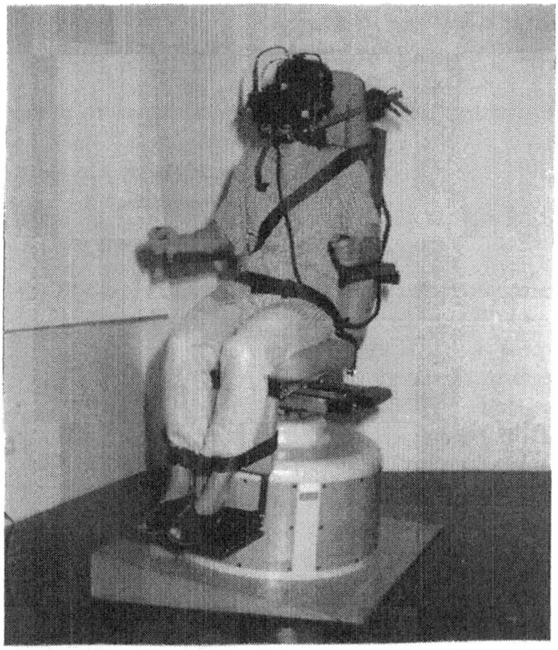

Figure 24.3: rotation chair for off-vertical-axis rotation. The chair is tilted laterally during constant velocity rotation by a motor device mounted underneath the floor. During the examination the patients have to wear a special head and neck restraint which is not shown here so that lateral tilt of head and neck during the examination is prevented (see Figure 24.4).

- development of atraumatic exposure of the footplate which is crossed by a bony bridge in the guinea pig;
- laser dosimetry by means of intraoperative monitoring of cochlear and facial nerve function;
- postoperative monitoring of cochlear and vestibular function;
- postoperative histological investigation of the maculae

Material and methods

Because of German regulations concerning animal studies we had to perform the microsurgical treatment in guinea pigs. The body weight of the female animals ranged from 130 g to 180 g.

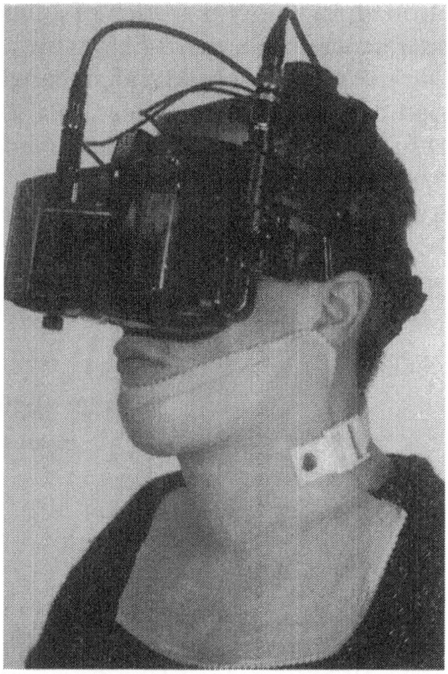

Figure 24.4: video-oculographic equipment consisting of a mask with semi-transparent glasses, infrared illumination of the eyes and lipstick video cameras for both eyes (SMI, Teltow, Germany). Head and neck restraint for prevention of spontaneous lateral head and neck tilt during registration of torsional eye movement.

Vestibular function was examined before and after the operation during the first postoperative week. Lateral tilt (x-axis) of the whole animal was performed on a platform. The statomotor response was observed by means of video documentation. The dynamic vestibulo-spinal response was evaluated by single picture analysis.

Intra- and postoperative monitoring of hearing was performed with auditory brainstem response (ABR) audiometry by use of bone conduction.

Operations were done under general anaesthesia. The bulla was exposed after retroauricular incision. The footplate was exposed by removing the ossified stapedial artery after interruption of the ossicular chain. The stapes was not touched. The footplate was irradiated by argon laser beam in non-contact mode via operation microscope using a micromanipulator (Figure 24.5). Laser impulses had a duration of 150 ms and intervals of 1 s. Argon laser power was 0.8 watts; diameter of the laser point was 0.28 mm. A pause of 30s for cooling was made after every 20 shots. The oval niche was sealed with connective tissue after the irradiation. Twenty hours after surgery the first vestibular tests were performed. After two weeks (group a) and after four weeks (group b), hearing was controlled for the two groups of animals respectively by means of ABR under general anaesthesia before the animals were killed. Intravital perfusion with glutaraldehyde and preparation of the temporal bones were carried out under deep anaesthesia. Decalcification was carried out mildly so that the otoliths were preserved. Semithin sections and electron microscopy were photodocumented.

The histological alterations of the labyrinth, especially the otolith organs, were related to the laser dose, which was applied intraoperatively. Latencies and amplitudes of the ABR potential V were compared pre- and postoperatively at the end of the observation time as well as intraoperatively.

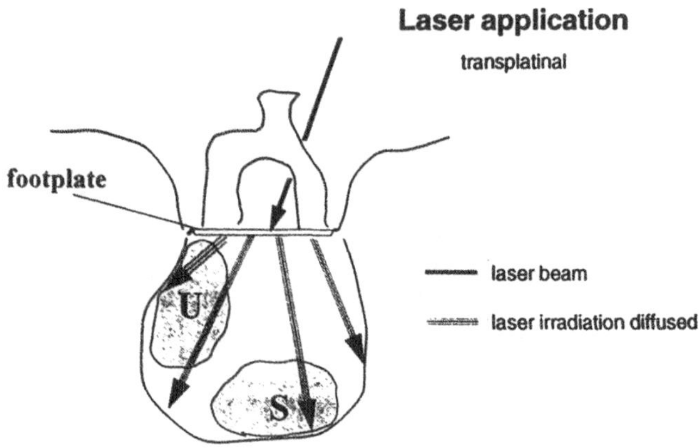

Figure 24.5: transplatinal laser irradiation of the vestibule. Diffraction of the laser beam in the footplate.

Results

All of the 35 animals had normal ABR thresholds and latencies of the potential V (3-3,2 ms) preoperatively. Lateral tilting of the upright standing animals resulted in stretching the legs on the side that was tilted downward and flexing contralaterally according to the vestibulospinal reflex response.

Intraoperative ABR recording showed increased latencies after application of more than 1400 laser shots. After more than 1800 pulses no ABRs could be recorded. After more than 2200 pulses, alterations of the foot plate and perforations were seen in the operation microscopic view.

The animals that had been treated with laser doses lower than 350 pulses showed no alterations in spontaneous motion or vestibulospinal reflexes during lateral tilting postoperatively. Animals with laser doses of more than 950 pulses had strong impairment of their body axis orientation. They moved in a curved direction to the operated ear. Body orientation during tilt in the yaw plane was strongly impaired – stand and gait were not possible on the first postoperative day. The animals with laser doses higher than 350 and lower than 700 pulses exhibited no obvious alteration of spontaneous motion. Vestibulospinal responses to lateral tilting were asymmetrical when the platform was tilted to the operated ear compared to tilting in the contralateral direction. The animals tended to hold the operated ear downward during the tilting procedure whereas tilting to the opposite side was tolerated as well as preoperatively. This side-dependent reflex asymmetry disappeared during the third postoperative day. The state of body orientation and the pattern of motion did not change after the fourth postoperative day. Later onset of vestibular impairment was not observed.

The ABR control findings, before the animals were killed, were identical to the previous findings at the end of laser surgery. Cochlear damage was not observed after application of less than 950 laser shots.

The histopathological study of the temporal bones revealed no abnormalities up to laser doses of 350 pulses. Specimens of labyrinths, irradiated with more than 950 pulses, showed perforation of the otolith vesicles by disruption of the membrane, separating perilymphatic and endolymphatic space. Resorptive granulation tissue and fibrocytes covered the vestibular neuroepithelium. No otoliths were seen in these specimens.

The effect of laser doses ranging from 350 to 700 shots was observed in the structure of the otoconial layer and the otolithic vesicle. The endolymph space of the saccule had collapsed and was adhesive to the otoconial layer. The otoconia were abnormally configured, their size and form was not uniform (Figure 24.6). The macula sacculi showed regular structure in semithin sections and electron microscopy. The type I and type II cells, as well as supporting cells, were well preserved (Figure 24.7). No cochlear or cristar damage could be found. The stapes footplate and the annular ligament were intact.

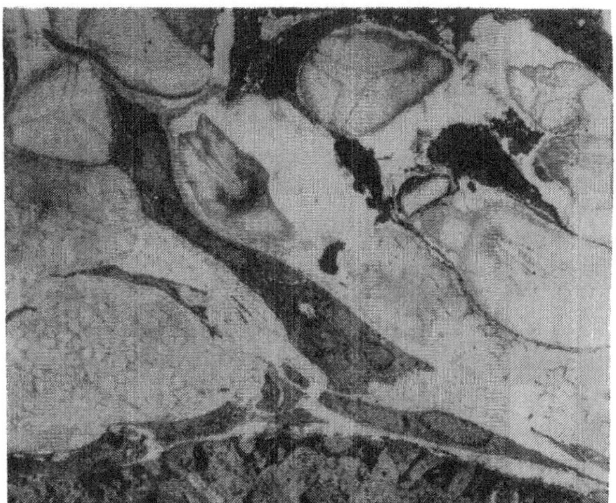

Figure 24.6: destruction of otoconial structures after laser irradiation (two weeks later).

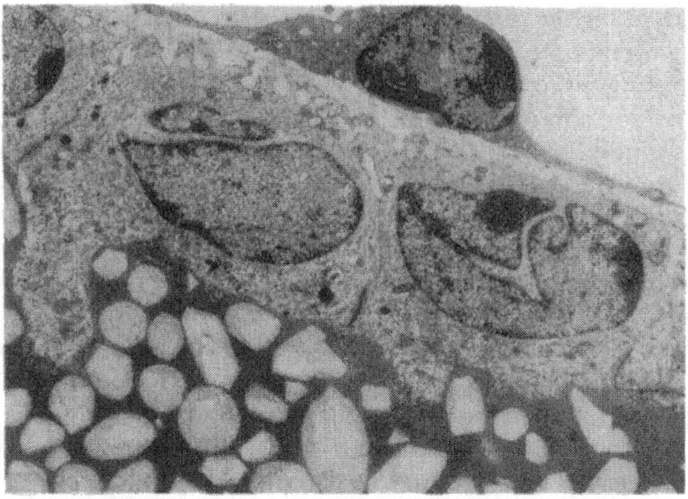

Figure 24.7: collapse of the otolithic vesicle. The otoconia are packed by the cellular layer (two weeks after laser irradiation).

Discussion

Abnormalities of inner ear fluids causing labyrinthine dysfunction of the maculae are well known (Gulya and Schuknecht 1995; Westhofen 1996). Endolymphatic hypertension is generally summarized under Ménière's disease. Often perilymphatic hypertension is looked upon as spontaneous perilymph fistula. Macular involvement in perilymphatic and endolymphatic hypertension is documented by others in previous investigations (Gulya and Schuknecht 1995). Clinical observations have shown that macular dysfunction is frequently part of acute and recurrent peripheral labyrinthine dysfunction.

Labyrinth ablation therapy can be complicated by impaired vestibular compensation in these patients, a consequence of their age distribution, with a maximum of about 50 years. Therefore, minimal surgical trauma and functional deficits will be expected when macular ablation can be performed without alteration of cristar and cochlear function. Opening of the footplate with direct laser irradiation of the saccule was reported by Nomura (1993) for the treatment of benign paroxysmal positional vertigo. Hearing was preserved. Nevertheless, the risk of hearing impairment by stapedotomy or stapedectomy in patients without ancylosis of the footplate is high. Therefore, transplatinal laser irradiation of the maculae was investigated in an animal model in the last few years. Depending on the wavelength and application technique of the laser beam the footplate is not opened during the laser application. Argon and Nd:YAG lasers were evaluated to determine a reasonable relationship between reflexion at the footplate and penetration through the footplate. The degree of laser diffusion cannot be calculated or exactly predicted because of the influence of footplate surface. The histological findings and the narrow window of application doses in guinea pigs show the limited influence of this effect. Data about laser effects on the inner ear were published in publications about laser stapedotomy. Thermic side effects were not described using CO_2, argon, Er:YAG lasers. Laser power and cumulative laser doses were lower than in our animal study with argon laser.

The guinea pig was chosen to conform with German animal protection rules. The species however is not easy to examine for loss of vestibular function. Guinea pigs tend to avoid movement and to keep in contact with the ground with their trunk. Registering vestibulo-ocular reflex responses was not possible in the free-moving animals. Therefore, a response was elicited, which expressed the dynamic properties of the vestibular system for the control of stand and gait. The normal vestibulospinal responses under comparable conditions were checked in animals who underwent a simulated operation but without irradiation of the footplate.

Prediction of the human vestibular responses after selective macula laser irradiation cannot yet be made. Morphological findings in guinea pigs can be extrapolated to human vestibular topography. However, laser doses in guinea pigs have to be re-evaluated for the human stapes because of its thicker footplate. Further long-term follow-up and observation of animals with preoperatively damaged canal function will be the next steps to be undertaken.

Conclusion

Selective lasersurgical ablation of otolith organs in guinea pigs can be accomplished by use of a non-contact argon laser without perforation of the footplate. Observation of the vestibulospinal responses and postoperative histomorphological findings show macular function loss in the first days after surgery and quick vestibular compensation process after the fourth day. Further studies have to evaluate whether this experimental microsurgical procedure can be refined to treat inner-ear diseases with chronic recurrent vestibular dysfunction and abnormal regulation of perilymphatic and endolymphatic pressure.

References

Bronstein AM, Gresty MA (1988) Short latency compensatory eye movement responses to transient linear head acceleration: a specific function of the otolith-ocular reflex. Exp Brain Res 71: 406–10.

Fernandez C, Goldberg J (1976a) Physiology of peripheral neurons innervating otolith organs of the squirrel monkey. I. Response to static tilts and to long duration centrifugal force. J Neurophysiol 39: 970–84.

Fernandez C, Goldberg J (1976b) Physiology of peripheral neurons innervating otolith organs of the squirrel monkey. II. Directional selectivity and force-response relations. J Neurophysiol 39: 985–95.

Gulya AJ, Schuknecht HF (1995) Anatomy of the temporal bone with surgical implications. New York/London: The Parthenon Publishing Group.

Nomura Y, Okuno T, Mizuno M (1993) Treatment of vertigo using laser labyrinthectomy. Acta Otolaryngol (Stockh) 113: 261–2.

Westhofen M (1994a) Objektivierung von Störungen des Otolithenapparates. In Stoll W (ed.) Schwindel und schwindelbegleitende Symptome. Wien/New York: Springer.

Westhofen M (1994b) Klinisch seitengetrennte Otolithenfunktionsprüfung mittels exzentrischer Rotation. Eur Arch Oto-Rhino-Laryngol Suppl II: 185.

Westhofen M, Mangold U, Sanchez M, Woellmer W (1995) Selektive Laser Labyrinthektomie der Otolithenorgane. Erste tierexperimentelle Ergebnisse. HNO Inf: 49.

Westhofen M (1996) Otolith disease – experimental findings and clinical implications. In: Ernst A, Marchbanks R, Samii M (eds) Intracranial and Intralabyrinthine Fluids: Basic Aspects and Clinical Applications. Berlin/Heidelberg: Springer Verlag: 263–78.

Correspondence to Professor M Westhofen, Department of ENT and Head and Neck Surgery, University Hospital Aachen, Pauwelsstr 30, D-52057 Aachen, Germany. Tel (+49) 241 8089360-1. Fax (+49) 241 8888365

Chapter 25
Posturographic correlates of peripheral and central vestibular disorders, as assessed by electronystagmography (ENG) and the Tetrax Interactive Balance System

R KOHEN-RAZ, A SOKOLOV, A KOHEN-RAZ, M DEMMER and M HAR-EL, Department of Otololaryngology, Wolfson Medical Center, Tel Aviv and Department of Special Education, Hebrew University, Jerusalem

ABSTRACT. One hundred and twenty patients of the ENT outpatient clinic of the Wolfson Medical Center, Holon, Israel, all suffering from vertiginous disorders of varying aetiology and symptomatology, were systematically evaluated by ENG examinations, according to which they were classified as peripheral, central, mixed central-peripheral pathology or else as possessing normal ENG patterns. The ENG findings were correlated with results of posturographic tests, using the innovative method of the Tetrax Interactive Balance System (IBS), based on the evaluation of postural sway on four different footplates, each measuring separately the pressure fluctuations produced by each heel and toe part. In contrast to previous studies carried out with other posturographic systems, which focus on displacement patterns of the point of gravity, results of this investigation showed that specific posturographic parameters, which can only be obtained with the IBS, differentiate significantly a) between patients with peripheral, central and normal ENG, b) between patients with normal and abnormal caloric tests, c) between left Vs right peripheral disturbances. These parameters were a) synchronization of sway between heel and toe, b) the amount of evenness of weight distribution over the four footplates (weight distribution harmony), c) the intensity of low frequency sway. An unexpected and unexplained finding was significantly more severe postural aberrations in left as compared to right peripheral vestibular pathology.

Key words: ENG, posturography, peripheral, central, vestibular disorders, caloric test, canal paresis.

Introduction

The potential contribution of posturography to elucidate the source of vertigo and dizziness as a complementary tool to ENG examinations has been well documented by research using different techniques of assessing postural control. The general trend of hitherto available results can be summarized as follows:

- In clinical samples of patients suffering from vertiginous disorders with varying symptoms and aetiology, there is a considerable percentage of subjects with negative (normal) ENG findings whereas results of posturography reflect pathology. On the other hand, cases with normal posturography and deviant ENG are rare. Thus, posturography is probably sensitive to an array of vestibular-spinal and vestibular-cervical disturbances, which affect the oculo-vestibular loops, tested by the ENG examination to a lesser extent (Furman 1994; Furman 1995; Lipp and Longridge 1994; Marison et al. 1994; Norre 1994).
- 'Dynamic' posturography has no unanimously documented advantage over 'static' posturography as to its diagnostic power to differentiate between the various sources and dynamics of vertiginous disorders (Furman 1994, 1995; Lipp and Longridge 1994; Marison et al. 1994; Norre 1994).
- Neither 'dynamic' nor 'static' posturography has hitherto detected reliable criteria of differential postural diagnosis between central and peripheral vestibular pathology (Furman 1995; Marison et al. 1994).
- Until now, no reliable signs reflecting the lateralisation of the vestibular pathology have been reported in the literature.

In the present study we report results obtained by an innovative posturographic method (Tetrax Interactive Balance System) based on the differential measurement of vertical forces on four different platforms, each bearing separately the weight of each heel and each toe-part respectively. This method yields postural parameters that cannot be obtained by traditional static or dynamic posturography, namely synchronization between the separate sway patterns on individual platforms, categories of Fourier sway frequency bands, weight distribution harmony etc. For a detailed description of the method see the article by Kohen-Raz (1991).

Subjects and method

The subjects of the study were 120 outpatients of the Otolaryngological Clinic of the Wolfson Medical Center, Tel-Aviv. Their age range was between 18 to 70 years. All suffered from vertiginous disorders of varying aetiology. They were given posturographic tests prior to undergoing routine clinical and ENG examinations. The main criterion variable was the general evaluation of the ENG record, summarizing the diagnosis as non-pathological, peripheral, central or mixed. In addition more specific ENG markers were used, such as absence versus presence of vertical gaze on the Hallpike and Gaze tests, percentage of response to caloric stimulation etc. The posturographic scores were statistically correlated with the ENG variables.

Results

As shown in Table 25.1 and Figure 25.1, general stability, the main parameter used by traditional posturography, does not differentiate between central and peripheral pathology, neither between the latter and subjects with normal ENG. On the other hand, the entire clinical population, including patients with normal ENG, is significantly less stable than healthy controls. These findings are identical with results of numerous other studies demonstrating that posturography is more sensitive that the ENG to disturbances of postural control mechanisms.

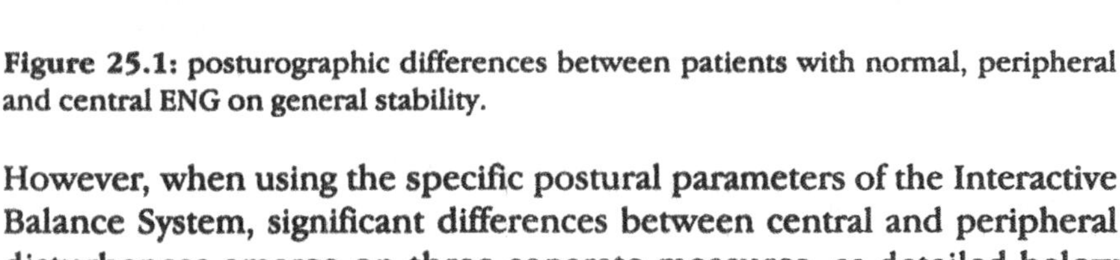

Figure 25.1: posturographic differences between patients with normal, peripheral and central ENG on general stability.

However, when using the specific postural parameters of the Interactive Balance System, significant differences between central and peripheral disturbances emerge on three separate measures, as detailed below (Table 25.2, Figure 25.2).

- In patients with central pathology synchronizations are significantly weaker on positions involving a) vestibular stress, b) changes in head positions.
- In cases of central pathology, low frequency sway (0,1 Hz) increases significantly when standing on solid surface with eyes closed (Romberg test) as compared to the performance with eyes open.
- Under vestibular stress (eyes closed on pads) weight distribution is significantly worse in patients with peripheral disorders.

Table 25.1: general stability in normal subjects and patients with normal peripheral and central ENG findings

Position	Diagnosis	N	M	SD
Romberg Eyes open	ENG central	33	15·3	8·18
	ENG peripheral	41	16·9	10·76
	ENG normal	10	17·4	8·07
	Normal Ss	30	11·7	4·70
Romberg Eyes closed On pads	ENG central	33	31·6	13·01
	ENG peripheral	41	30·7	16·42
	ENG normal	10	31·6	12·94
	Normal Ss	30	21·9**	8·21

p < 0·001 between normal Ss and patient population

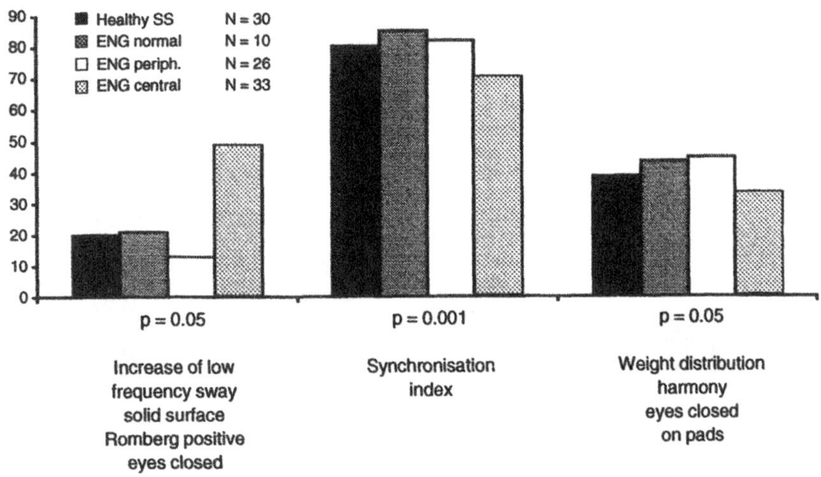

Figure 25.2: posturographic differences between patients with central versus peripheral ENG symptomatology.

To further investigate posturographic characteristics of peripheral vestibular pathology, patients with unilateral weakness over 23%, which is considered to be a reliable criterion of peripheral disturbance, were compared with those with unilateral weakness under 23% (Figure 25.3). Interactive posturography differentiated at a high level of significance between these two groups, the patients with weakness over 23% showing a significantly increased intensity of low frequency sway when standing on pads, eyes closed (vestibular stress) and turning the head in the saggital and lateral plane. This congruence between the postural and ENG parameters represents a valuable cross validation of the empirically derived ENG criterion of 23% weakness, besides demonstrating the validity and reliability of the Interactive Balance System.

Table 25.2: posturographic differences between patients with peripheral versus central pathology

Diagnosis	Synchronizations			Low frequency sway eyes closed		Weight distribution harmony	
	N	M	SD	M	SD	M	SD
Central	33	706***	(169)	49*	(87·6)	5·6*	(3·29)
Peripheral	41	823	(127)	13	(53·6)	7·5	(4·82)
ENG Norm.	10	855	(91·3)	21	(40·1)	7·3	(5·34)
Norm. Ss	20	804	(158)	20	(63·3)	6·5	(3·31)

Asterisks indicate significance of differences between central and peripheral pathology.
*** $p < 0{\cdot}001$
* $p < 0{\cdot}05$

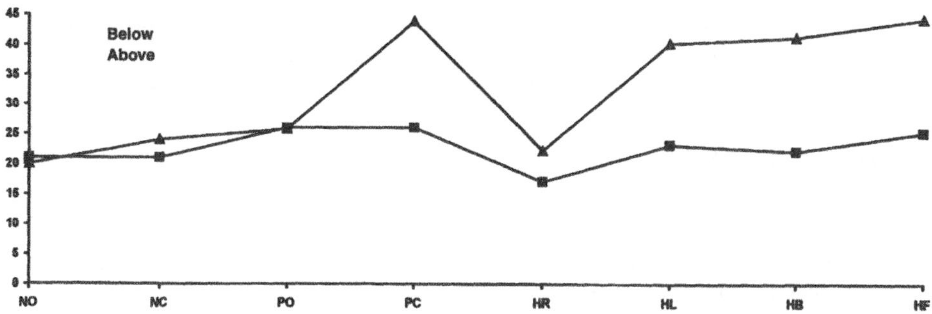

Figure 25.3: posturographic differences between patients with unilateral weakness below and above 23% intensity of low-frequency sway on eight standard positions.

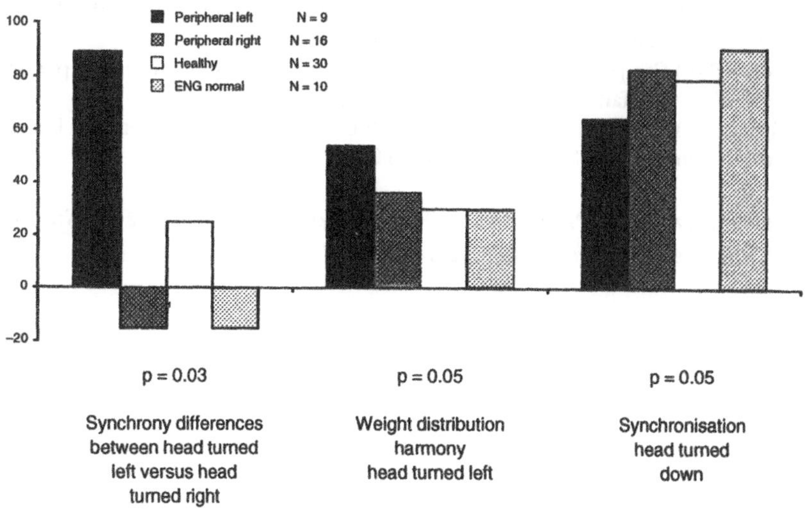

Figure 25.4: posturographic differences between patients with left versus right peripheral vestibular pathology.

Finally, results presented in Figure 25.4 demonstrate the sensitivity of the IBS to differentiating between left and right peripheral vestibular pathology. Again only the 'IBS – specific measures' of synchronization and weight distribution turned out to be valid predictors. The rather unexpected finding, however, is the fact that *left* peripheral pathology is characterized by typical deviant postural scores whereas patients with right peripheral pathology perform on these 'laterality sensitive parameters' within normal range. Closer inspection of the data reveals that on these same parameters patients with a left peripheral pathology tend to show responses similar to subjects with central vestibular symptoms, whilst patients with a right peripheral pathology perform within the normal range. These findings might be tentatively explained on the basis of Previc's theory (Previc, 1991) according to which the *left* vestibular system is dominant. Consequently, in comparison to a right peripheral pathology, a left vestibular disturbance would be more pervasive, i.e. more generally handicapping to the sensori-motor (and cognitive-motor) functions of the organism, controlled by central functions. Obviously these finding need to be validated by further studies.

References

Furman JM (1994) Posturography, uses and limitations. Baillières Clin Neurol 3: 501–3.

Furman JM (1995) Role of posturography in the management of vestibular patients. Otolaryngol Head Neck Surg 112: 8–15.

Kohen-Raz R (1991) Application of tetraxiametric posturography in clinical and developmental diagnosis. Percep Mot Skills 73: 635–56.

Lipp M, Longridge (1994) NS Computerized dynamic posturography: its place in the evaluation of patients with dizziness and imbalance. J Otolaryngol 23: 177–83.

Marison C, Hawken MM, Kennard C, Kenyon G (1994) Kenposturo dynamic platform measurement in Ménière's disease. J Vestib Res 4: 409–19.

Norre ME (1994) Can posturography contribute to the diagnosis of vertigo in patients where other tests fail to do so? Acta Otolaryngol (Stockh) 114: 465–72.

Previc FH (1991) A general theory concerning the prenatal origins of cerebral lateralisation in humans. Psychol Rev 98: 299–334.

Correspondence to Professor R Kohen-Raz, Department of Special Education, Hebrew University, Jerusalem, Israel.

Chapter 26 Vertigo and progressive sensori-neural hearing loss – peripheral cause or vascular compression of CN VIII?

MARGARETA B MØLLER, Department of Neurological Surgery, University of Pittsburgh School of Medicine, Pittsburgh PA, USA

ABSTRACT. Vestibular symptoms can have several causes. Effective treatment of pathologies of the inner ear is different from that of patients in whom the pathology is in the central portion of the vestibular nerve. The diagnostic methods that can differentiate between these two causes of balance disorders are described.

Key words: disabling positional vertigo, microvascular decompression operations, vestibular nerve.

Introduction

Vascular compression of cranial nerves in the posterior fossa (cerebello-pontine angle) can cause symptoms of hyperactivity and loss of function. The most common disorders caused by vascular compression are trigeminal neuralgia (TGN) and hemifacial spasm (HFS). Less common are glossopharyngeal neuralgia (GPN) and geniculate neuralgia.

Microvascular decompression (MVD) offers a high cure rate with moderate complications: 81% for TGN and 85% for HFS (Apfelbaum 1988; Jannetta 1990; Sindou and Mertens 1993; Barker et al. 1995, 1996). Vascular compression of CN VIII may cause symptoms of vertigo, tinnitus and often slowly progressive unilateral hearing loss. In a study of 207 consecutive patients with vertigo who underwent MVD of CN VIII, the cure rate was 79% for unilateral compression and 77% for patients with bilateral compression (Møller et al. 1993). Nine patients who previously had vestibular nerve section were excluded, as were three patients who had several procedures. Together, these 207 patients underwent 254 MVDs and 1.6% (four patients) lost hearing on the operated side and 1.6% (four patients) suffered a hearing loss.

Vascular compression of the vestibular portion of the eighth nerve causes specific symptoms of vertigo and/or disequilibrium together with nausea. We have labelled this form of vestibular entity disabling positional vertigo (DPV).

Patients with DPV have a more or less constant spinning sensation or true vertigo that is always associated with a varying degree of nausea, occasionally leading to vomiting. In most patients the onset of symptoms is subtle. Nevertheless, the symptoms increase steadily over time, similar to HFS, which starts with twitching of the muscles under one eye and then slowly progresses to involve the entire half face, including the platysma. Some patients with DPV have a sudden onset of severe vertigo, nausea or vomiting thus similar to the onset of TGN. Such a sudden onset of vestibular symptoms may resemble vestibular neuronitis, but the patient's symptoms fail to resolve over time and instead slowly get worse.

All these symptoms increase by activity and decrease or abate with bed rest. The patients therefore, over time, become more and more limited with regard to activities such as driving a car, working in general. Many patients over time become essentially confined to their homes. Vestibular suppressant medications such as Antivert or Dramamine is of no help in these patients, but a low dosage of Valium (2–4 mg/day) gives relief in many of the patients with DPV. A patient with DPV typically walks in a way that is often interpreted as ataxia, but the way the patient with DPV walks is more similar to that of an intoxicated person, with the exception that a patient with DPV does not stagger randomly, but instead slowly drifts to one side, that is the affected side.

The patients often summarize their symptoms as: 'I feel like I have had too much to drink' and 'I feel as if I walk like a drunk.' Some patients complain of disequilibrium – 'on a boat feeling' – associated with nausea and they are unsure of where the floor is under their feet constantly. Patients with DPV often have symptoms from adjacent cranial nerves making it easier to find which side is affected by vascular compression.

Clinical findings

Deep sharp ear pain (geniculate neuralgia) was present in 33 patients with unilateral DPV. Tinnitus was also often present but usually mild in 33 patients. Seventeen of our 207 patients had a mild but persistent twitching of the muscles under one eye and 22 patients had facial pain or numbness, symptoms from CN VII and CN V respectively.

When the blood vessel is also in contact with the cochlear portion of the eighth nerve, the patients suffer a slowly progressive but never fluctuating hearing loss on the affected side. The hearing loss is usually a flat sensorial-neural loss ranging from 40–50 dB to an advanced loss around 75 dB PTA and occurred in 33 patents. In these patients who had a marked, flat hearing loss, the onset was always in the high frequency range and it progressed rather rapidly to 60–75 dB. The discrimination scores were disproportionally low, 8–20% (Figure 26.1 and 26.2). Twelve patients had a marked high-frequency hearing loss, more pronounced on the affected side, while nine patients had a low-frequency up-sloping loss but with no fluctuations of the hearing. Nineteen patients had a 'notch' in the mid-frequency range.

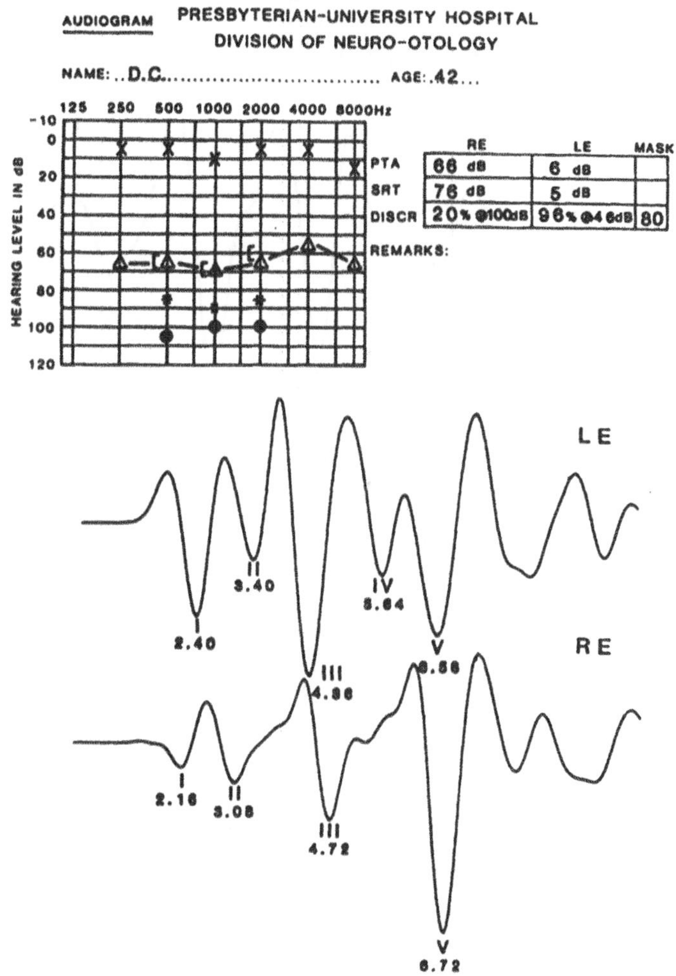

Figure 26.1: patient with DPV on the right side, three years' duration. PTA was 66 dB but with only 20% speech discrimination score on the right ear. Acoustic reflexes were elevated when elicited from the right ear. The reflex decay test was positive. Brainstem auditory evoked potential (BAEP) showed increased IPL I-III on the right side and increased IPL III-V when elicited from the left ear.

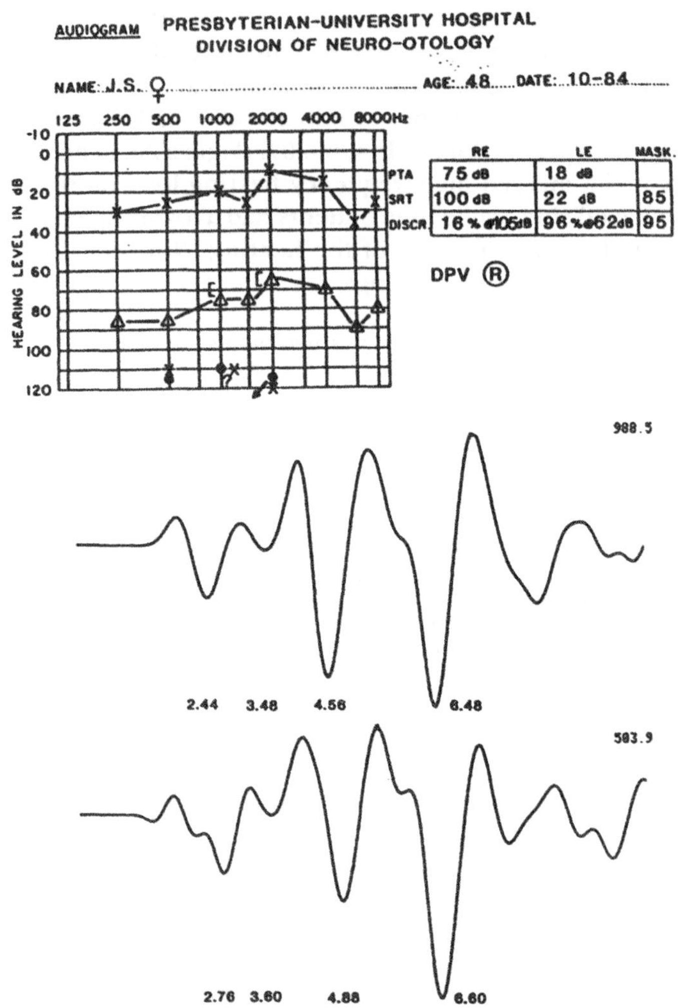

Figure 26.2: A. Patient with DPV on the right side, 15 years' duration. The patient had marked hearing loss, low speech discrimination score on the right ear. The acoustic reflex was elevated–absent. The BAEP elicited from the right ear was abnormal.

Twenty nine patients who were diagnosed with DPV had a low-frequency or flat hearing loss may suggest the diagnosis of Ménière's disease. However, the symptoms that our patients exhibit are very different from the description given by Prosper Ménière of the disease that carries his name. Patients with Ménière's disease have attacks, lasting from 1–24 hours, with severe vertigo, nausea and occasionally vomiting. The attacks are preceded by increasing aural fullness and an increase in the patient's tinnitus together with a decrease in hearing, initially in the low frequency range. When the attack is over, the patients have no more vertigo, the hearing improves and the patient can function normally

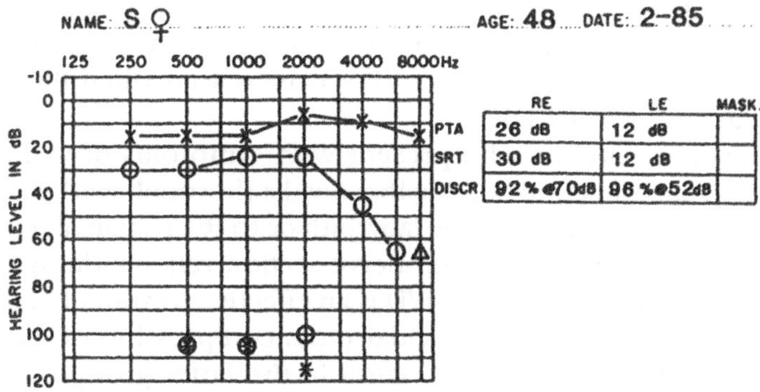

Figure 26.2: B. The same patient four months after MVD. Note the improvement of hearing. One year after the operation, the patient's hearing had returned to its value before the operation (A).

until the next attack. However, over time, the hearing loss progresses from involving only the low-frequency range to become a flat sensorial-neural loss. Speech discrimination of patients with Ménière's disease is within the range of patients with cochlear injuries.

Stahle (1984) followed some patients with Ménière's disease for as long as 32 years. He states that after 5–10 years of Ménière's disease, the patient reaches the 'end stage,' which is characterized by a flat audiogram and a pure tone average of 50-60 dB. The discrimination score is 50-60% and the patient has a 50% reduction of the caloric response in the affected ear.

Twenty nine of 107 patients (14%) had a flat or low frequency hearing loss and their hearing did not fluctuate. The discrimination scores of patients with DPV are much lower than normally found in individuals with similar hearing loss of cochlear origin. Patients with DPV also have a reduced caloric response of 25–35% on average.

It is important to be able to differentiate between peripheral cause (hydrops) of vestibular disturbances and vascular compression of cranial nerve VIII. In our group of 207 patients, nine patients had undergone a vestibular nerve section at another institution after which they became much worse.

We feel that a careful case history is the most important means to differentiate between these two groups of patients. Special tests such as the acoustic middle-ear reflex response (AR) and recording of the brainstem auditory evoked potential (BAEP) can help to differentiate a peripheral lesion from a more central (eighth nerve) lesion. The thresholds for the acoustic reflex response are normal in patients with labyrinthine lesions but they are elevated in patients with eighth nerve lesions or they may not be possible to elicit in such patients.

The latencies of the various peaks of the BAEP in patients with inner ear disorders is normal and the peak pattern is normal whereas the BAEP is abnormal in patients with eighth nerve lesions. In such patients the interpeak latencies (IPL) of peak I–III are increased when elicited from the ipsilateral side and/or the IPL III–V is increased when the BAEP is elicited from the normal side.

In summary: patients who present with symptoms of vertigo and progressive sensori-neural hearing loss need a thorough work-up to find out if the lesion is localized to the labyrinth or to the proximal portion of the eighth nerve. While vestibular nerve section relieves the symptoms effectively in the former group, MVD is an effective treatment of patients in the group that represents vascular cross compression of the eighth cranial nerve. Vestibular nerve section in the latter group causes the symptoms to get much worse and MVD will not relieve the symptoms of vertigo after a destructive procedure has been carried out.

References

Apfelbaum RI (1988) Surgical management of disorders of the lower cranial nerves. In Schmidek HH, Sweet WH, (eds) Operative Neurosurgical Techniques, Indications, Methods and Results (Vol 2). Orlando: Grune & Stratton, pp. 1063–82.

Barker FG, Jannetta PJ, Bissonette DJ, Shields PT, Larkins MV (1995) Microvascular decompression for hemifacial spasm. J Neurosurg 82: 201–10.

Barker FG, Jannetta PJ, Bissonette DJ, Larkins MV, Jho HD (1996) The long-term outcome of microvascular decompression for trigeminal neurologic. N Engl J Med 334: 1077–83.

Jannetta PJ (1990) Treatment of trigeminal neuralgia by micro-operative decompression. In Yomans JR (ed.) Neurological Surgery. Philadelphia: WB Saunders, pp. 3928–42.

Møller MB, Møller AR, Jannetta PJ, Jho HD, Sekhar LN (1993) Microvascular decompression of the eighth nerve in patients with disabling positional vertigo: selection criteria and operative results in 207 patients. Acta Neurochir 125: 75–82 .

Sindou M, Mertens P (1993) Microvascular decompression (MVD) in trigeminal and glosso-vago-pharyngeal neuralgias. A twenty year experience. Acta Neurochir 58: 168–70.

Stahle J, Friberg U, Svedberg A (1991) Long-term progression of Ménière's Disease. Acta Otolaryngol (Stockh) 111 (Suppl 485): 78–83.

Correspondence to Dr MB Møller, University of Texas at Dallas, Callier Center for Communication Disorders, 1966 Inwood Rd, Dallas TX 75235-7298, US.

Chapter 27
Acetazolamide therapy of endolymphatic hydrops audiologically monitored by low-frequency masking – a case report

EA KABUDWAND, D ADLER ENT Department, Klinikum Buch, Berlin, Germany; G SCHOLZ, D MROWINSKI, ENT-Department, Virchow-Klinikum (Charité), Berlin, Germany

ABSTRACT. Carbonic anhydrase (CA) is an enzyme widely distributed in the human body, for example in the choroid plexus of the brain, the eye, and in high concentrations in the inner ear. It is believed to play a role in the production or resorption of the inner-ear fluid. Acetazolamide is an inhibitor of CA and is thus believed to reduce endolymphatic hydrops (EH) and also intraocular pressure in glaucoma, and has been successfully used in patients with Ménière's disease. We administered acetazolmide to a 40-year-old female referred to our clinic with the second episode of unilateral low-frequency hearing loss and tinnitus within one year and who was unsuccessfully treated with plasma expanders and steroids. Using low-frequency masking (LFM) we showed a reduced phase-dependent modulation of the masked thresholds in the affected ear as seen in EH and an improvement by three times of the modulation depth during acetazolamide therapy. Pure-tone thresholds also improved.

The diagnosis and monitoring of endolymphatic hydrops using LFM provides a non-invasive means to evaluate the risks and benefits of a pathophysiologically directed treatment of EH, in this case with acetazolamide.

Key words: endolymphatic hydrops, drug treatment, low-frequency biasing.

Introduction

Current clinical diagnostic tests of endolymphatic hydrops (EH) such as the glycerol test and electrocochleography are either of doubtful clinical value or invasive and unpleasant for the patient. Low-frequency masking (LFM) is devised to show EH in a non-invasive manner and has been clinically used in patients with Ménière's disease (Mrowinski et al. 1996). The masked threshold of a short stimulus depends on the phase of a superimposed low-frequency masker tone. In patients with suspected EH a reduced modulation is observed. Using this method we monitored the modulation depths in a 40-year-old patient with fluctuating sensorineural hearing loss, tinnitus and aural pressure, without vertigo during the therapy with acetazolamide. This drug is supposed to reduce fluid pressure in the inner ear and has already been used in animal models of EH. The results of clinical studies in patients with Ménière's disease are controversial.

Methods

We suppose that during the application of a low-frequency tone the basilar membrane moves synchronously over its total length (Figure 27.1). The sensitivity of the organ of Corti is varied periodically by this movement, in the apical region more than in the base. This alteration can be tested by the recording the masked threshold of a short test stimulus at different phases of a masker tone. A short stimulus (tone burst 2 kHz, duration 2 ms) and a low-frequency masker tone (30 Hz, 115 dB SPL) are produced by separate transducers and applied via a small ear probe (etymotics). The level of the masker and the phase correlations are controlled by a microphone. The masked threshold of the stimulus is recorded in phase steps of 30 degrees (Mrowinski et al. 1995).

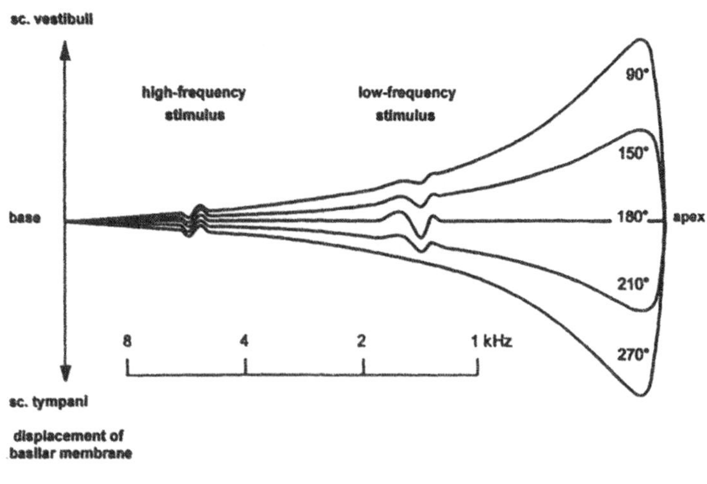

Figure 27.1: phase-dependent displacement of the basilar membrane for test stimuli and low-frequency masker (schematically).

The results of 12 subjects with normal hearing are summarised in Figure 27.2. Two masking maxima of different amplitudes can be recognised at phase 90° and 270°. The higher maximum, at 270°, corresponds to the rarefaction maximum of the tympanic membrane, causing a movement of the basilar membrane towards the scala tympani. The difference in the masked threshold between maximal and minimal masking is called the modulation depth (MD).

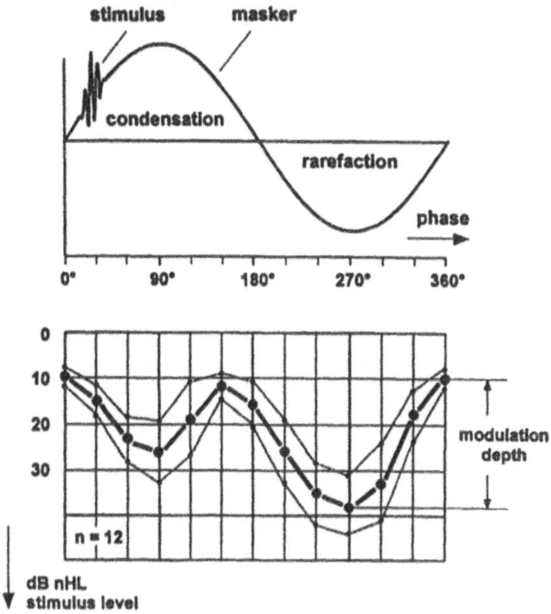

Figure 27.2: top: masker tone with superposed test stimulus; bottom: masked threshold of 12 normal-hearing subjects (mean and SD) for a tone burst with a masker tone. Hearing threshold without masker = 0 dB; difference between maximal and minimal masking = modulation depth.

Patient's history

A 41-year-old female patient presented with a three-week history of unilateral sensorineural hearing loss, tinnitus and aural fullness without vertigo. This was the second episode and she had already been under care of the otolaryngologist with the same symptoms a year earlier. The audiometric tests showed a sensorineural hearing loss at low and middle frequencies in the right ear and normal hearing in the left ear. The vestibular bithermal calorics showed a decreased sensitivity in the affected ear.

We treated the patient with a daily dose of 500 mg acetazolamide (Diamox) for five days and reduced this to 250 mg for another 10 days and 250 mg every other day for one month. Afterwards we lost contact with the patient due to change of residence. After three months we examined the patient again. The treatment had meanwhile been interrupted due to colics in the region of both kidneys, 10 minutes after each drug intake.

Results

The initial pure-tone and phase audiograms before the drug treatment are presented in Figure 27.3. The pure tone audiogram of the affected right ear showed a middle- and low-tone hearing loss of 40 dB. The modulation depth of the right ear (MD = 7 dB) was very reduced as seen in patients with Ménière's disease. The normal hearing contralateral ear showed a near normal modulation (MD = 19).

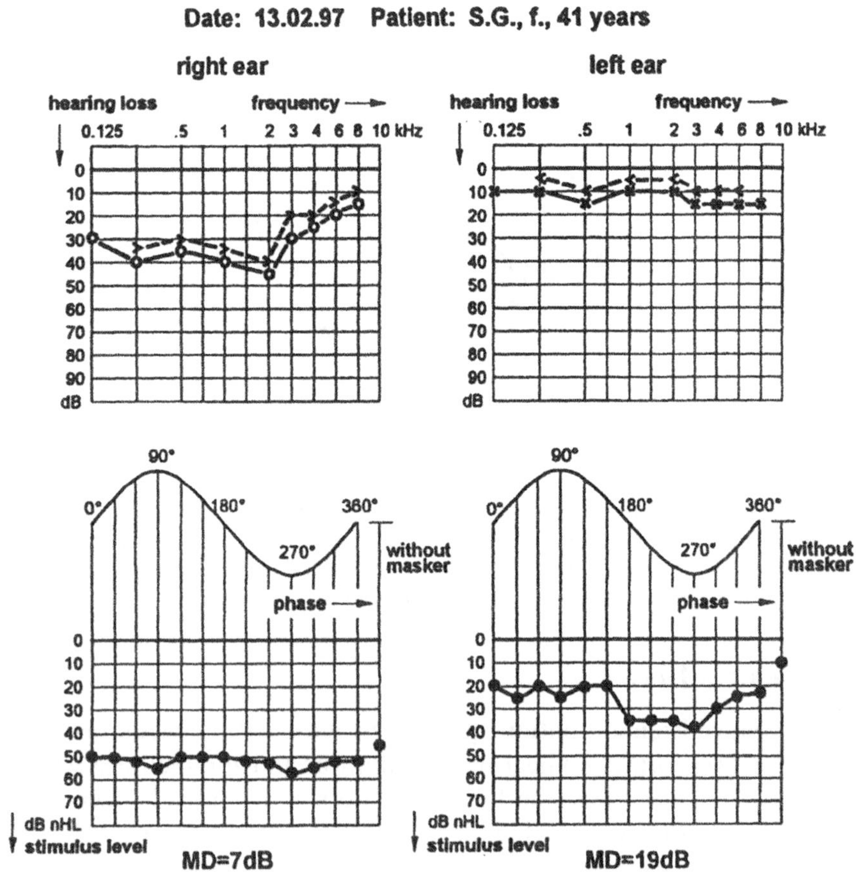

Figure 27.3: pure-tone audiogram (top) and phase audiogram (bottom) of patient SG at admission.

After six days of acetazolamide treatment the patient reported a hearing improvement and a reduction of tinnitus. The hearing threshold of the right ear reached almost normal values (Figure 27.4), MD increased up to 19 dB. MD of the contralateral left ear also increased to 25 dB, the tone audiogram remaining unchanged. At the beginning of the treatment the patient reported a paraesthesia of the fingers, which disappeared after four days. Three months later after the interruption of the therapy the patient showed the same symptoms with almost identical audiometric findings as seen before the therapy.

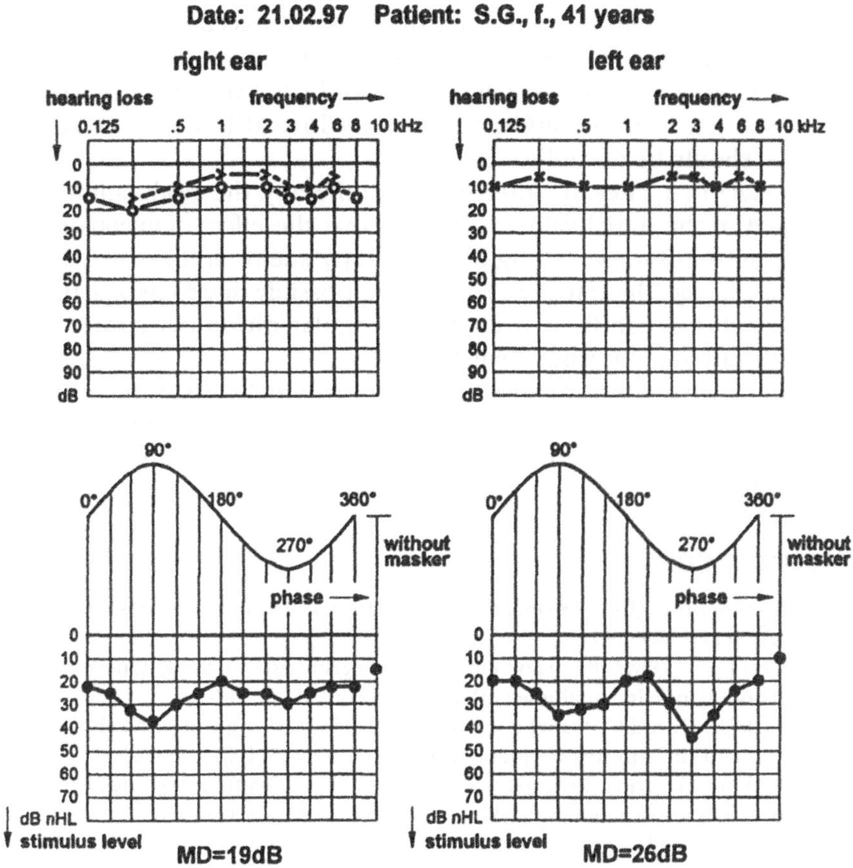

Figure 27.4: pure tone audiogram (top) and phase audiogram (bottom) of patient SG six days after therapy with acetazolamide.

Discussion

During the therapy, acetazolamide led to an improvement of hearing and tinnitus in a patient with unilateral low-tone sensorineural hearing loss. A corresponding reduction of EH was monitored by low-frequency masking. We suppose that the inhibition of carbonic anhydrase (CA) in the inner ear causes a decrease of fluid pressure in the cochlea.

CA is an enzyme widely distributed in the human body and is found in many tissues where secretory functions exist, for example in the choroid plexus of the brain, the eye and in high concentrations in the inner ear (Maren 1977). It catalyses the hydration of carbon dioxide and is believed to play an important role in the secretion of K^+ and H^+ in the endolymph, formation of HCO_3^- and overall production or resorption of the inner-ear fluid (Watanabe and Ogawa 1984).

In experimentally induced hydrops in animals, Shinkawa and Kimura (1986) showed a reduction of endolymphatic hydrops by acetazolamide. Brookes et al. (1984) found a useful effect with acetazolamide in only four of 14 patients with Ménière's disease and a high incidence of side effects such as paraesthesia, headache and drowsiness. Norris (1988) suggests that acetazolamide can be paradoxical in its effect and the therapy should be carefully monitored.

The fact that, after the interruption of therapy, the modulation depth declined to the initial values, indicates that the effect of acetazolamide might be temporary. A long-term therapy seems to be restricted by adverse effects. The transient colics were possibly due to hypercalciuria. This condition can be prevented by regular blood pH controls and administration of bicarbonate if necessary.

The diagnosis and monitoring of endolymphatic hydrops using LFM provides a non-invasive means for the evaluation of risks and benefits of a pathophysiologically directed treatment of EH, in this case with acetazolamide.

References

Brookes GB, Booth JB (1984) Oral acetazolamide in Ménière's disease. J Laryngol Otol 98(11): 1087–95.

Maren TH (1977) Physiology and chemistry of cerebrospinal fluid, aqueous humor and endolymph in Squalus acanthias. J Exp Zool 1993: 317–24.

Mrowinski D, Gerull G, Nubel K, Scholz G (1995) Masking and pitch shift of tone bursts and clicks by low-frequency tones. Hear Res 85(1–2): 95–102.

Mrowinski D, Scholz G, Krompass S, Nubel K (1996) Diagnosis of endolymphatic hydrops by low-frequency masking. Audiol Neurootol 1:125–34.

Norris CH (1988) Drugs affecting the inner ear. A review of their clinical efficacy, mechanisms of action, toxicity, and place in therapy. Drugs 36(6): 754–72.

Shinkawa H, Kimura RS (1986) Effect of diuretics on endolymphatic hydrops. Acta Otolaryngol 101(1–2): 43–52.

Watanabe K, Ogawa A (1984) Carbonic anhydrase activity in stria vascularis and dark cells in vestibular labyrinth. Ann Otol Rhinol Laryngol 93: 262–6.

Correspondence to Dr E A Kabudwand, Klinikum Buch, HNO, Karowerstr 11, D13125 Berlin. Fax (+49) 30-94014505.

Chapter 28
Effects of middle ear pressure changes on the inner ear function in Ménière's disease: a review

B DENSERT, S ARLINGER, O DENSERT, K SASS and L ÖDKVIST, Department of Otolaryngology, University of Lund, Sweden and Department of Otolaryngology, University Hospital, Linköping, Sweden

ABSTRACT. A review of experimental and clinical experience concerning effects of middle ear pressure changes on inner ear function was undertaken. The relationships between external pressure changes and inner ear function in Ménière's disease were demonstrated for the first time when patients with acute vestibular and cochlear symptoms caused by this disorder were treated with reduced pressure in a pressure chamber. Transmission of various pressure changes from the middle-ear to the inner ear fluids has been studied in a series of animal experiments in order to clarify the inner ear pressure-regulating mechanisms. It was shown that the cochlear aqueduct plays the main role as pressure-release factor. It was also found that a complex pressure wave applied in the middle ear was the most effective way to induce a positive pressure change in the perilymphatic compartment. It was postulated that such pressure changes could influence the inner ear fluid pressure relations in endolymphatic hydrops and improve inner ear symptoms. Effects of middle ear pressure changes have been investigated in a series of clinical studies. The results indicated an improvement in vestibular and cochlear symptoms in patients with definite Ménière's disease. Transtympanic electrocochleography (TTECochG) has enabled the objective study of the effects of middle ear pressure changes on cochlear electrophysiology. Clinical, placebo-controlled studies have shown statistically significant increase in the amplitudes of the action potential (AP) and a decrease in the summating potential (SP) as well as in the AP/SP ratio in the active group. The findings indicated an improvement in inner-ear electrophysiology after exposure to pressure applications. Two weeks' treatment with placebo control has shown statistically significant improvement in the active group in hearing function as well as subjective symptoms such as vertigo, tinnitus and functionality profile.

Key words: inner-ear pressure, cochlear aqueduct, Ménière's disease, electrocochleography, pressure treatment.

Introduction

Ménière's disease is an inner-ear disorder that, despite extensive research, still constitutes a major clinical challenge. Attacks of rotatory vertigo and dizziness dramatically affect the patient's daily life. Cochlear damage caused by a long-standing endolymphatic hydrops presents a difficult rehabilitation problem. Tinnitus can be particularly resistant to known forms of treatment and frequently causes a debilitating condition in patients. Permanent damage to sensory organs of the vestibular and cochlear systems occurs in most cases after a few years. Independently of large individual variability in symptomatology, functional disability is a frequent outcome and a major problem in these patients.

A variety of medical and surgical methods, aimed at reduction of the endolymphatic volume, have been tried in order to control the symptoms. The results of the function preserving methods have not been fully satisfactory and recurrence of vertigo is a most common problem. Destructive methods show a higher success rate at the cost of damage to the inner-ear sensory organs. Apart from the vestibular nerve section, which permanently deprives the patient of the peripheral balance function on the affected side, no surgical method has, so far, been fully effective in preventing recurrence of vertigo.

Local overpressure treatment is a non-destructive method that has been developed from long-term research of the relationships between external pressure changes and the inner-ear function in Ménière's disease.

The aim of this paper is to present a review of experimental and clinical experience gained over the years concerning the effects of middle-ear pressure changes on the inner-ear function in Ménière's disease, and to present results of the latest clinical, studies of middle-ear pressure changes on cochlear electrophysiology and clinical symptoms in this disorder.

Pressure chamber treatment

The first clinical experiments, using a pressure chamber, demonstrated that patients with acute cochlear symptoms improved when exposed to hypobaric pressure, which created a relative overpressure in the patient's middle ear (Densert et al. 1975; Ingelstedt et al. 1976). It was also shown that symptoms deteriorated when pressure was increased in the pressure chamber – creating a negative relative pressure in the middle ear. Figure 28.1 illustrates changes in hearing levels in relation to alterations of pressure in a pressure chamber. Improvement in cochlear symptoms was accompanied by relief of vertigo when exposed to negative pressure in the pressure chamber. Since then a hypobaric pressure chamber has been used in patients with Ménière's disease for treatment of inner ear symptoms (Tjernström et al. 1979; Van Deelen et al. 1987; Kitahara et al. 1994).

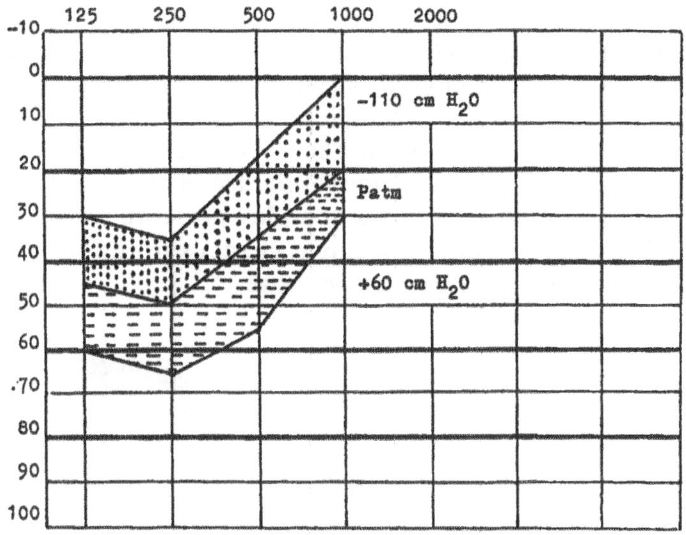

Figure 28.1: changes in hearing levels depending on alterations in chamber pressure. Improvement after hypobaric pressure, while hearing deteriorates due to an increase in chamber pressure.

A pressure chamber is considered to have a limited value as a treatment modality, but its use has had a significant impact on the development of research of the effects of pressure changes on the inner-ear function.

Animal experiments

Pressure-regulating mechanisms have been studied by Carlborg et al. (1981) in a series of animal experiments. It was demonstrated that the cochlear aqueduct was the main pressure release point responsible for stabilisation of changes induced in the inner ear fluid pressure via a communication route to the cerebrospinal fluid (CSF). It was found that other communication routes, such as the endolymphatic duct and sac, had a lesser role with regard to pressure stabilisation. Carlborg et al. (1990) have also demonstrated that in the presence of a patent cochlear aqueduct, the inducement of a pressure change in the inner-ear fluids was dependent on the rise and/or fall time of the pressure change. Effects of a hypobaric pressure chamber seem thus to be limited by these pressure regulating mechanisms.

In order to induce a positive pressure change in the inner-ear fluids, transmission of various pressure pulses from the middle ear to the perilymphatic fluid were studied in the cat by Densert et al. (1986). It was shown that a complex pressure wave consisting of a sine pressure wave superimposed on a static pressure component had longer time constants as compared to a square pressure wave, which was equalised almost instantly in the presence of a patent cochlear aqueduct. Figure 28.2 shows recordings of pressure changes applied in the middle ear of a cat and the corresponding changes in the perilymphatic fluid. A complex pressure wave induced a more pronounced positive pressure change in the inner ear compared with a square pressure wave, which is equalised via the cochlear aqueduct within a few seconds. A negative pressure change caused by the rebound flow of perilymphatic fluid through the cochlear aqueduct, initiated by a complex pressure wave, is smaller compared with the rebound negative changes in the perilymphatic pressure induced by a square pressure wave.

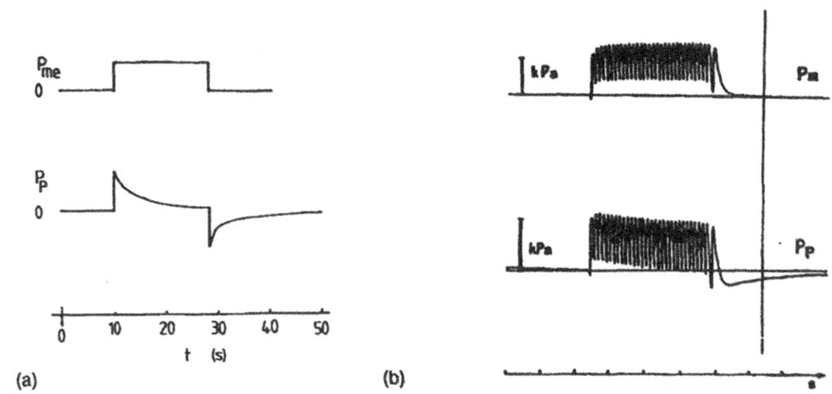

Figure 28.2: a change in the perilymphatic fluid induced by a square pressure wave is shown in (a) and a change induced by a complex pressure wave in (b). Note the difference in pattern of inner-ear pressure stabilisation. The cochlear aqueduct is open.

Feasibility studies of the local overpressure method in patients with Ménière's disease

Effects of hypobaric pressure changes have been explained in the light of the hypothesis that the inducement of a positive pressure change in the middle ear, transmitted to the inner-ear fluids, may influence the inner-ear fluid pressure relations in cases of endolymphatic hydrops. It was postulated that a positive pressure change induced in the perilymphatic compartment might force an excess of the endolymphatic fluid out of the endolymphatic compartment via the endolympahtic duct and other pressure communicating routes.

Effects of transmission of complex pressure waves applied in the middle ear have been studied by Densert et al. (1986) in various groups of patients with well-defined diagnosis of Ménière's disease. It was shown that patients could be relieved from vestibular symptoms after a period of exposure to complex pressure pulses. Improvement in hearing thresholds and speech discrimination scores was also shown in these groups of patients. There was also an improvement in the inner-ear frequency selectivity indicating a change in inner-ear hydrodynamics. These clinical studies have been conducted as feasibility studies aimed at investigating the therapeutic potential and safety of the local overpressure method in the control of symptoms in Ménière's disease.

It was also found that the use of local overpressure over longer periods of time could give a satisfactory control of vertigo and had a preserving effect on hearing function. No adverse effects on cochlear or vestibular function have been observed in connection with application of middle-ear pressure pulses.

Effects of middle ear pressure changes on cochlear electrophysiology

Previous clinical research of effects of middle ear pressure changes on clinical symptoms in Ménière's disease has indicated a therapeutic potential of this treatment modality. An improvement in inner ear fluid balance could be considered responsible for the improvement of hearing and symptoms.

Despite clear indications of an improved inner ear function observed in patients with Ménière's disease in relation to local overpressure, there has been still no direct evidence that the inducement of a positive pressure change into the inner ear fluids could in fact alter inner ear hydrodynamic conditions. Our current knowledge of the behaviour of electrocochleographic recordings in endolymphatic hydrops is to a great extent formed by observations of changes in the electric potentials in response to glycerol dehydration or other agents likely to influence the inner ear fluid balance. Since there is no true animal model of Ménière's disease and no direct method of measuring the endolymphatic volume/pressure changes in patients, the assessment of electrophysiological potentials using transtympanic electrocochleography (TTECochG) seems to have an unquestioned value in monitoring hydrops in relation to pressure applications.

Transtympanic electrocochleography is a well-documented method of recording the electrical activity of the cochlea and the first order eight nerve fibres in response to acoustic stimulation. It is recognised as a powerful diagnostic tool in evaluating cochlear function in various inner ear disorders, endolymphatic hydrops in particular. Acoustic stimuli such as an alternating click and a long tone burst evoked stimulus dependent electric cochlear potentials – the action potential (AP) and the summating potential (SP). The click evoked AP is the sum of electric activity of the individual cochlear nerve fibres, firing in synchrony along the length of the basilar membrane. The SP is known as a multifactorial electric potential originating from various non-linear processes within the cochlea. The SP arises from the outer hair cells and is described as a DC shift that occurs at high intensities of the stimulus. Numerous electrophysiological studies (Gibson et al. 1987; Daumon et al. 1986) have shown that electrocochleographic changes associated with endolymphatic hydrops are characterised by an increase in the SP. The SP is therefore considered to reflect a degree of deflection of the cochlear partition caused by hydrops.

Application of pressure pulses

Pressure pulses transmitted from the ear canal through the middle ear to the inner ear fluids is shown in Figure 28.3. Pressure pulses are applied in the ear canal by a close fitting cuff. The pressure is distributed via the ventilation tube in the middle ear, and through the round window membrane to the inner ear fluids. Since the inner ear fluids are not compressible and the membranous labyrinth is enclosed in a bony capsule, an increase in perilymphatic pressure may direct an excess of endolymphatic fluid out of the endolymphatic compartment through the pressure release points such as the endolymphatic duct and sac – thus activating the longitudinal flow. Circulatory flow of the endolymphatic fluid through the inner ear membranes and connecting blood vessels or other mechanisms, which may activate the turn over of the endolymph, could be other feasible systems contributing to a decrease in endolymphatic fluid.

The effects of middle ear pressure changes on the cochlear electric potentials have been investigated in patients with well-defined Ménière's disease. The first step was to investigate whether a few minutes' exposure to pressure pulses might influence the hydropic ear. TTECochG was performed immediately before and after the exposure.

An example of TTECochG recordings is shown in Figure 28.4. There was an increase in the AP amplitude associated with an improvement in the click response morphology. The burst evoked responses showed a decrease in the SP potential at 1 and 2 kHz. The increase in the AP potential and the decrease in the SP potential were statistically significant (Densert et al. 1997).

Changes in TTECochG recordings obtained in patients after the pressure applications show an increase in the synchronised cochlear response, which, along with a decrease in the AP/SP ratio, indicates an improvement in inner ear electrophysiology. Changes in the cochlear responses support the hypothesis that transmission of middle ear pressure changes alleviates the effects of the endolymphatic hydrops on the sensory organs of the cochlea. The results can be considered as a proof of mechanisms indicating that pressure pulses directly affect the underlying pathology responsible for symptoms in Ménière's disease.

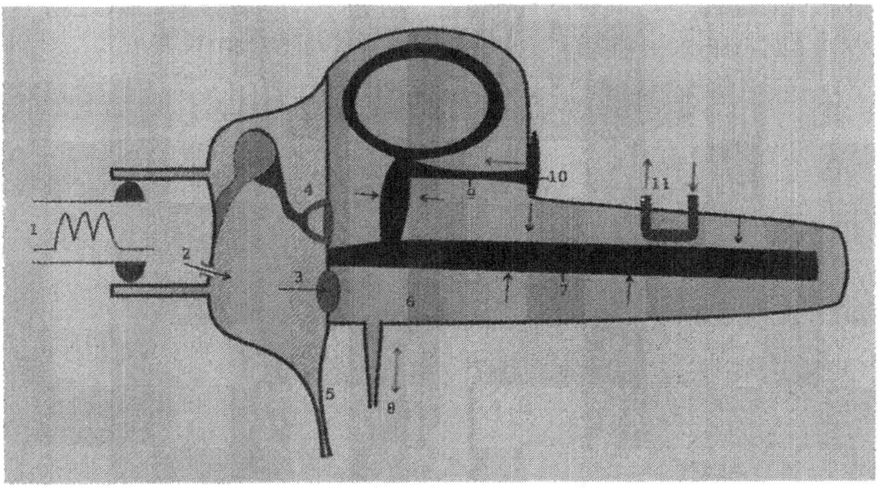

Figure 28.3: a complex pressure wave (1) is applied in the ear canal and transmitted to the middle ear via the ventilation tube (2). A pressure change is then mainly transmitted through the round window membrane (3) to the inner ear fluids. Important pressure regulation points are denoted; the Eustachian tube (5), scala tympani (6), scala media (7), cochlear aqueduct (8), endolymphatic duct (9) and sac (10), blood vessels (11).

Effects of middle ear pressure changes on clinical symptoms

Evaluation of the effects of pressure pulses on the inner ear symptoms in Ménière's disease has been the next stage of assessment of efficacy of the local overpressure method. The aim was thus to estimate the effects of middle ear pressure pulses, applied over a longer period on rotatory vertigo, dizziness, fullness of the ear, tinnitus and hearing function. Assessment of vertigo, dizziness, aural pressure and tinnitus was made using visual analogue scales (VAS).

A questionnaire, concerning the functionality profile, was given to gather information about the patient's ability to perform in daily life and at work. Functions such as ability to function at work, drive a car, participate in social functions, free time activities, enjoy leisure and rest was evaluated. Transtympanic electrocochleography was conducted before and after the treatment period.

The treatment consisted of repeated pressure pulses applied in the middle ear via the ear canal and through the ventilation tube. Pressure applications were of the amplitude of 12 cm of water and consisted of complex pressure pulses composed of static pressure and a low

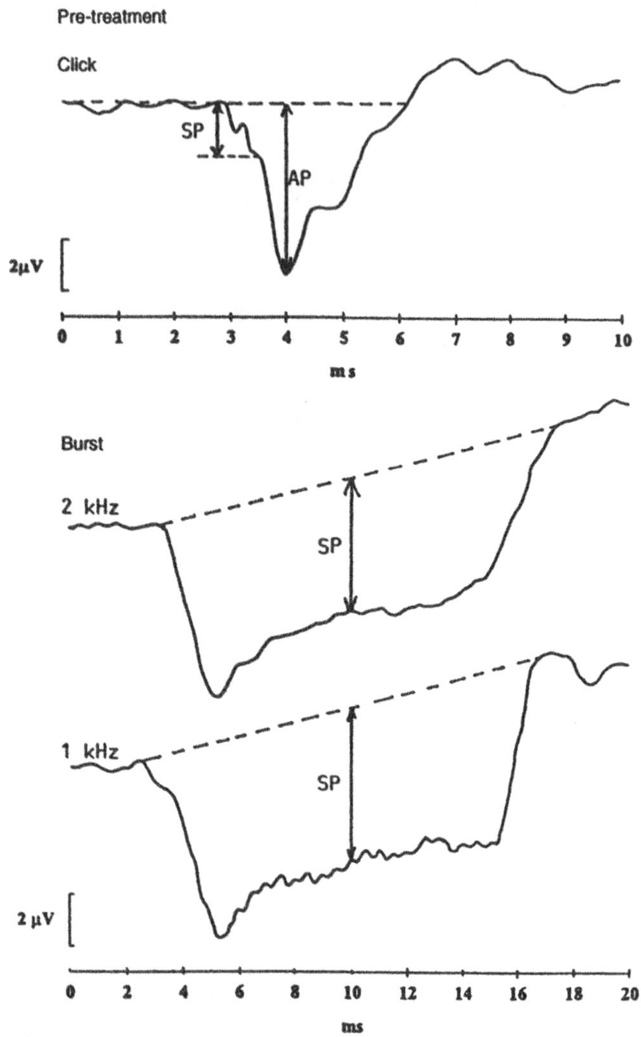

(a)

Figure 28.4: shows changes in the electrocochleographic recordings in a patient before (a) and after (b) three minutes' exposure to the pressure pulses. There is an improvement in the click response pattern and an increase in the AP amplitude. The SP component is decreased in the burst evoked responses.

frequency sinus modulation. A special pressure generator constructed for this purpose delivered controlled pressure pulses of predetermined parameters. A built-in software computer system ensured stable and repeatable pressure levels.

The patients were given the pressure treatment for two weeks. Patients randomised to receive placebo were instructed to use a placebo device for the same period of time. A preliminary analysis of the results is presented here.

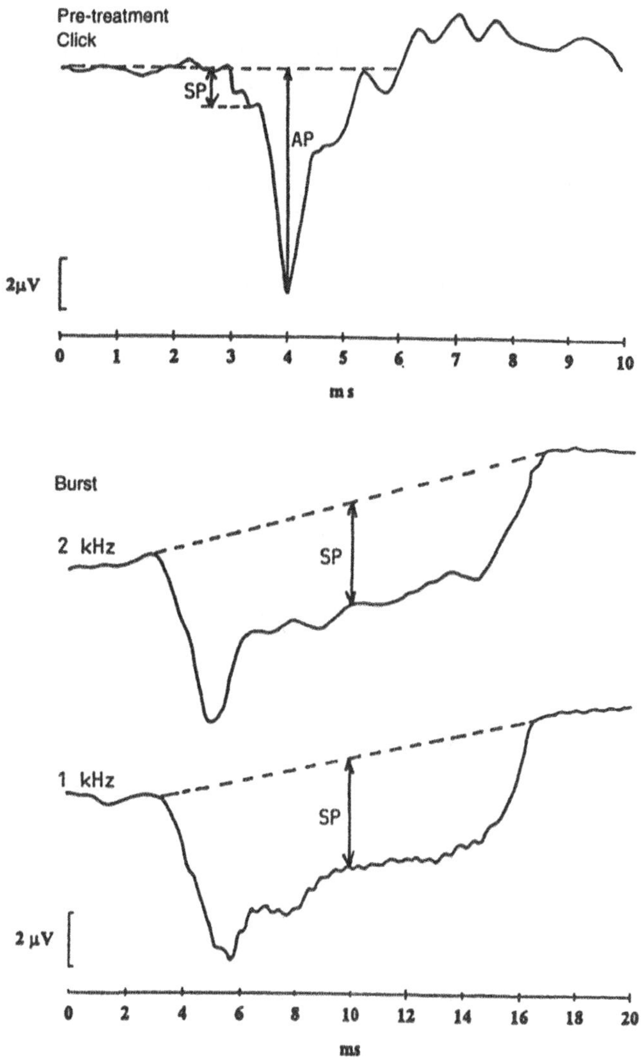

(b)

Figure 28.4: (cont)

A significant improvement in frequency and intensity of vertigo, dizziness, aural pressure and tinnitus was found in the active group. Functionality profile was also significantly improved. For all these parameters the $p < 0.0005$.

In many patients in the active group, changes in functionality profile have been quite dramatic and shown a substantial improvement in key functions such as work, rest, social life and driving. No patients in the active group became worse during the period of treatment. No significant improvement was found in the placebo group. A few patients deteriorated during placebo treatment.

Hearing improvement and/or improvement in speech discrimination scores of more than 12% was found in several patients in the active group. No improvement in hearing function was found in the placebo group.

Transtympanic electrocochleography recordings showed a significant decrease in the SP potentials at 1 and 2 kHz in the active group.

Changes in TTECochG recordings are shown in Figure 28.5 in a patient before and after two weeks of treatment. An improvement in pure tone audiometry is shown in the same patient.

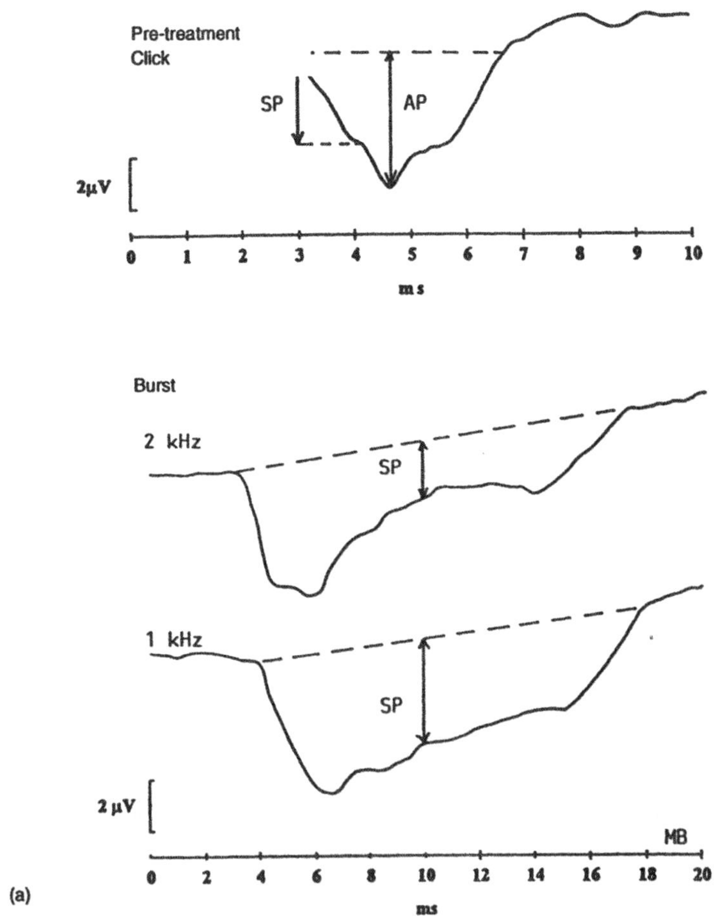

Figure 28.5: changes in the electrocochleographic recordings before (a) and after (b) two weeks of local pressure treatment. A corresponding improvement in the audiometry (c and d) is shown in the same patient. Speech audiometry was also improved.

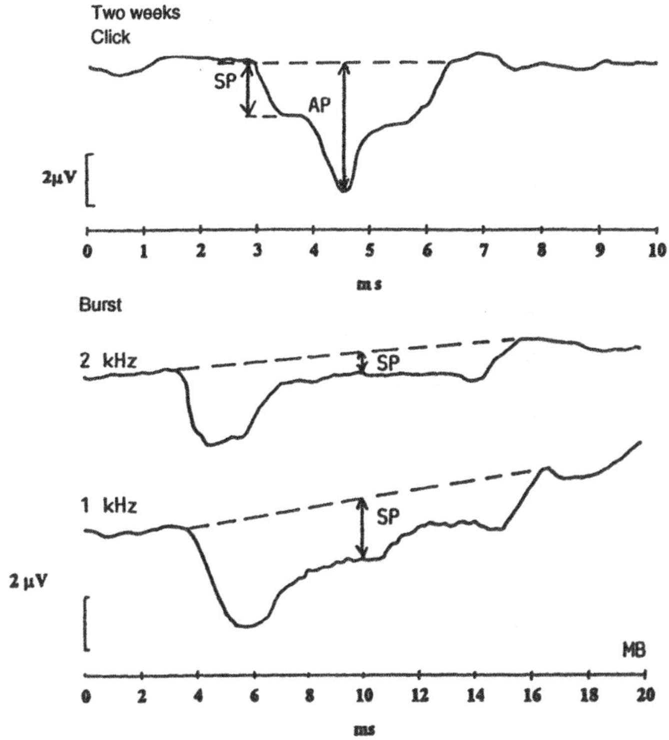
Two weeks
Click
SP
AP
2μV
0 1 2 3 4 5 6 7 8 9 10
ms
Burst
2 kHz
SP
1 kHz
SP
2 μV
MB
0 2 4 6 8 10 12 14 16 18 20
ms

(b)

(c)

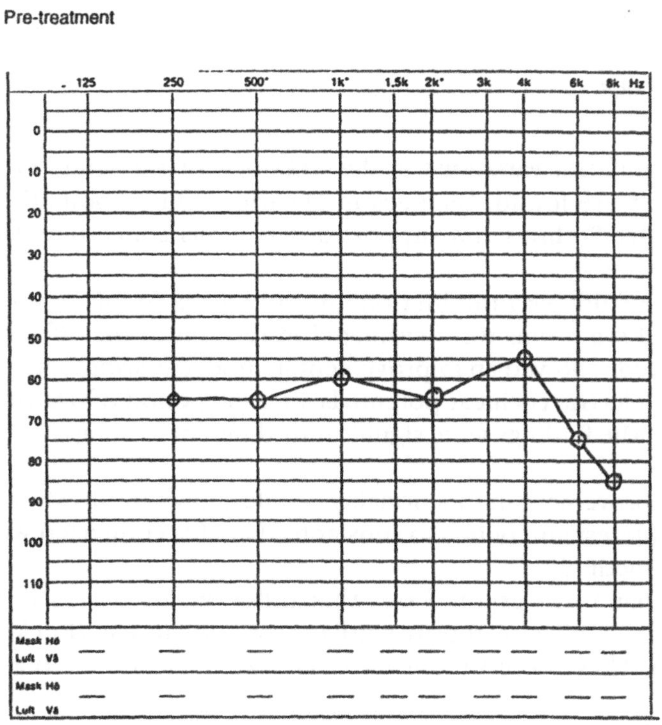
Pre-treatment
125 250 500 1k 1.5k 2k 3k 4k 6k 8k Hz
0 10 20 30 40 50 60 70 80 90 100 110
Mask Hö
Luft Vä
Mask Hö
Luft Vä

Two weeks

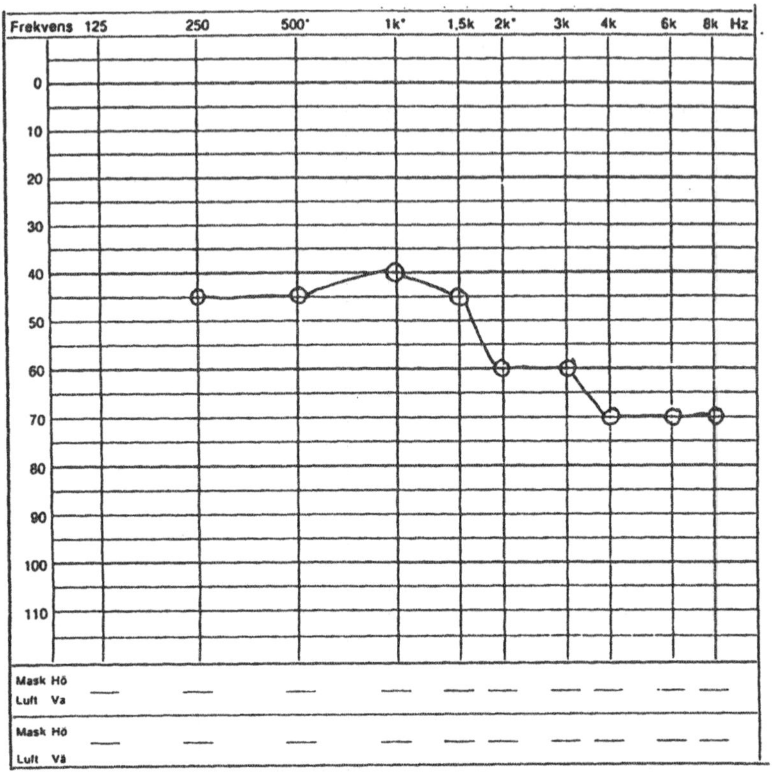

(d)

Figure 28.5: Contd

No significant variations in the TTECochG parameters in the placebo group were found. Changes in electric cochlear potentials represent objective findings and support observations concerning the inner ear symptoms.

The study was the first one where effects of local overpressure on clinical symptoms were evaluated in relation to the effects of placebo. The study showed an improvement in inner ear symptoms and a change in the electric cochlear potentials after the period of overpressure treatment. The findings were significantly different as compared to a placebo group.

Improvement in functionality profile shown in this study is of particular interest and points to substantial benefits connected with this treatment method.

It has been pointed out that different methods of treatment in Ménière's disease report similar rates of success (Ruckensain et al. 1991). These observations have often been attributed to a non-specific effect of placebo, which is believed to have an influence on symptoms in Ménière's disease, particularly on vertigo. There are only a few placebo-controlled studies undertaken with patients with Ménière's disease, probably because of problems concerning feasibility of these studies. The role of placebo in this disorder has so far been difficult to assess. The results presented here indicate that the effects of placebo in a group of patients with active vestibular symptoms may have previously been overestimated.

Conclusions

The method of local overpressure is based on long experimental and clinical experience. The pivotal, controlled clinical studies indicating clinical efficacy of this method have only recently been conducted. The results point out that the local pressure method may be useful in relieving subjective symptoms and restoring the functional disability which is considered to be a major complaint in this disorder. The effects are significantly superior to those obtained by a placebo device, an observation that puts a new light on the role of placebo in Ménière's disease. More effort is needed to assess the long-term efficacy of the pressure method on the function of the inner ear.

References

Carlborg B, Densert B, Densert O (1982) Functional patency of the cochlear aqueduct. Ann Otol Rhinol Laryngol 91: 209–15.

Carlborg B, Farmer J, Carlborg A (1990) Effect of hypobaric pressure on the labyrinth. Acta Otolaryngol (Stockh) 110: 386–93.

Dauman R, Aran JM, Portmann M (1986) Summating potential and water balance in Ménière's disease. Ann Otol Rhinol Laryngol 95: 389–95.

Densert O, Ingelstedt S, Ivarsson A (1975) Immediate restoration of basal sensorineural hearing (Mb Ménière) using a pressure chamber. Acta Otolaryngol (Stockh) 80: 93–100.

Densert B, Densert O, Erlandsson B, Shepard H (1986) Transmission of complex pressure waves through the perilymphatic fluid in cats. Acta Otolaryngol 102: 403–9.

Densert B (1986) Effects of overpressure on hearing function in Ménière's disease. Acta Otolaryngol 103: 32–42.

Densert B, Arlinger S, Densert O (1987) Air-bone gap in Ménière's disease after exposure to overpressure. Audiology 26: 339–47.

Densert B, Arlinger S, Densert O, Sass K, Ödquist L (1997) Effects of middle ear pressure changes on the cochlear electric potentials in patients with Ménière's disease. Am J Otol 18: 726–33.

Densert O, Carlborg B, Stagg J (1981) Transmission of low frequency pressure steps to the perilymphatic fluid. Acta Otolaryngol (Stockh) 91: 55–64.

Ingelstedt S, Ivarsson A, Tjernström Ö (1976) Immediate relief of symptoms during acute attacks of Ménière's disease using a pressure chamber. Acta Otolaryngol (Stockh) 82: 368–78.

Kitahara M, Kodama A, Ozawa H, Izukura H (1994) Pressure test for diagnosis of Ménière's disease. Acta Otolaryngol (Stockh) Suppl 510: 107–10.

Kugamani H, Nishida H, Baba M (1982) Electrocochleographic study of Ménière's disease. Arch Otolaryngol 108: 284–8.

Moffat DA, Gibson WPR, Ramsden RT, Morrison AW, Booth JB (1978) Transtympanic electrocochleography during glycerol dehydration. Acta Otolaryngol (Stockh) 85: 156–66.

Ruckenstain MJ, Rutka JA, Hawke M (1991) The treatment of Ménière's disease: Torok revisited. Laryngoscope 101: 211–16.

Tjernström Ö (1974) Alternobaric vertigo. An experimental study in man of vertigo due to atmospheric pressure changes. Malmö General Hospital. Thesis.

Tjernström Ö, Casselbrandt M, Harris S (1979) Hearing improvement in attacks of Ménière's disease treated with pressure chamber. Adv Otolaryngol 25: 54–60.

Van Deelen GW, Hulk J, Hiuzing EH (1978) The use of the underpressure chamber in the treatment of patients with Ménière's disease. J Laryngol Otol 101: 229–35.

Correspondence and requests for further information about the preliminary analysis or other aspects of this paper to Dr Barbara B Densert, Department of Otolaryngology, General Hospital, 301 85 Halmstad, Sweden. Fax (+46) 35 131497.

Chapter 29
Tullio phenomenon: disturbed inner ear homeostasis?

CLIODNA F O'MAHONEY, Department of Audiology and Audiological Medicine, Great Ormond Street Hospital for Children, London UK, and LINDA M LUXON, Institute of Laryngology and Otology, and National Hospital for Neurology and Neurosurgery, London UK

ABSTRACT. The Tullio phenomenon denotes vestibular activation in response to sound. This activity may arise from the semicircular canals (canal Tullio) or from the otolith organs (otolith Tullio). At high sound intensities (>125–130 dBSPL), such activation is considered to be normal in both animals and humans. However there are many patients who experience vestibular symptoms, and in whom vestibular signs can be demonstrated, in association with commonplace sounds of much lower intensities, such as the ringing tone of the telephone, traffic, humming etc., and as such cause considerable distress. Various vestibular responses in association with sound have been reported in the literature, including vestibulo-ocular, vestibulo-spinal and vestibulo-collic activity. The Tullio phenomenon has been associated with several different aetiologies including perilymphatic fistulae, ossicular dislocation or subluxation, hypermobility of the stapes footplate, Ménière's disease, cochlear dysplasia and psychological disturbance. However many patients do not fit into any of these categories and the precise pathophysiological mechanism or mechanisms still remain obscure. Recent work has highlighted the role of new techniques such as tympanic membrane displacement analysis, video-oculography and threshold measurement of the click-evoked vestibulo-collic reflex in the assessment of these patients, the results of which suggest saccular hypersensitivity may be an important factor. The

many similarities between the Tullio phenomenon, Hennebert's sign (nystagmus provoked by non-acoustic positive or negative ear canal pressure) and dizziness induced by increasing intracranial pressure are outlined and the possible common pathophysiological mechanisms discussed.

Key words: Tullio phenomenon, inner ear, dizziness, homeostasis, pathophysiology, vestibule, Ménière's disease.

Introduction

The Tullio phenomenon denotes vestibular activation in response to sound. It is named after Tullio, who in 1929 demonstrated sound-induced nystagmus in pigeons with fenestrations in the bony horizontal semicircular canal – the so-called 'third window' between the middle and inner ears. It was first described in humans by Benjamins (1938).

It is thought that if the intensity of sound is loud enough (generally accepted to be >125–130 dBSPL), normal subjects and animals will show signs of vestibular excitation (Parker 1976; 1978), but there is in addition a relatively small subpopulation who develop vestibular signs and symptoms at intensities encountered in everyday life, and which cause significant distress as a result. Examples of the precipitants of these symptoms/signs include traffic noise, drilling, loud voices, children crying, and dial and ringing tones on the telephone. Interestingly, patients often complain that their own voices cause symptoms, particularly while uttering vowels or humming. Many professional singers/musicians (particularly wind instrument players) complain that certain notes cause them to feel dizzy, when they play themselves, which can clearly cause career difficulties. However they remain asymptomatic while listening to others playing the same notes at even higher intensities. This raises the question whether the vestibular activation is specific to sound or indeed if non-acoustic pressure changes may also be responsible.

Typically patients describe the symptoms as a sensation of the world moving, vibrating or tilting on exposure to the offending sound or sounds. Blurring/jumping of vision, often in time to the beat of music, and a feeling of dizziness/vertigo/unsteadiness/as if one is going to fall are other common complaints. As well as these symptoms, patients often describe associated bizarre-sounding symptoms, such as being able to 'hear their eyes move', an 'echoey' or reverberatory quality to sounds including their own voices or a pressure sensation in the affected ear.

Almost without exception, descriptions in the literature of the Tullio phenomenon relate to sound of frequency from 25 Hz up to 1000 Hz (De Vries and Bleeker 1949; Kacker and Hinchcliffe 1970; Kwee 1976; Vogel et al. 1986; Dieterich et al. 1989; Ishizaki et al. 1991; O'Mahoney et al. 1998). Whereas in terms of the range of sound frequencies this is considered low frequency, under normal conditions the functional range of frequency of head motion to which the vestibular system responds is up to approximately 10Hz. Thus in some way the pathology that gives rise to the Tullio phenomenon alters the frequency response of the vestibular labyrinth.

Various aspects of vestibular activation in response to acoustic stimulation have been objectively demonstrated, including vestibulo-ocular, vestibulo-collic and vestibulo-spinal activity. Until recently objective demonstration of vestibular stimulation has mainly relied on recording eye movements. Such recording has become more sophisticated and amenable to analysis with the advent of video-oculography and three dimensional search coil recording techniques. Eye movements suggestive of horizontal semicircular canal stimulation (canal Tullio) have been reported in both animals and humans, and eye movements consistent with stimulation of the otolith organs (Vogel et al. 1986), both utricle (Dieterich et al. 1989) and saccule (O'Mahoney et al. 1998) have also been reported, (Otolith Tullio). However, while ascribing the origin of eye movements to the horizontal semicircular canal is generally thought to be relatively straightforward – stimulation of which gives rise to horizontal nystagmus (however see Fluur and Melstrom 1970(a) and (b) below) – the situation is not as clearly defined for the other vestibular end-organs. In particular, research into eye movements evoked by stimulation of the otolith organs is scanty and contradictory. Suzuki et al. (1969a) reported predominately rotatory eye movements, but with in addition a vertical and horizontal component, on stimulation of the utricular nerve. Fluur and Melstrom (1970a) reported pure horizontal or pure vertical eye movements depending on which quadrant of the utricular macula of the cat was stimulated. These results were not reproduced by Curthoys (1987) who found that stimulation of any area of the saccular or utricular macula in guinea pigs leads to ipsilateral upward or upward/torsional eye movement, with some skew deviation evident in those guinea pigs investigated contralaterally. Pure horizontal movements were not observed. Similar upward or upward/torsional movements were observed by other workers in cats, monkeys and rabbits in response to otolith stimulation (Suzuki et al. 1969b; Gernandt 1970; Markham 1973; Chan 1979). Dieterich et al. (1989) reported eye movements in one patient with Tullio phenomenon, which they attributed to utricular stimulation but that are also consistent with saccular stimulation. O'Mahoney et al. (1998) have reported 14 patients with symptoms of the Tullio phenomenon, in whom eye movements were consistent with, although not pathognomic of, saccular stimulation. Results of click-evoked vestibulo-collic reflex testing on these patients, supports a saccular origin for these symptoms and signs (Young et al. 1977; Didier and Cazals 1989; McCue and Guinan 1994; Murofushi et al.

1995; Colebatch et al. 1994). Nevertheless, deducing the origin of torsional and/or vertical nystagmus remains somewhat speculative, and such nystagmus could arise from either otolith organ or indeed from a combination of otolith organs and/or posterior and/or superior semicircular canals.

As well as vestibulo-ocular activity, abnormal vestibulo-spinal and vestibulo-collic activity has been reported in response to sound stimulation. Vestibulo-spinal activity has been investigated functionally by measuring body sway using a force platform (posturography) and also electrophysiologically by measuring the electromyographic response to sound stimulation. Ishizaki et al. (1991) reported 152 patients with various inner-ear pathologies (including Ménière's disease, perilymphatic fistula, vertigo associated with chronic otitis media with or without cholesteatoma, and peripheral vestibular pathology of unknown origin) in whom increased postural sway was demonstrated in approximately 25% of these patients when they were exposed to low frequency sound (< 63 Hz). No increased sway was observed in 55 healthy volunteers.

Dieterich et al. (1989) also demonstrated sound-induced postural imbalance in their patient with Tullio phenomenon using a piezoelectric force platform. In addition the latencies of the EMG responses of the tibialis anterior, gastocnemius, neck and arm muscles were unusually short, which is thought to indicate direct stimulation by sound of the otolith-vestibulospinal reflex. Colebatch et al. (1994, 1998) have also shown short latency EMG responses to sound.

A significantly lower threshold of vestibulo-collic reflex activation has also been demonstrated in patients with symptoms and signs of the Tullio phenomenon (Colebatch et al. 1994, 1998; Bronstein et al. 1995; O'Mahoney et al. 1998). These data are thought to be indicative of pathological hypersensitivity of the saccular macula. Thus, from the above, various aspects of vestibular activation have been well demonstrated but the mechanism(s) whereby a relatively low intensity of sound can cause vestibular activation are still a source of much investigation. Several theories have been postulated.

Fistula between middle and inner ear

Arising from Tullio's original observations on fenestrated pigeons, it was thought that for the phenomenon to occur, it was necessary that there be a defect in the bone around the semicircular canals in order that the crista be stimulated by sound. Much of the subsequent animal work has been based on this assumption. In the 1980s Wit and Bleeker with various co-authors (Wit and Bleeker 1982; Wit et al. 1981; 1984, 1985) reported their observations on fenestrated pigeons, demonstrating vestibular nerve sensitivity to sound in the audio-frequency range even after cochlear extirpation. In humans, it used to be believed that the Tullio phenomenon was pathognomic of congenital syphilis (Mayer and Fraser 1936), the pathophysiology thought to be gummatous softening of the bone around the semicircular canal. More recently other causes of perilymphatic fistulae have been associated with the Tullio phenomenon, including fistulae associated with cholesteatoma (Ishizaki et al. 1991), 'spontaneous' fistulae of both round and oval windows (Fox et al. 1988) and an oval window fistula due to barotrauma (Ildiz and Dundar, 1994). Indeed Fox et al. (1988) suggested that the high degree of correlation between symptoms of the Tullio phenomenon and subsequent demonstration of an active fistula at surgery was such that the presence of sound-induced vestibular symptoms was a useful tool in establishing the diagnosis of a perilymphatic fistula, a diagnosis that is open to much controversy both pre- and peri-operatively.

Hypermobility/dislocation of the stapes footplate

Kaker and Hinchcliffe (1970) reported a patient with Tullio phenomenon who, during surgery, was shown to have subluxation of the stapes, causing the footplate to come into direct contact with the saccule, which they postulated to be the route of vestibular stimulation. Vogel et al. (1986) suggested a similar mechanism in their patient but as surgical exploration was refused, this could not be confirmed. Dieterich et al. (1989) described a patient with sound induced oscillopsia associated with skew deviation of the eyes (dysconjugate vertical eye movements and ocular torsion) in whom surgical exploration revealed a subluxed stapes footplate with a hypertrophic stapedius muscle. It was suggested that, for a given sound pressure, the amplitude of footplate motion was pathologically increased, and this led to overstimulation of the utricle.

Ménière's disease

Using a similar logic to that immediately above but in reverse, pathological contact between a normal stapedial footplate and the dilated saccule/utricle (as occurs in endolymphatic hydrops) could give rise to sound-induced vestibular activation. Although there are several reports in the literature of patients with Ménière's disease and Tullio phenomenon (Kacker and Hinchcliffe 1970; Ishizaki et al. 1991), in most instances the diagnostic criteria used to make the diagnosis of Ménière's disease are not recorded.This is an attractive explanation, but given the prevalence of Ménière's disease and the extent of research carried out into this condition, it is surprising that the association does not seem to be a commonly reported phenomenon in association with Ménière's disease and, indeed, is not mentioned in many large studies. This may, however be partly due to two extraneous factors. Firstly a label of 'Ménière's disease' is often ascribed somewhat arbitrarily to patients with dizziness of unknown aetiology, without applying accepted diagnostic criteria. Secondly, the symptoms of Tullio phenomenon can appear so bizarre that sometimes patients can be dismissed as being overanxious, or at best these symptoms are ignored and physicians' attention directed to the more established symptoms of Ménière's disease – episodic vertigo, tinnitus and hearing loss. For these reasons, Tullio phenomenon in relation to Ménière's disease may be more common than is generally appreciated. However in the series of 14 patients with Tullio phenomenon reported by O Mahoney et al. (1998) no patient was considered to have Ménière's disease in accordance with standard criteria (Alford 1972; Beasley 1996).

Vestibulo-fibrosis

This is another postulated mechanism by which abnormal stapes-to-vestibular end-organ coupling occurs. Cawthorne (1956) reported Tullio symptoms in patients with chronic adhesive otitis media – the symptoms being mediated via fibrotic connections between the stapes and vestibular labyrinth secondary to chronic inflammation. Ishizaki et al. (1991) suggested a similar aetiology in their patients with sound-induced postural instability associated with long-standing chronic otitis media.

Bony dysplasia of the inner ear

Kwee (1976) reported that 50 of 150 congenitally deaf children attending a deaf school complained of symptoms of the Tullio phenomenon and were found to have horizontal nystagmus on direct observation with Frenzl's glasses. Kwee concluded that the children who experienced sound-induced symptoms were those with Mondini type cochlear dysplasia, the enlarged saccule seen in this type of deformity, making pathological contact with the stapes footplate and so giving rise to the symptoms. However, no radiological/histological proof of this hypothesis was reported. There has been much interest recently in patients with hearing loss who have been shown to have pathologically widened vestibular aqueducts (one of the features of Mondini type dysplasia) but without the other radiological features of the Mondini dysplasia. In the now quite extensive literature on widened vestibular aqueducts, mention is made en passant of such patients experiencing the Tullio phenomenon (Schessel and Nedzelski 1992). However it has not been systematically studied.

Hennebert's sign (1909)

Similarities have been drawn between the Tullio phenomenon and Hennebert's sign, the latter being the provocation of nystagmus by non-acoustic pressure, both positive and negative, applied to the ear canal and thus transmitted to the middle ear. As with Tullio phenomenon, it was originally believed that Hennebert's sign was indicative of a bony fistula of the labyrinth due to syphilitic osteitis. Again, as in Tullio phenomenon, further work on Hennebert's sign suggested that patients with Ménière's disease may also exhibit a positive Hennebert's sign, and the unifying pathological process between both conditions (congenital syphilis and Ménière's disease) was suggested to be 'vestibulofibrosis' giving rise to membranous attachment and thus abnormal coupling between the stapes footplate and the vestibular labyrinth. Moreover, a patient with profound deafness, attributed to acute labyrinthitis in childhood, later developed vertigo and a positive Hennebert's sign. At labyrinthectomy for relief of vertigo, the patient was found to have fibrous adhesions between the stapes and membranous labyrinth (Nadol 1974). The range of aetiologies suggested by a positive Hennebert's sign was further broadened to include perilymphatic fistulae not associated with syphilis, again as has been suggested (see above) for the Tullio phenomenon.

Psychological causes

Given the unusual nature of the symptoms of the Tullio phenomenon, it is perhaps not surprising that it has on occasions been attributed, perhaps erroneously, to overanxiety or hysteria (Coats and McCrary 1979). On the other hand, although the symptoms may be explicable on an organic basis, it is quite likely, given the distressing nature of these symptoms and the possible consequences, for example in career terms, that secondary psychological factors may complicate the picture, as they do in many other vestibular pathologies (Eagger et al. 1992). In terms of management, these psychological factors may need to be addressed in their own right.

Transmission of pressure to the inner ear

The above theories of the pathophysiological mechanism of the Tullio phenomenon are dependent in one way or another on enhanced transmission of acoustic pressure, from the middle ear to the vestibular system (either canal and/or otolith end-organs) as the means of generation of the Tullio phenomenon. There is also the suggestion from the work on Hennebert's signs that this entity may be another manifestation of the same underlying processes, rather than an entirely separate condition. In addition, Dieterich et al. (1989) described eye movement in the opposite direction from that evoked by sound when their subject performed the Valsalva manoeuvre. The authors suggested that the increased intracranial pressure associated with the Valsalva manoeuvre was transmitted to the labyrinthine fluids and stimulated the utricle, thereby producing symptoms and signs. The utricular stimulation during Valsalva was in the opposite direction to that induced by sound and so also reversed the direction of eye movement.

Interestingly, in the literature (Dieterich et al. 1989; O'Mahoney et al. 1998) many of the patients who complained of sound-induced vestibular symptoms, found their symptoms were provoked by playing wind instruments. The possibility must be considered that these symptoms arise not from the sound pressure being conducted to the inner ear via the outer and middle ears, but that they arise secondary to increased intracranial pressure due to forceful blowing. O'Mahoney et al. (1998) describe patients who in addition to sound-induced symptoms experienced similar vestibular symptoms during childbirth and defecation, both of which increase intracranial pressure.

The existence of a functional pressure transfer mechanism between the CSF and the inner ear has been well characterised in animals and humans, both *in vivo* and *in vitro*.

Carlborg and co-workers (Carlborg et al. 1982, 1983) investigated pressure changes in the inner ear in relation to variation in CSF pressure before and after occlusion of the cochlear aqueduct. They concluded that this structure was of central importance for pressure regulation within the inner ear in the face of alterations in CSF pressure in cats, both alive and dead (Carlborg, Densert and Densert, 1982; Carlborg and Farmer 1983). Kitihara et al. (1994) described how, in guinea pigs, the cochlear and vestibular aqueducts as well as the eustachian tube, via their connections to the perilymph, endolymph and middle ear respectively, maintain, by means of valve-like mechanisms, middle- and inner-ear pressure within limits despite changes in atmospheric pressure on the one hand and CSF pressure on the other. In humans, direct measurement of perilymphatic pressure and its change in relation to changes in intracranial pressure is not possible *in vivo* because of the inaccessibility of the inner ear. However using tympanic membrane displacement techniques (Densert et al. 1977; Marchbanks 1980, 1984) perilymphatic pressure can be measured indirectly and non-invasively. This technique has been further developed in order to measure perilymphatic pressure changes which occur with changes in intracranial pressure (Marchbanks et al. 1987; Reid et al. 1990) and thus to confirm the existence of CSF to perilymph pressure transmission in humans, thought to be via the cochlear aqueduct. Such measurements are relevant to understanding the Tullio phenomenon and vestibular symptoms induced by the Valsalva manoeuvre, as they provide confirmation of the presence or absence of a pressure transmission route from

CSF to inner ear. Clearly if such a route is not patent, then the vestibular symptoms/signs reported in association with the Valsalva manoeuvre cannot be explained on the basis of pressure transmission. Tympanic membrane displacement measurements have recently been carried out in patients with Tullio phenomenon many of whom had in addition Valsalva-induced vestibular signs and symptoms (O'Mahoney et al. 1998).

The association between vestibular symptoms and signs provoked by sound, non-acoustic positive and negative meatal/middle ear-pressure changes and changes in intracranial pressure is further explored by O'Mahoney et al. (1998), and it is suggested that they may all be different manifestations of the same process.

From the above it seems clear that the Tullio phenomenon and the related conditions of Hennebert's and Valsalva-induced vestibular symptoms and signs are not disease-specific entities but are manifestations of vestibular dysfunction caused by a variety of different pathologies and are thus seen in a heterogeneous group of patients. Careful analysis of the precise symptoms and signs, using recently described investigative tools such as tympanic membrane displacement analysis, video-oculography and click-evoked vestibulo-collic reflex testing, may be able to provide detailed information on the functional integrity of specific parts of the vestibular system – canal versus otolith, saccule versus utricle. Further research in this area will help expand the current body of knowledge on vestibular function (both otolith and canal), characterise the inter-relationships between the brain and the inner ear fluids, allow an understanding of the precise pathophysiological mechanisms subserving the Tullio phenomenon and hopefully point the way towards effective management of patients who experience the Tullio and related phenomena.

References

Alford BR (1972) Ménière's disease: criteria for diagnosis and evaluation of therapy for reporting. Report of subcommittee on equilibrium and its measurement. Trans Am Acad Ophthalmol Otolaryngol 76: 1462–4.

Benjamins CE (1938) Les Réactions acoustiques de Tullio chez l'homme. Acta Otolaryngol 26: 213–16.

Bleeker JD, de Vries HL (1949) The microphonic activity of the labyrinth of the pigeon. Acta Otolaryngol (Stockh) 37: 289.

Bleeker JD, Wit HP, Segenhout JH (1980) Evidence for sound perception with the labyrinth. Acta Otolaryngol (Stockh) 89:76–84.

Bronstein AM, Faldon M, Rothwell J, Gresty MA, Colebatch J, Ludman H (1995) Clinical and Electrophysiological Findings in the Tullio Phenomenon. Acta Otolaryngol (Stockh) Suppl 520: 209–11.

Carlborg B, Densert B, Densert O (1982) Functional patency of the cochlear aqueduct. Ann Otol 91: 209–15.

Carlborg BIR, Farmer JC (1983) Transmission of cerebrospinal fluid pressure via the cochlear aqueduct and endolymphatic sac. Am J Otolaryngol 4(4): 273–82.

Cawthorne T (1956) Chronic adhesive otitis. J Laryngol Otol 70:559–64.

Chan YS, Hwang JC, Cheung YM (1979) Vestibular function of saccule in cats as indicated by the response of Deiters' nucleus to static tilts. Exp Brain Res 35: 591–4.

Coats JAC, McCrary JA (1979) Voluntary nystagmus masquerading as Tullio's phenomenon. Arch Otolaryngol 105: 288–9.

Colebatch JG, Day BL, Bronstein AM, Davies RA, Gresty MA, Luxon LM, Rothwell JC (in press) Vestibular hypersensitivity to click stimulation in the Tullio phenomenon. J Neurol Neurosurg Psychiat.

Colebatch JG, Rothwell JC, Bronstein A, Ludman H (1994) Click-evoked vestibular activation in the Tullio phenomenon. J Neurol Neurosurg Psychiat 57: 1538–40.

Curthoys IS (1987) Eye movements produced by utricular and saccular stimulation. Aviat Space Environ Med 58: Supplement, A192–197.

Densert O, Ivarsson A, Pedersen K (1978) The influence of perilymphatic pressure on the displacement of the tympanic membrane. Acta Otolaryngol (Stockh) 84–220.

De Vries HL, Bleeker JDJW (1949) The microphonic activity of the labyrinth of the pigeon. II. The response of the cristae in the semicircular canals. Acta Otolaryngol 37: 298–306.

Didier A, Cazals Y (1989) Acoustic responses recorded from the saccular bundle on the eighth nerve of the guinea pig. Hear Res 37: 123–8.

Dieterich M, Brandt TH, Fries W (1989) Otolith function in man. Results from a case of otolith Tullio phenomenon. Brain 112: 1377–92.

Eagger S, Luxon LM, Davies RA, Coehlo A, Ron MA (1992) Psychiatric morbidity in patients with peripheral vestibular disorder: a clinical and neuro-otological study. J Neurol Neurosurg Psychiatry 55: 383–7.

Fluur E, Mellstrom A (1970a) Utricular stimulation and oculomotor reactions. Laryngoscope 80: 1701–12.

Fluur E, Mellstrom A (1970b) Saccular stimulation and oculomotor reactions. Laryngoscope 80: 1713–21.

Fox EJ, Balkany TJ, Arenberg IK (1988) The Tullio phenomenon and perilymph fistula. Otolaryngol Head Neck Surg 98: 88–9.

Gernandt BE (1970) Nystagmus evoked by utricular stimulation. Exp Neurol 27: 90–100.

Hennebert C (1909) Vestibularreaktion auf die pneumotische. Intern. Zblat Ohr Rhinolaryngol 8: 532.

Ildiz F, Dundar A (1994) A case of Tullio phenomenon in a subject with oval window fistula due to barotrauma. Aviat Space Environ Med 65: 67–9.

Ishizaki H, Pyykko I, Aalto H, Starck J (1991) Tullio phenomenon and postural stability: Experimental study in normal subjects and patients with vertigo. Ann Otol Rhinol Laryngol 100: 976–83.

Kacker S K, Hinchcliffe R (1970) Unusual Tullio phenomena. J Laryngol Otol 84: 155–66.

Kitahara M, Suzuki M, Kodama A (1994) Equilibrium of inner and middle ear pressure. Acta Otolaryngol (Stockh) (Suppl 510): 113–15.

Kwee HL (1976) The occurrence of the Tullio phenomenon in congenitally deaf children. J Laryngol Otol 90: 501–7.

Marchbanks RJ (1984) Measurement of tympanic membrane displacement arising from aural cardiovascular activity, swallowing and intra-aural muscle reflex. Acta Otolaryngol (Stockh) 98: 119–29.

Marchbanks RJ, Reid A, Martin AM, Brightwell AP, Bateman D (1987) The effect of raised intracranial pressure on intracochlear fluid pressure: three case studies. Brit J Audiol 21: 127–30.

Markham CH, Estes MS, Blanks RHI (1973) Vestibular influences on ocular accommodation in cats. Equilibrium Res 3: 102–15.

Mayer O, Fraser JS (1936) Pathological changes in the ear in late congenital syphilis. J Laryngol Otol 51: 683–714.

McCue MP, Guinan JJ (1994) Acoustically responsive fibers in the vestibular nerve of the cat. J Neurosci 14: 6058–70.

Murofushi T, Curthoys IS, Topple AN, Colebatch JG, Halmagyi GM (1995) Responses of guinea pig primary vestibular neurons to clicks. Exp Brain Res 103: 174–8.

Nadol JB Jr (1974) Positive fistula sign with an intact tympanic membrane. Arch Otolaryngol 100: 273–8.

O Mahoney CF, Luxon LM, Bronstein A, Colebatch JG, Davies R, Gresty M, Ludman H, Marchbank RJ, Rothwell JC (submitted) Further insights into the origin and aetiology of the Tullio phenomenon: Findings in fourteen patients. Brain.

Parker DE et al. (1976) Effects of sound on the vestibular system. Aerospace Med Res Lab TR 75–89.

Parker DE, Tubbs RL, Littlefield VM (1978) Visual-field displacements in human beings evoked by acoustical transients. J Acoust Soc Am 63: 1912–18.

Pillsbury HC, Postma DS (1983) The Tullio phenomenon, fistula test, and Hennebert's sign: clinical significance. Otolaryngol Clin North Am 16(1): 205–7.

Reid A, Marchbanks RJ, Burge DM, Martin AM, Bateman DE, Pickard JD, Brightwell AP (1900) The relationship between intracranial pressure and tympanic membrane displacement. Brit J of Audiol 24: 123–9.

Schessel DA, Nedzelski JM (1992) Presentation of large vestibular aqueduct syndrome to a dizziness unit. J Otolaryngol 21(4): 265–9.

Suzuki JI, Tokumasu K, Goto K (1969a) Eye movements from single utricular nerve stimulation in the cat. Acta Otolaryngol 68: 350–62.

Suzuki JI, Goto K, Tokumasu K, Cohen B (1969b) Implantation of electrodes near individual vestibular nerve branches in mammals. Ann Otol Rhinol Laryngol 78: 815–26.

Tullio P (1929) Das Ohr und die Entstehung der Sprache und Schrift. Berlin: Urban & Schwarzenberg.

Vogel P, Tackmann W, Schmidt FJ (1986) Observations on the Tullio phenomenon. J Neurol 233: 136–9.

Wit HP, Bleeker JD (1982) Sensitivity of the vestibular system to acoustic stimuli. Arch Otorhinolaryngol 234: 205–11.

Wit HP, Bleeker JD, Mulder HH (1984) Responses of pigeon vestibular nerve fibers to sound and vibration with audiofrequencies. J Acoust Soc Am 75(1): 202–8.

Wit HP, Bleeker JD, Segenhout JH (1981) Vestibular and cochlear responses to acoustic transients. Acta Otolaryngol (Stockh) 92: 409–22.

Wit HP, Scheurink AJW, Bleeker JD (1985) Hearing thresholds of normal and fenestrated deaf pigeons. Acta Otolaryngol (Stockh) 100: 36–41.

Young ED, Fernandez C, Goldberg M (1977) Responses of squirrel monkey vestibular neurons to audio-frequency sound and head vibration. Acta Otolaryngol 84: 352–60.

Correspondence to Dr Cliodna F O'Mahoney, Department of Audiology and Audiological Medicine, Great Ormond Street Hospital for Children, London, UK.

Chapter 30 Traumatic sudden sensorineural hearing loss in children

BRENDA KOU and ROBERT MACDONALD, Hospital For Sick Children, Toronto, Canada

ABSTRACT. The authors studied all cases of traumatic sudden sensorineural hearing loss at the Hospital for Sick Children, Toronto, Canada, between 1980 and 1995. Traumatic sudden sensorineural hearing loss is by definition a hearing loss of greater than or equal to 30 dB occurring within three days of trauma. A case study of a child with bilateral enlarged vestibular aqueducts, is discussed. Up to one-third of cases were found to have an inner-ear malformation on high resolution CT scan. The majority (two-thirds) of cases resulted in unilateral sudden hearing loss. The presence of bilateral traumatic sudden sensorineural hearing loss, moreover, was found to be strongly correlated with the presence of a congenital inner-ear malformation.

Key words: traumatic, sudden, sensorineural, hearing loss, children, enlarged vestibular aqueduct.

Introduction

Sudden sensorineural hearing loss may be defined as a greater than or equal to 30 dB sensorineural hearing loss occurring in at least three contiguous frequencies that develops over a period of less than three days. Its aetiology may be idiopathic, infectious/inflammatory, vascular or traumatic. When sudden sensorineural hearing loss occurs in association with head trauma, the clinical suspicion that a perilymphatic fistula is involved is heightened. Moreover, it is well known that ears with a congenital middle- or inner-ear malformation may be predisposed to the development of a perilymphatic fistula.

In this article we present an illustrative case of a child with previously normal hearing who suffered traumatic bilateral sudden sensorineural hearing loss and was subsequently found to have bilateral enlarged vestibular aqueducts on high resolution CT scan of the temporal bones. This case prompted a 15-year review at the Hospital for Sick Children in Toronto to answer the following question: 'In traumatic sensorineural hearing loss, what is the likelihood that the child has a pre-existing congenital inner-ear malformation?'

Case report

NM is a previously healthy five-year-old child with previously normal hearing. She was sitting in a chair, tipped the chair backwards, and fell hitting her head against the floor. As a result of this minor head injury, the child became suddenly profoundly deaf bilaterally. She had no loss of consciousness, no vertigo and no other injuries.

Her audiogram showed a right-ear pure-tone threshold of 110 dB and a left ear pure tone threshold of 96 dB. Auditory brainstem response testing confirmed the profound loss. A high resolution CT scan of the temporal bones demonstrated bilateral enlarged vestibular aqueducts. She was diagnosed with possible bilateral perilymphatic fistula and was taken to the operating room for a left tympanotomy within three days of the accident, and two months later a right tympanotomy. The left tympanotomy demonstrated a perilymphatic fistula at the round window niche and was repaired with a fat graft and tissue.

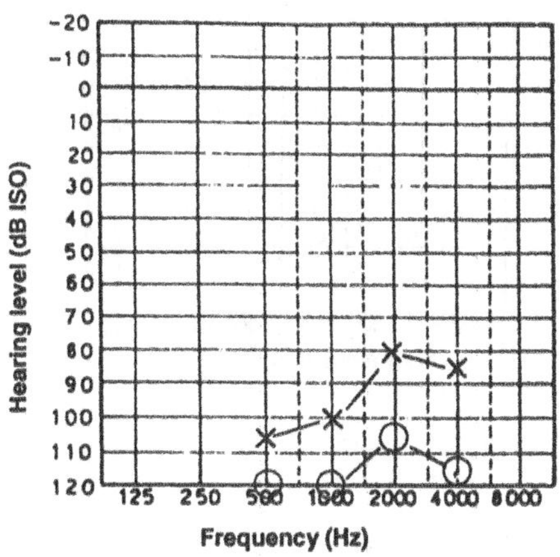

Figure 30.1: pre-operative audiogram.

Two months later she underwent a right tympanotomy. At this second operation, however, no perilymphatic fistula was found, and the oval and round windows were both packed with fat graft prophylactically.

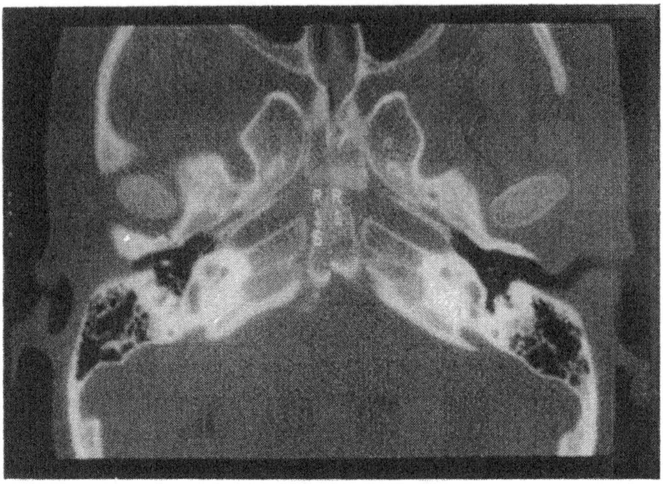

Figure 30.2: bilateral enlarged vestibular aqueducts.

A follow-up audiogram demonstrated little change in the left-ear pure-tone thresholds, but some improved upsloping was noticed in the right-ear thresholds, especially in the high frequencies up to 60 dB at 8000 Hz.

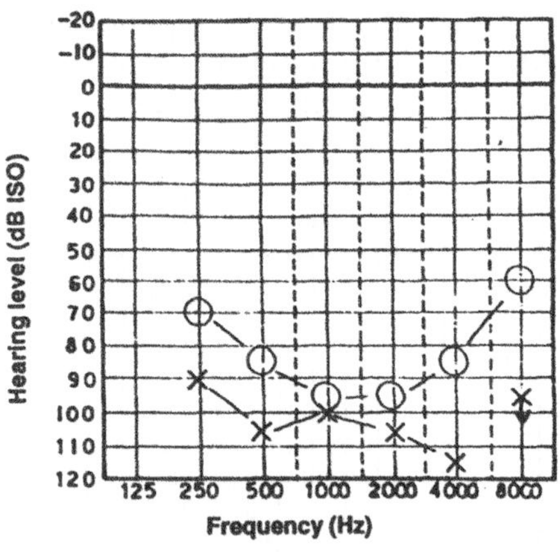

Figure 30.3: post-operative audiogram.

Materials and methods

A 15-year retrospective chart review was conducted at Toronto's Hospital for Sick Children between the years 1980 to 1995. All cases must have had an identifiable preceding traumatic event and have had at least a 30 dB sensorineural hearing loss in three contiguous frequencies occurring within three days of trauma.

High resolution CT scans of the temporal bones with bone

algorithms and coronal/axial views were performed on all children evaluated after 1988 as part of the diagnostic work-up of the hearing assessment program at the Hospital for Sick Children.

Results

Patient demographics

Twelve children (16 ears) were identified as meeting our specifications for traumatic sudden sensorineural hearing loss. The male to female breakdown was as follows: eight males (66%) and four females (44%). The average age of presentation was 9.7 years (range two to 15 years).

Aetiology of trauma

Most of the trauma was blunt secondary to falls. There was one penetrating injury from a twig, which penetrated through the tympanic membrane and round window causing a traumatic perilymphatic fistula. There was also one case of pneumatic-induced sudden sensorineural hearing loss secondary to nose blowing. Most of the cases of traumatic sensorineural hearing loss were unilateral (eight cases = 66%), whereas only four cases (44%) were bilateral.

CT diagnosis

Of the 12 children studied, nine children (75%) had a high resolution CT scan of the temporal bones with bone algorithms: one-third of these children who had CT scans had an inner-ear malformation, one-third had temporal bone fractures, and one-third had 'other' miscellaneous CT findings (one had an air fluid level in the mastoid, one had a soft tissue swelling in the external auditory canal, and one was normal). Among the three cases of inner-ear malformation, two had bilateral enlarged vestibular aqueducts and one had a bilateral Mondini malformation.

Management

Eight (50%) of the 16 ears affected by traumatic sudden sensorineural hearing loss were explored. Of the eight ears explored, five (63 %) had an inner-ear malformation, and three (38 %) did not. Only in one of the five ears with inner-ear malformations was a perilymphatic fistula identified at surgery (20%). In contrast, two of the three ears (66%) that did not have an inner-ear malformation on CT scan were found at surgery to have a perilymphatic fistula.

Discussion

The enlarged vestibular aqueduct syndrome was first described by Valvassori in 1978 and is characterized by an abnormally patent vestibular aqueduct. It is the single most common inner-ear malformation on CT scan and is associated with congenital, progressive, or sudden sensorineural hearing loss. Vestibular disturbances have also been described as occurring as part of the syndrome. It occurs due to an arrest in development of the inner ear at the fifth week of gestation.

Table 30.1: CT diagnosis and management strategies

Patient	CT diagnosis	OR	Perilymphatic fistula
1	Bilateral Mondini	Left and right Tympanotomy	None
2	Bilateral enlarged vestibular aqueducts	Left and right tympanotomy	Left PLF at round window
3	Air fluid level Right mastoid	–	–
4	Right longitudinal temporal bone fracture	–	–
5	Right longitudinal temporal bone fracture	–	–
6	Normal	–	–
7	–	Left tympanotomy	–
8	Soft tissue swelling external ear canal (left)	Left tympanotomy	–
9	Bilateral transverse temporal bone fractures	–	–
10	–	–	–
11	–	Right tympanotomy	Right PLF at round window
12	Bilateral enlarged vestibular aqueducts	Left tympanotomy	–

PLF = perilymphatic fistula

Another name for the vestibular aqueduct is the endolymphatic duct. It contains endolymph and connects the utricle and the saccule of the inner ear. It should not be confused with the cochlear aqueduct, which contains perilymph and connects the posterior cranial fossa with the scala tympani. The cochlear aqueduct has long been postulated as the mechanism by which infection is transmitted between the inner ear and the central nervous system. To date, however, there has been no convincing radiologic evidence for an 'enlarged cochlear aqueduct' syndrome.

The relationship between sudden sensorineural hearing loss and trauma has traditionally been thought to be due to the presence of a perilymphatic fistula in which an abnormal communication exists between the middle and the inner ears. The concept of a labyrinthine membrane rupture via either the implosive or the explosive routes has been described by Goodhill (1971). The implosive route results from transmission of increased middle ear pressure due to Valsalva manoeuvres to the perilymphatic space through the oval and round windows. The explosive route, on the other hand, results from transmission of increased intracranial cerebrospinal fluid pressure to the perilymph via either the cochlear aqueduct or the internal auditory canal.

The predisposition of children with Mondini malformations to recurrent meningitis and perilymphatic fistula is well documented in the literature. It is felt to be related to defects in the tegmen, Hyrtl's fissure, oval window and the stapes footplate.

Recent attention has been focused on the enlarged vestibular aqueduct syndrome and its associated sensorineural hearing loss and vestibular disturbances, which may be precipitated by minor head trauma. Four possible aetiologic mechanisms for sensorineural hearing loss have been described in the literature. The first theory hypothesizes that ears with enlarged vestibular aqueducts are predisposed to the development of perilymphatic fistula. Another theory is that trauma may compress the dural envelope of the endolymphatic sac, allowing the reflux of high protein, hyperosmolar fluid through an abnormally patent vestibular aqueduct. A third theory postulates the presence of traumatic intracochlear membrane ruptures of either the basilar or Reissner's membrane which may produce sudden stepwise hearing loss. The final theory hypothesizes the presence of endolymphatic hydrops resulting in the accumulation of excess volumes of endolymph and progressive sensorineural hearing loss.

Regardless of the mechanism of traumatic hearing loss in congenital inner ear malformations, our study has demonstrated that up to one-third of children who present with sudden sensorineural hearing loss and receive a high resolution CT scan of the temporal bones will have a congenital inner-ear malformation. This is especially true if the hearing loss is bilateral, as 75% of the children in this category had an inner-ear malformation.

An interesting result of our study is the fact that the perilymphatic fistula rate in the five ears with congenital inner-ear malformations was only 20%, compared to 66% in the three ears who did not have an inner-ear malformation. This emphasizes the fact that perilymphatic fistula are not the only mechanism for traumatic sudden sensorineural hearing loss, and perhaps reflux of hyperosmolar fluids, or intracochlear membrane ruptures may be just as important in the pathogenesis of the hearing loss.

Conclusions

In conclusion, our study demonstrates that sudden sensorineural hearing loss in association with trauma is a rare event (12 cases, 16 ears in 15 years' experience at the Hospital for Sick Children). The standard investigation protocol of a high resolution CT scan of the temporal bones in this scenario will show a congenital inner-ear malformation in one-third of cases overall, and in up to 75% of cases with bilateral traumatic sudden sensorineural hearing loss.

References

Arcand P (1991) The large vestibular aqueduct syndrome and sensorineural hearing loss in the pediatric population. J Otolaryngol 20(4): 247–50.

Goodhill V (1971) Sudden deafness and round window rupture. Laryngoscope 81(9): 1462–74.

Herther C, Schindler RA (1985) Mondini's Dysplasia with recurrent meningitis. Laryngoscope 95: 655–8.

House JW, Morris MS, Kramer SJ, Shasky GL, Coggan BB, Putter JS (1991) Perilymphatic fistula: surgical experience in the United States. Otolaryngol Head Neck Surg 105(1): 51–61.

Jackler R, De la Cruz A (1989) The large vestibular aqueduct syndrome. Laryngoscope 99(12): 1238–43.

Kimmelman C (1993) Sudden Hearing Loss, Self Instructional Package (SIPAC) of the American Academy of Otolaryngology, Head and Neck Surgery.

Levenson M (1989) The large vestibular aqueduct syndrome in children. Arch Otolaryngol Head Neck Surg 115: 54–8.

Schessel D, Nedzelski JM (1992) Presentation of large vestibular aqueduct syndrome to a dizziness unit. J Otolaryngol 21(4): 265–9.

Valvassori GE, Clemis JD (1978) The large vestibular aqueduct syndrome. Laryngoscope 88: 723–48.

Weber P, Perez BA, Bluestone CD (1993) Congenital perilymphatic fistula and associated middle ear anomalies. Laryngoscope 103(2): 160–4.

Weissman J, Weber P, Bluestone CD (1994) Congenital perilymphatic fistula: computed tomography appearance of middle ear and inner ear anomalies. Otolaryngol Head Neck Surg 111(3): 243–9.

Correspondence to Dr Brenda Kou, 1 Glentworth Road, North York, Ontario, Canada M2J 2E4.

Chapter 31
Fleeting hearing loss: reversible dysfunction of the cochlea during acute bacterial meningitis

MP RICHARDSON, PT RUDD, Bath Unit for Research into Paediatrics; A REID, Department of Audiology, Royal United Hospital, Bath; MJ TARLOW, Department of Paediatric Infectious Disease, Birmingham Heartlands Hospital, Birmingham, UK

ABSTRACT. Sensorineural hearing loss has long been recognised as a serious complication of bacterial meningitis. However, until recently, the pathophysiology of this condition was unclear. We recently completed a multicentre study in which 124 children with newly diagnosed bacterial meningitis underwent repeated audiological testing with transient evoked otoacoustic emissions (OAEs), auditory brainstem responses and tympanometry. The first tests were performed within six hours of diagnosis. Three types of hearing loss were detected: permanent sensorineural deafness, conductive deafness and a rapidly reversible form of sensorineural hearing loss. Thirteen children (10.5%) exhibited the latter type of impairment. All children with this 'fleeting hearing loss' had absent otoacoustic emissions (OAEs) soon after admission to hospital but recovered normal hearing within one to five days. To the best of our knowledge, this phenomenon had not been described before. The absence of OAEs in children with fleeting hearing loss indicates that the phenomenon represents a reversible dysfunction of the cochlea. During meningitis, bacteria and inflammatory cells from the subarachnoid space probably invade the inner ear through the cochlear aqueduct or the modiolus. Permanent post-meningitic deafness appears to be caused by the resulting structural damage to the sensory cells of the organ of Corti. However, this mechanism is not thought to be the cause of fleeting hearing loss because structural damage to the sensory apparatus is probably irreparable. Possible causes of fleeting hearing loss include disruption of the endocochlear potential or reversible dysfunction of the hair cells. It is likely that these effects are mediated by bacterial toxins.

Key words: bacterial meningitis, hearing loss, cochlea, cochlear aqueduct, otoacoustic emissions.

Introduction

Sensorineural deafness is the commonest neurological complication of bacterial meningitis in childhood. Around 10% of survivors are affected (Baraff et al. 1993). Until recently, the natural history and pathophysiology of post-meningitic hearing loss were poorly understood. However it is now clear that hearing loss develops during the acute stage of the infection and that the cochlea is the site of the auditory lesion in some, if not all, cases of post-meningitic deafness (Nadol 1978; Dodge et al. 1984; Wald et al. 1995; Richardson et al. 1997).

In order to find out more about the development of post-meningitic hearing loss we recently carried out a detailed prospective study of auditory function in children with acute bacterial meningitis. In this study we made extensive use of otoacoustic emissions (OAEs). OAEs are minute soundwaves that can be recorded from nearly all healthy ears but are absent in most cases of hearing impairment (Kemp 1978). OAE testing was particularly useful for this study because the method is quick, highly sensitive and non-invasive. Moreover, as emissions are abolished by lesions of the cochlea but are preserved in more central lesions, OAEs can be used to distinguish between the sensory and neural components of sensorineural deafness (Richardson 1995). This feature of OAEs allowed us to determine the extent to which the cochlea is affected by bacterial meningitis.

The main findings of this study have already been published (Richardson et al. 1997). The aim of this chapter is to discuss in detail our discovery of fleeting hearing loss during acute bacterial meningitis. This phenomenon, which appears to represent a rapidly reversible dysfunction of the cochlea, is almost certainly due to the spread of an inflammatory process from the intracranial to the intralabyrinthine fluids.

Methods

The methodology of this study has already been described (Richardson et al. 1997). Essentially, we recruited children with newly diagnosed bacterial meningitis from 21 hospitals in the South and West of England and South Wales. We asked to be informed of all cases of meningitis within one hour of the diagnosis being suspected so that the first hearing tests could be performed as early as possible. The study protocol called for transient evoked otoacoustic emissions to be recorded at 0–6, 6–12, 12–24, and 36–48 hours after diagnosis. Otoacoustic emissions were also recorded at the time of discharge from hospital. For all recordings we used the ILO88 System (Otodynamics, Hatfield, UK) as previously described (Richardson et al. 1995). The patients' hearing was also tested by auditory brainstem responses (ABR) and tympanometry before discharge from hospital. Whenever possible these two techniques were used to confirm and characterise hearing loss in children who failed to produce OAEs at any stage of their illness.

Hearing impairment was classified as sensorineural if the child had an ABR threshold ≥ 30 dB HL and a normal tympanogram. Sensorineural hearing loss was further classified as cochlear (sensory) if OAEs were absent, or retrocochlear (neural) if OAEs were preserved. Children with an ABR threshold ≥ 30 dB HL and an abnormal (type B) tympanogram were classified as having a conductive hearing loss. We performed audiological follow-up using OAEs, brainstem responses and tympanometry on children who had hearing loss of any degree at the time of discharge from hospital. All children enrolled in the study were referred for local audiological follow-up and we were informed of the results of these assessments.

Results

One hundred and twenty-four children with acute bacterial meningitis were studied. Their ages ranged from six weeks to 15 years (median 2.1 years). The commonest causative organisms were *Neisseria meningitidis* (74%) and *Streptococcus pneumoniae* (15%).

Three types of hearing loss were detected in this study: permanent sensorineural deafness, conductive hearing loss, and a rapidly reversible sensorineural hearing loss. Three children (2.4%) developed permanent sensorineural deafness. All three had absent OAEs at the time of their first assessment, indicating a lesion of the cochlea. Bilaterally raised hearing thresholds of 70 to 110 dB HL were confirmed by ABR. At follow-up all three of these children have persistent severe to profound sensorineural hearing loss. Five children (4.0%) were diagnosed as having conductive hearing loss. All but two of these children have since regained normal hearing.

Thirteen children (10.5%) failed to produce OAEs at the time of their first assessment but produced normal emissions later in their illness (see Table 31.1). This phenomenon was bilateral in eight cases and unilateral in five. All of these children had normal tympanograms, which excludes middle ear dysfunction and suggests that they had a rapidly reversible pathological process in the inner ear. We were able to detect a moderate-to-severe hearing impairment in eight of these children by using ABR or clinical examination. The remaining children regained normal OAEs before a hearing impairment could be confirmed. All the children with reversible hearing loss had normal auditory assessments at discharge from hospital and at follow-up.

Table 31.1: details of children with fleeting hearing loss during bacterial meningitis

Patient No.	Sex	Age (years)	Causative organism	Time from diagnosis to: detection of hearing loss by OAEs	recovery of normal hearing
1	male	0·3	*N. meningitidis*	3 hours*	5 days
2	female	0·4	*N. meningitidis*	12 hours*	40 hours
3	female	0·4	*N. meningitidis*	8 hours*	20 hours
4	male	0·4	*N. meningitidis*	2 hours	42 hours
5	female	0·4	*S. pneumoniae*	3 hours	3 days
6	male	0·8	*N. meningitidis*	7 hours*	18 hours
7	female	0·9	*N. meningitidis*	2 hours*	42 hours
8	female	2·5	*N. meningitidis*	2 hours	5 days
9	male	3·8	*N. meningitidis*	8 hours	44 hours
10	male	4·2	*N. meningitidis*	5 hours*	21 hours
11	male	7·4	*N. meningitidis*	12 hours	42 hours
12	male	10·5	*N. meningitidis*	1 hour**	48 hours
13	male	12·1	*N. meningitidis*	2 hours**	14 hours

*Moderate to severe hearing loss confirmed by auditory brainstem responses,
** complained of deafness on admission to hospital

Otoacoustic emission recordings from one child with rapidly reversible hearing loss are shown in Figure 31.1. This child (patient no. 7 in Table 31.1) was admitted to hospital at the age of nine months with a short history of fever and severe drowsiness. A clinical diagnosis of meningococcal meningitis was made and this was later confirmed on microscopy of her cerebrospinal fluid. Otoacoustic emission recordings were first attempted two hours after the clinical diagnosis of meningitis had been made. At this stage OAEs could not be recorded from either ear (Figure 31.1, upper recording). Otoacoustic emissions were also absent on two other occasions during the first day of admission and a bilaterally raised hearing threshold of 60 dB HL was detected by ABR. On the following day robust OAEs were recorded from both ears (lower recording). She recovered quickly from her meningitis and was found to have normal auditory function at discharge from hospital and on two subsequent occasions.

Discussion

To the best of our knowledge this study was the first to detect rapidly reversible sensorineural hearing loss in patients with acute bacterial meningitis. This form of hearing loss is remarkable because it was found to be present at the time of admission to hospital and usually resolved within 48 hours. In order to distinguish it from the slowly resolving

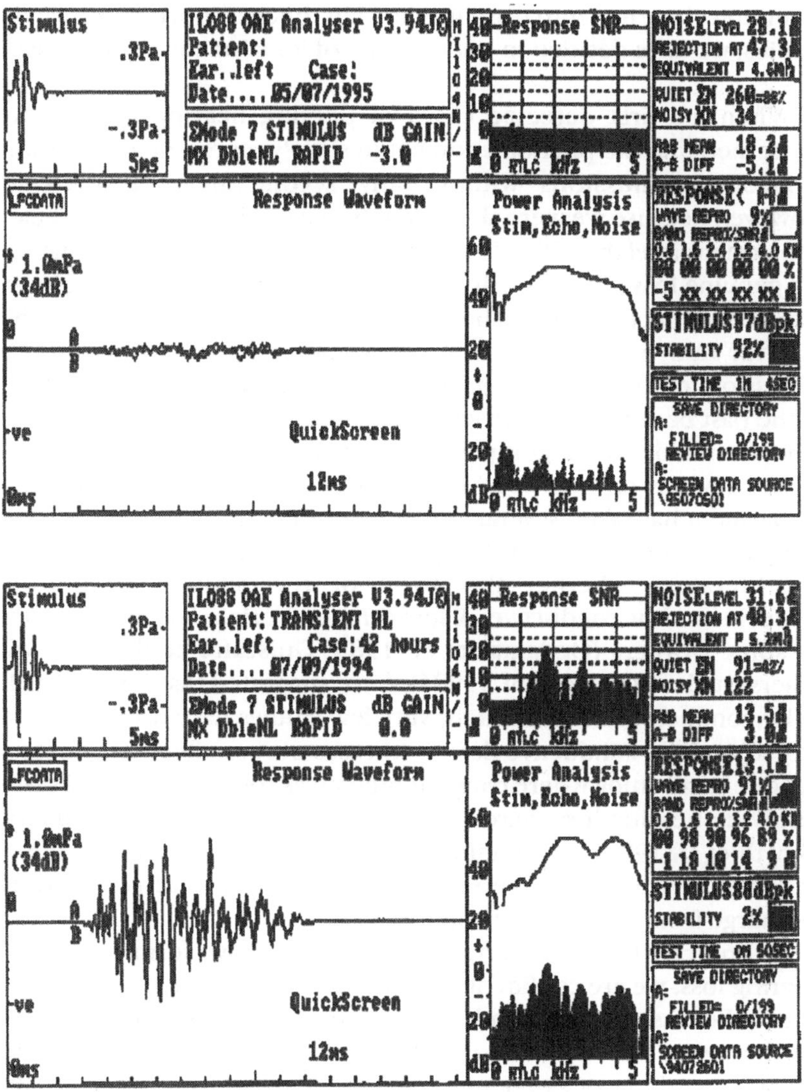

Figure 31.1: transient evoked OAE recordings in a child with fleeting hearing loss. The lower left panel of each display (labelled 'Response Waveform') shows a visual representation of soundwaves recordings (OAEs) in the external ear. The adjacent panel ('Power Analysis') shows Fast Fourier Transformations of various acoustic parameters across the auditory frequencies (0–5 kHz). This gives information about the strength of the following sounds: i) the eliciting stimulus (black line), ii) the OAE signal (black area), iii) an estimate of contaminating noise (hatched area). Above this panel is another Fast Fourier Transformation ('Response SNR') showing the signal-to-noise ratio of the recording. The remaining panels give statistical analyses of stimulus, OAE signal, and noise. The upper recording shows absent OAEs two hours after the diagnosis of meningitis. No waveform can be seen and the Fast Fourier Transformations show no OAE signal above the background noise. The lower recording shows a strong OAE signal 40 hours later.

reversible hearing impairment that has been reported in previous studies, we have decided to call this type of impairment 'fleeting hearing loss'.

When we first detected cases of fleeting hearing loss we were concerned that the phenomenon could simply represent an abolition of otoacoustic emissions without any accompanying hearing loss. However, because we were able to confirm a moderate or severe hearing impairment in eight of the 13 children, we are now confident that there is a genuine loss of hearing. It is perhaps surprising that fleeting hearing loss has not been described before. The most probable reason for this is that our study was the first to examine children immediately after their admission to hospital. Patients may not have reported the phenomenon in the past because most are very young and the hearing loss occurs at a time when most meningitis patients have diminished consciousness.

In previous reports of reversible hearing loss after meningitis the impairment has taken several weeks, or even months, to resolve (Dodge et al. 1984; Guiscafré et al. 1984; Vienny et al. 1984). It is unclear from the literature whether this form of reversible hearing loss is sensorineural or conductive. A sensorineural origin has been proposed by some workers (Özdamar et al. 1983; Guiscafré et al. 1984) while others, such as Dodge et al. (1984), have suggested that it is more likely to be conductive. Unfortunately, very few studies have used tympanometry to exclude conductive defects in cases of reversible hearing loss. In the one small study where tympanometry was routinely used, all cases of resolving hearing loss were associated with abnormal tympanograms (Smyth et al. 1988). Our findings agree with this report in that the only children in our study who had a hearing loss that resolved after discharge from hospital had conductive defects.

In contrast to the uncertainty about the above form of reversible hearing loss, we are confident that fleeting hearing loss is sensorineural. This is because all cases had absent OAEs and normal tympanograms. Furthermore, because OAEs are abolished by lesions of the inner ear but are preserved in cases of neural or central deafness, our findings clearly suggest that the cochlea is the site of the auditory lesion in fleeting hearing loss. From the results of our study it is impossible to determine the aetiology and pathophysiology of fleeting hearing loss. On the other hand, the pathological processes involved in permanent post-meningitic deafness have been well described. This knowledge may be used to discuss possible mechanisms of fleeting hearing loss because both forms of hearing loss appear to result from events within the cochlea in the early stages of meningitis (Richardson et al. 1997).

Lesions of the cochlea have often been reported or suspected in post-meningitic deafness. Post mortem studies have frequently described a form of purulent labyrinthitis in meningitis. The lesions observed have ranged from collections of inflammatory cells in the perilymphatic space (scala tympani and scala vestibuli in Figure 31.2), to destruction of the entire membranous labyrinth (Fraser and Dickie 1920; Crowe 1930; Perlman and Lindsay 1939; Merchant and Gopen 1996). Gross inflammation of the cochlea with bacterial infiltration has also been found in animals with experimental meningitis (Moxon et al. 1974; Kay 1991; Bhatt et al. 1993). In some experiments ultrastructural damage to the organ of Corti has also been reported. Typical lesions include intracellular disorganisation of hair cells, fusion and stripping of hair cell cilia, and rupture of the reticular lamina (Tarlow et al. 1991; Osborne et al. 1995; Winter et al. 1996).

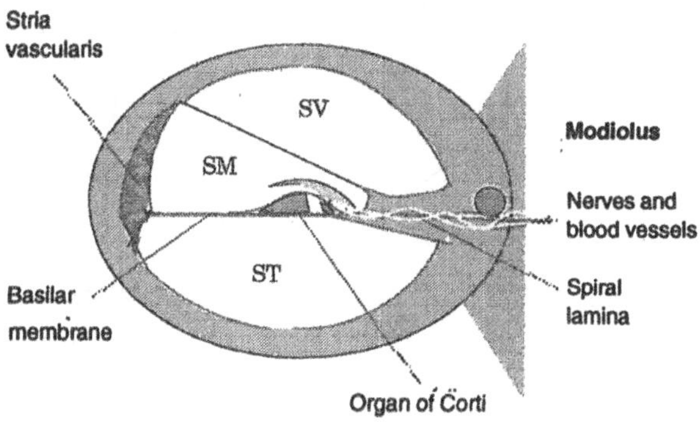

Figure 31.2: schematic cross-section of one turn of the cochlea (SV: scala vestibuli; SM: scala media; ST: scala tympani).

Post mortem studies and animal experiments have also been used to find out how bacteria and inflammatory cells enter the cochlea. Meningitis is primarily an infection of the subarachnoid space. However, as Figure 31.3 shows, the inner ear is closely related to this compartment. The figure suggests that there are two main routes for the spread of inflammation from the cerebrospinal fluid to the inner ear. The first route is through the cochlear aqueduct. In adult humans this channel is often closed but in children it appears to be wider and is more frequently patent (Perlman and Lindsay 1939; Palva 1970). This route of invasion is supported by human and experimental pathological studies in which the earliest stages of inflammation have been noted in the cochlear aqueduct and in the basal turn of the scala tympani, where the aqueduct opens into the inner ear (Perlman and Lindsay 1939; Bhatt et al. 1993). In experimental meningitis, electron microscopy studies have suggested that, after entering the scala tympani, bacteria and inflammatory cells penetrate the scala media by creating pores in the basilar membrane (Osborne et al. 1995). As expected with invasion via the cochlear aqueduct, this penetration is most pronounced in the first turn of the cochlea (Winter et al. 1996). In animal experiments deafness to high frequencies develops before deafness to low frequencies (Bhatt et al. 1993). This finding also supports the aqueduct as the route of invasion because high frequency sounds are detected in the basal part of the cochlea.

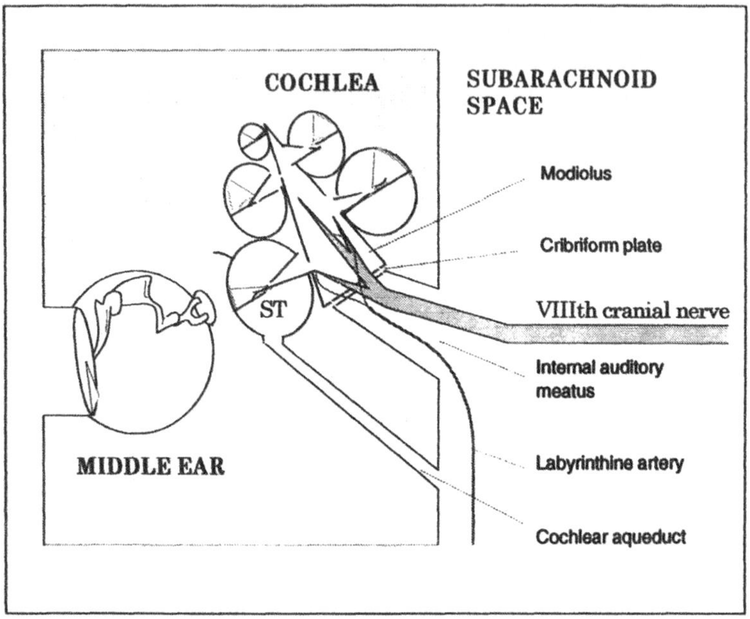

Figure 31.3: schematic section of the temporal bone showing relationship of the cochlea to the subarachnoid space (ST: scala tympani of the first turn of the cochlea).

The second proposed route of cochlear invasion is through the cribriform plate and the modiolus. Post mortem findings have been described that are consistent with this route (Fraser and Dickie 1920; Perlman and Lindsay, 1939). In a recent pathological study, Merchant and Gopen (1996) proposed that the modiolus was the major route of invasion because inflammmation of the modiolus was strongly associated with purulent labyrinthitis whereas inflammation of the cochlear aqueduct was not. On the other hand, there is also evidence that the modiolus is not the main route of entry to the cochlea. Firstly, as Merchant and Gopen suggest, inflammation spreading from the modiolus should cause neural hearing loss before producing sensory (cochlear) hearing loss. In our study at least, sensory hearing loss appears to be the first to develop. The absence of neural deafness in meninigits is also suggested by the finding that the auditory nerves of children deafened by meningitis can be stimulated by electrical cochlear implants (Eisenberg et al. 1984). Secondly, the easiest way for inflammation to spread from the modiolus to the membranous labyrinth is through the neural and vascular channels of the spiral lamina (Kay 1991; see Figure 31.2).

Bacteria and inflammatory cells would then be expected to emerge in the endolymph of the scala media. In pathological studies, however, this compartment is rarely affected by purulent labyrinthitis (Perlman and Lindsay 1939; Merchant and Gopen 1996). Finally, invasion of the cochlea via the modiolus has never been documented in experimental meningitis. However, it should be noted that the cochlear aqueducts in experimental animals (which are usually rodents) are shorter and wider than those in humans (Bhatt et al. 1993; Merchant and Gopen 1996). Their modioli are also smaller and more compact. Thus the modiolar route is naturally less likely in these animals.

Besides the cochlear aqueduct and modiolus, other routes of cochlear invasion have also been proposed. These include septic embolism via the blood supply to the scala media (Fraser and Dickie 1920; Igarishi et al. 1974) and through the middle ear in cases of concurrent otitis media and meningitis (Crowe 1930; Djeric et al. 1994). However, most of the evidence to date suggests that the cochlear aqueduct or the modiolus is the main route of invasion (Merchant and Gopen 1996). We believe that, in children especially, the cochlear aqueduct is likely to be the most important. Indeed, the fact that the aqueduct is more likely to be patent in children than adults may explain why post-meningitic deafness is more common in children.

Although the audiological findings are similar in permanent and fleeting hearing loss, it is likely that the pathophysiological mechanisms are different. This is because the contents of the membranous labyrinth are very delicate and purulent infiltration would almost certainly produce permanent damage. One possibility is that fleeting hearing loss is caused by raised intralabyrinthine pressure secondary to the raised intracranial pressure of meningitis. Alternatively, the phenomenon could be explained by a sudden reduction in pressure at the time of lumbar puncture. Michel and Brusis (1992) reported nine cases of hearing loss after this procedure. However, this is unlikely to be the whole explanation of fleeting hearing loss because three of our patients had evidence of hearing loss before lumbar puncture was performed. Another patient did not undergo the procedure. Furthermore, all of the cases reported by Michel and Brusis occurred in adults, most of whom had been subjected to invasive procedures such as myelography or spinal anaesthesia. The fluid shifts through a paediatric needle in diagnostic lumbar puncture are likely to be much less. The patients described by Michel and Brusis also experienced a longer lasting form of deafness than that observed in our patients. Indeed, three of their patients did not regain normal hearing. Further evidence that alterations in pressure are not responsible for fleeting hearing loss comes from experiments with OAEs. Büki et al. (1996) found that OAEs are preserved even during large changes in intracranial pressure.

Serous or non-purulent labyrinthitis has often been proposed as a cause of reversible sensorineural hearing loss after meningitis (Trolle 1950; Nadol 1978). Nadol (1978) illustrated the condition in a patient who experienced reversible deafness and then died of another complication of meningitis. The pathological specimens showed deposits of an unidentified eosinophilic material in the turns of the cochlea. There was no evidence of bacterial invasion or an inflammatory infiltrate. Most authors suggest that serous labyrinthitis is caused by bacterial toxins (Trolle 1950; Nadol 1978). In recent years animal experiments have confirmed that bacterial toxins, or the host's response to them, can disrupt the organ of Corti and the stria vascularis (Tarlow et al. 1991; Comis et al. 1993; Guo et al. 1994). Some of the damage observed, such as stripping of the stereocilia of hair cells is likely to be irreversible. Other changes may resolve quickly. These include swelling and surface cratering of hair cells and supporting cells (Huang et al. 1990; Clerici et al. 1995; Osborne et al. 1995; Winter et al. 1996); disruption of the stria vascularis (Gao et al. 1994; Osborne et al. 1995); and the creation of pores in the basilar membrane and supporting cells of the organ of Corti (Osborne et al. 1995; Winter et al. 1996). The latter two processes could cause deafness by diminishing the endocochlear potential that provides energy for the organ of Corti. (The endocochlear potential is produced by metabolic activity in the stria vascularis and is dependent upon the structural integrity of the borders of the scala media.) Further evidence for the role of toxins in fleeting hearing loss comes from ABR studies in experimental meningitis. In one such study, Comis et al. (1993) found that the early stage of hearing loss produced by toxin was reversible. It is therefore quite possible that toxic labyrinthitis is the cause of fleeting hearing loss.

From the results of our study it is tempting to propose that fleeting hearing loss represents a form of cochlear dysfunction that would progress to permanent damage if the meningitis were not treated. This is certainly possible, given that the onset of deafness and the site of the lesion appear to be identical in fleeting and permanent sensorineural hearing loss. There is also some evidence from animal experiments to support this hypothesis. In their rabbit model of meningitis, Bhatt et al. (1993) found that animals invariably showed evidence of hearing loss around 12 hours after the induction of meningitis. All the animals went on to develop severe sensorineural deafness. In a subsequent experiment the animals were given antibiotics at the time of the onset of hearing loss (Bhatt et al. 1995). In this case, nearly all of the animals were found to have normal hearing two days later.

In summary, we have described the discovery of fleeting hearing loss in children with acute bacterial meningitis. This phenomenon appears to be caused by a rapidly reversible dysfunction of the cochlea. The pathological processes involved are unclear but they almost certainly involve the transmission of an inflammatory reaction from the subarachnoid space to the inner ear.

References

Baraff LJ, Lee SI, Schriger DL (1993) Outcomes of bacterial meningitis in children: a meta-analysis. Pediatr Infect Dis J 12: 389–94.

Bhatt SM, Lauretano A, Cabellos C, Halpin C, Levine RA, Xu WZ, Nadol JB Jr, Tuomanen E (1993) Progression of hearing loss in experimental pneumococcal meningitis: correlation with cerebrospinal fluid cytochemistry. J Infect Dis 167: 675–83.

Bhatt SM, Cabellos C, Nadol JB, Halpin C, Lauretano A, Xu WZ, Tuomanen E (1995) The impact of dexamethasone on hearing loss in experimental pneumococcal meningitis. Pediatr Infect Dis J 14: 93–6.

Büki B, Avan P, Lemaire JJ, Dordain M, Chazal J, Ribari O (1996) Otoacoustic emissions: a new tool for monitoring intracranial pressure through stapes displacements. Hear Res 94: 125–39.

Clerici WJ, DiMartino DL, Prasad MR (1995) Direct effects of reactive oxygen species on cochlear outer hair cell shape in vitro. Hear Res 84: 30–40.

Comis SD, Osborne MP, Stephen J, Tarlow MJ, Hayward TL, Mitchell TJ, Andrew PW, Boulnois GJ (1993) Cytotoxic effects on hair cells of guinea pig cochlea produced by pneumolysin, the thiol activated toxin of Streptococcus pneumoniae. Acta Otolaryngol (Stockh) 113: 152–9.

Crowe SJ (1930) Pathologic changes in meningitis of the internal ear. Arch Otolaryngol 11: 537–68.

Djeric DR, Schachern PA, Paparella MM, Jaramillo M, Haruna S, Bassioni M (1994) Otitis media (silent): a potential cause of childhood meningitis. Laryngoscope 104: 1453–60.

Dodge PR, Davis H, Feigin RD, Kaplan SL, Jubelirer DP, Stechenberg BW, Hirsh SK (1984) Prospective evaluation of hearing impairment as a sequela of acute bacterial meningitis. N Engl J Med 311: 869–74.

Eisenberg LS, Luxford WM, Becker TS, House WF (1984) Electrical stimulation of the auditory system in children deafened by meningitis. Otolaryngol Head Neck Surg 92: 700–5.

Fraser JS, Dickie JKM (1920) Meningitic neuro-labyrinthitis. Proc Roy Soc Med 13: 23–62.

Guiscafré H, Benitez-Díaz L, Martínez MC, Muñoz O (1984) Reversible hearing loss after meningitis: prospective assessment using auditory evoked responses. Ann Otol Rhinol Laryngol 93: 229–32.

Guo Y, Wu Y, Chen W, Lin J (1994) Endotoxic damage to the stria vascularis: the pathogenesis of sensorineural hearing loss secondary to otitis media? J Laryngol Otol 108: 310–3.

Huang MH, Dulon D, Schacht JS (1990) Outer hair cells as potential targets of inflammatory mediators. Ann Otol Rhinol Laryngol 99 (Suppl 148): 35–8.

Igarashi M, Saito R, Alford BR, Fillipone MV, Smith JA (1974) Temporal bone findings in pneumococcal meningitis. Arch Otolaryngol 99: 79–83.

Kay R (1991) The site of the lesion causing hearing loss in bacterial meningitis: a study of experimental meningitis in guinea-pigs. Neuropathol Appl Neurobiol 17: 485–93.

Kemp DT (1978) Stimulated acoustic emissions from within the human auditory system. J Acoust Soc Am 64: 1386–91.

Merchant SN, Gopen Q (1996) A human temporal bone study of acute bacterial meningogenic labyrinthitis. Am J Otol 17: 375–85.

Michel O, Brusis T (1992) Hearing loss as a sequel of lumbar puncture. Ann Otol Rhinol Laryngol 101: 390–4.

Moxon ER, Smith AL, Averill DR, Smith DH (1974) Haemophilus influenzae meningitis in infant rats after intranasal inoculation. J Infect Dis 129: 154–62.

Nadol JB (1978) Hearing loss as a sequela of meningitis. Laryngoscope 88: 739–55.

Osborne MP, Comis SD, Tarlow MJ, Stephen J (1995) The cochlear lesion in experimental bacterial meningitis of the rabbit. Int J Exp Path 76: 317–30.

Özdamar Ö, Kraus N, Stein L (1983) Auditory brainstem responses in infants recovering from bacterial meningitis. Arch Otolaryngol 109: 13–8.

Palva T (1970) Cochlear aqueduct in infants. Acta Otolaryngol 70: 83–94.

Perlman HB, Lindsay JR (1939) Relation of the internal ear spaces to the meninges. Arch Otolaryngol 29: 12–23.

Richardson M (1995) Otoacoustic emissions. Arch Dis Child 73: 284–6.

Richardson MP, Williamson TJ, Lenton SW, Tarlow MJ, Rudd PT (1995) Otoacoustic emissions as a screening test for hearing impairment in children. Arch Dis Child 72: 294–7.

Richardson MP, Reid A, Tarlow MJ, Rudd PT (1997) Hearing loss during bacterial meningitis. Arch Dis Child 76: 134–8.

Smyth V, O'Connell B, Pitt R, O'Callaghan M, Scott J (1988) Audiological management in the recovery phase of bacterial meningitis. Int J Pediatr Otorhinolaryngol 15: 79–86.

Tarlow MJ, Comis SD, Osborne MP (1991) Endotoxin induced damage to the cochlea in guinea-pigs. Arch Dis Child 66: 181–4.

Trolle E (1950) Defective hearing after meningococcal meningitis. Acta Otolaryngol 38: 384-402.

Vienny H, Despland PA, Lütschg J, Deonna T, Dutoit-Marco ML, Gander C (1984) Early diagnosis and evolution of deafness in childhood bacterial meningitis: a study using brainstem auditory evoked potentials. Pediatrics 73: 579–86.

Wald ER, Kaplan SL, Mason EO, Sabo D, Ross L, Arditi M et al. (1995) Dexamethasone therapy for children with bacterial meningitis. Pediatrics 95: 21–8.

Winter A, Marwick S, Osborne M, Comis S, Stephen J, Tarlow MJ (1996) Ultrastructural damage to the organ of Corti during acute experimental Escherichia coli and pneumococcal meningitis in guinea pigs. Acta Otolaryngol (Stockh) 116: 401–7.

Correspondence to Dr Martin Richardson, Paediatric Infectious Diseases Unit, Lanesborough Wing, St George's Hospital, London SW17 OQT. Tel. 0181 725 3922. Fax 0181 725 3262. E-mail: mrichard@sgbms.ac.uk.

Chapter 32
Recurrent perilymphatic fistula in patients with possible patent cochlear aqueduct: a role for cochlear aqueduct blockade – report of five cases

DUDLEY J WEIDER, FRANK E MUSIEK, Section of Otolaryngology, Department of Surgery Dartmouth-Hitchcock Medical Center, Lebanon, New Hampshire, US

ABSTRACT. A subgroup of patients who present with symptoms and findings of perilymphatic fistula do not respond with permanency to simple reinforcement of the oval or round windows. Symptoms of vertigo, tinnitus, aural fullness and hearing loss may remit or stabilize for varying periods, only to return upon physical straining or, on occasion, in the absence of any obvious precipitating event. It is our theory that these patients most probably possess a patent cochlear aqueduct to a varying degree.

Five such cases are presented in which there is good evidence for cochlear aqueduct patency. Initially their perilymphatic fistulae were treated by simple reinforcement of their oval and round windows with areolar tissue. Secondary surgeries included endolymphatic shunt, lumbar and ventriculo-peritoneal shunt, vestibular nerve section and cochlear aqueduct blockade.

Key words: perilymphatic fistula, cochlear aqueduct, blockade, vertigo, tinnitus, hearing loss, surgery.

Introduction

During the past 22 years the first author (DJW) has encountered a small subgroup of patients who present with symptoms and findings of perilymphatic fistula (PLF) but do not respond permanently to simple reinforcement of the oval window (OW) and/or round window (RW). Each patient has experienced stable postoperative hearing and freedom from vertigo, only to have a normal life occurrence, such as a sneeze or the act of lifting an object, bring on previous symptoms, including vertigo, hearing loss, pressure and often tinnitus. Some success in treating this group of patients has been achieved by Weider et al. (1988, 1991, 1992, 1996) by using secondary measures designed to reduce perilymphatic pressure at the site of the PLF leak. These methods include endolymphatic shunt (ELS), lumbar-peritoneal shunt, ventriculo-peritoneal shunt and cochlear aqueduct (CA) blockade.

Presented are five patients believed to have classic symptoms and findings of PLF and who responded to ordinary OW and RW reinforcement for varying periods of time only to have symptoms of PLF return. Methods of clinical evaluation and clinical treatment are presented.

Case reports

Case 1

This now 44-year-old man presented to the Dartmouth-Hitchcock Medical Center (DHMC) in July 1986 with a seven-year history of fluctuating hearing in his right and only hearing ear (Weider 1991). A five-month history of wide daily fluctuation and deterioration in hearing prompted his initial visit. Oval window and RW PLFs were repaired in August 1986, and an ELS was performed. In March 1987 an OW PLF revision was performed, and a lumbar drain left in place for five days postoperatively. Both PLF repairs resulted in stabilization of hearing at improved levels for about four weeks, but hearing fluctuation, deterioration and tinnitus returned coincident with minor physical straining. Cochlear aqueduct patency was suspected when indium 111 DTPA was collected on a cerebral spinal fluid (CSF) saturated cotton ball (placed in the ear canal) 12 hours after introduction into the spinal fluid by way of lumbar puncture. A ventilation tube had been placed in the right tympanic membrane permitting egress of fluid collecting in the middle ear.

Taking the lead from Farrior and Endicott (1971), the patient's CA was exposed in the posterior cranial fossa and occluded with a muscle plug. This stopped drainage of fluid (presumably CSF) from the middle ear through the ventilation tube, leading to a mild decrease in high-pitched tinnitus and to relatively stable hearing thresholds; results have lasted for 10 years (see Figure 32.1). In May 1997 the patient's SRT was 60 dB with a PTA of 56.25 dB and a speech discrimination score (SDS) of 96%. Compared to a previous audiogram (Weider, 1991) there has been 5 dB to 10 dB improvement at 250 and 500 Hz, stability at 1000 Hz, and 10 dB to 15 dB decline in frequencies at 2000 dB and above.

The CA is found in the posterior fossa, just below the internal auditory meatus and just above the ninth nerve. The lumen is located with a 4-mm right-angle pick. Drilling exposes the proximal funnel shaped lumen and muscle is placed over it, suspended from the dura with a microclip (Weider, 1981). Recently the lumen has been filled with bone dust covered with fibrin glue; the muscle is held in place by the glue. The brain expands against the muscle, further stabilizing it. Despite the intracranial approach the cochlea is at less risk than with the transcanal approach of Farrior and Endicott (1971). Brainstem audiometry is used to monitor hearing during these procedures.

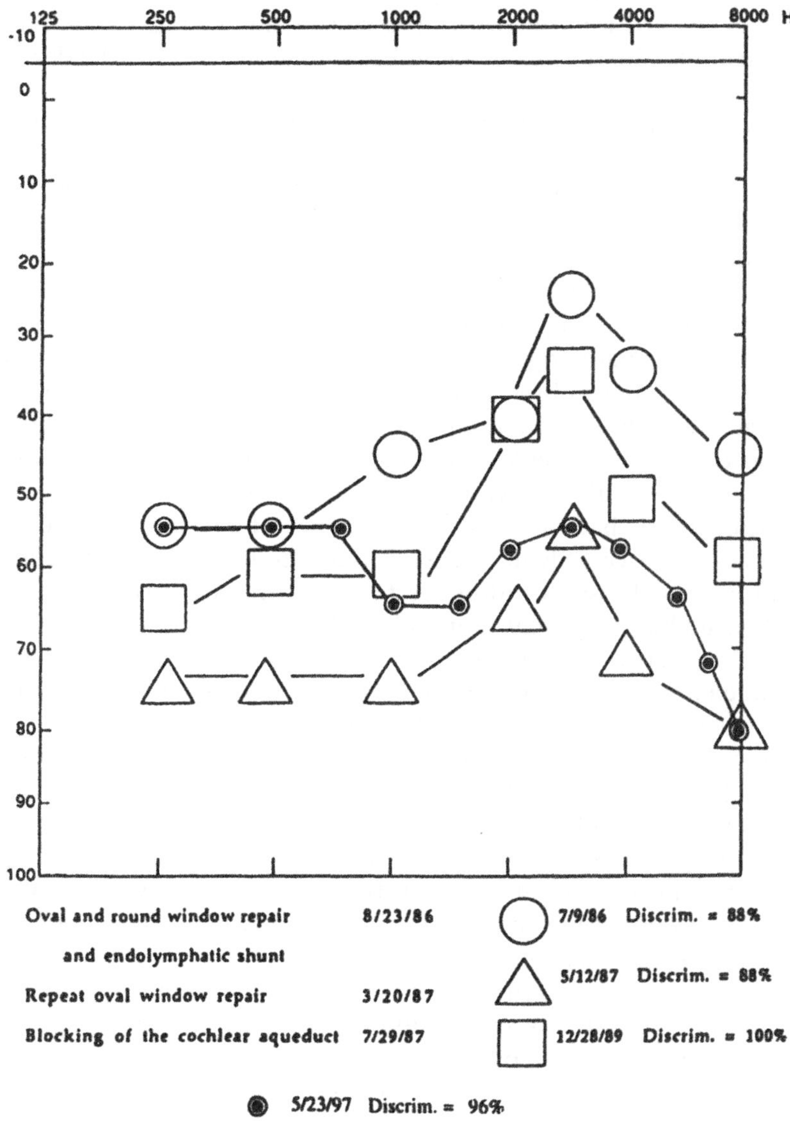

Figure 32.1: serial audiograms from 9 July 1986 to 23 May 1997 along with documentation of surgical events undergone by case 1.

Case 2

This now 50-year-old man presented to the DHMC in January 1985 with a six-year history of disequilibrium which had become constant and was accentuated by physical stress (Weider 1996). Hearing was normal bilaterally. During his Quix test (Lehrer et al. 1980), eyes-closed turning test (ECTT) (Singleton 1978) and eyes-closed walking test (ECWT) (walking 20 feet toward a target with eyes closed) he veered to the left but normalized 1 hour after being given 40 mg of IM Lasix (Weider 1988, 1992).

The patient responded positively (total resolution of vertigo) to three PLF repairs and an ELS (left ear) over a period of two years. Within months of each repair relapse occurred in response to minor physical straining. A vestibular nerve section was performed in June 1987, resulting in the absence of vertigo for five years.

In December 1992 the patient's symptoms returned. His Quix and ECWT veered to the right (normalizing after 40 mg of IM Lasix). Bed rest and two PLF repairs (right ear) stopped the vertigo for several weeks to a few months. Minor physical stress or exercise triggered reappearance of vertigo and mild nausea. Interestingly, he complained of bobbing oscillopsia despite measurable vestibular function (electronystagmogram) in the right ear. Head (vertex) pressure became a constant.

The patient had elevated CSF pressure of 280 mm of water. After removal of 15 cc CSF, he was symptom-free for 36 hours (absence of bobbing oscillopsia, tinnitus, vertex pressure and vertigo). In March 1994 a lumbar-peritoneal shunt relieved his symptoms for several weeks but they returned coincident to his dentist placing him in the Trendelenburg position. The shunt was eventually removed because of the development of 'low-pressure' headaches upon standing.

A right suboccipital craniotomy with CA blockade was performed in June 1996 because of continued symptoms of stress-induced vertigo and constant low-grade nausea. Postoperatively his stability improved, bobbing oscillopsia and nausea stopped and hearing remained stable. Tinnitus in both ears diminished. He continued to have some trouble walking on uneven ground and felt mildly nauseated when exposed to fluorescent lighting.

This significant improvement lasted for five months when, with increased exercise, his symptoms returned. He currently experiences pressure in his head, some bobbing oscillopsia and vertigo with any exertion. An LP in February 1997 demonstrated normal CSF pressure. A recent transtympanic electrocochleogram (EcoG) demonstrated a summating potential (SP) of 80% which does not change on straining (as in Case 5). His Quix test and ECWT continue to veer right (normalizing after 40 mg of IM Lasix, frusemide). Platform posturography in April 1997 demonstrates diminished vestibular function, which also improved markedly one hour following 40 mg of IM Lasix. Hearing has never fluctuated (SRT of 10 dB and an SDS of 100%).

This man did quite well for 5 months after CA blockade. Recent events indicate that our repair or technique for achieving occlusion was inadequate to hold up over time. More permanent methods for CA occlusion are under consideration. An ELS is being considered for this patient, as his SP is elevated (diagnostic of hydrops); several authors have found ELS applicable in this situation (Lehrer et al. 1987; Seltzer and McCabe 1986; Weider 1988, 1991, 1992). I (DJW) have found a very high incidence of PLF in patients with stress-induced vertigo when Quix and related tests (ECTT, ECWT, platform posturography) return to normal following glycerin (Lehrer et al. 1980) or Lasix administration (Weider and Johnson 1988; Weider 1992) (see Appendix 1).

Case 3

This now 34-year-old woman first presented at age 17 in January 1983 (Weider and Johnson 1988; Weider 1992). She had been doing cartwheels when she became vertiginous and nauseated. She presented with a mild low-tone hearing loss in her right ear, a mild right canal paresis and a subjectively positive fistula test in the right ear. In April 1983 a right exploratory tympanotomy revealed PLFs of the OW and RW, which were repaired with ear-lobe fat. After a three-week rest period the patient returned to her sport of track and won the district hurdles championship. One year later upon joining an aerobics class she felt a pop in her right ear and a return of all the symptoms of stress-induced vertigo and some high-pitched tinnitus. Hearing remained little affected. Bed rest was not effective in relieving her symptoms. She was re-explored in October 1984 and both windows were reinforced. Postoperatively she had complete resolution of symptoms. She remained asymptomatic until February 1986 when coincident with the delivery of her first child all her symptoms returned. This time immediate bed rest led to resolution of symptoms. Her next two children were delivered by Caesarean section without event.

Early in 1995 she again became symptomatic. Bed rest did not lead to symptom resolution. In April 1996 an endolymphatic sac decompression coupled with RW and OW reinforcement was performed; her symptoms resolved for three weeks only to return after her dentist put her in the Trendelenburg position (as in Case 2). Again bed rest was ineffective. Any slight physical effort led to vertigo and nausea. Tinnitus became constant and high pitched. Her low-frequency hearing had deteriorated only slightly. Spinal fluid pressure was normal.

In June 1996 the patient underwent a right suboccipital craniotomy and CA blockade. Stress-induced vertigo ceased immediately. Her hearing improved slightly to its previous baseline. She remained asymptomatic for six months until December 1996. These periods remained relatively minor until July of 1997 when her stress-induced vertigo again reached preoperative dimensions. A posterior fossa vestibular nerve section was performed in September of 1997 in conjunction with a posterior fossa endolymphatic shunt. At the time of this writing she is 5 months postoperative and experiencing no further stress-induced vertigo. Her sensorineural hearing has declined about 15 dB through all frequencies, and her SDS remains at 96%. Hopefully this procedure will afford her lasting resolution to a 17-year problem.

Case 4

In January 1995 a 64-year-old farmer and general maintenance man experienced a sudden increase in tinnitus in his left ear. He described the tinnitus as having a machinery-like hum. Commensurate with the onset of his tinnitus he started feeling mildly off balance with a definite increase in vertigo with physical activity, especially lifting. He also felt the hearing in his left ear to be slightly diminished. The following month his left-ear tinnitus increased and hearing decreased as he struck a nail with a hammer, and within minutes he experienced nausea and vertigo. He feels his hearing never improved and has fluctuated noticeably since this event. When first seen by an otolaryngologist, his left-ear SRT and PTA (250-2000 Hz) were 30 dB, with a SDS of 84%.

He was explored for a PLF in June 1995, but reinforcement of an OW PLF did not improve his symptoms. The patient was referred to the DHMC in August 1995. His audiology had not changed. His Quix test and ECWT veered to the left but normalized one hour after IV Lasix and lasted about 16 hours. Interestingly the tinnitus in his left ear also diminished but did not disappear.

In November 1995 the patient underwent an exploratory tympanotomy with revision reinforcement of the OW and RW. Prior to administering general anesthesia a lumbar puncture revealed a CSF pressure of 270 mm of water. A diagnosis of benign intracranial hypertension (or pseudotumour cerebri) was made and a lumbar drain was placed and left in for four postoperative days. Postoperatively the patient's vertigo stopped completely and tinnitus diminished. Upon return to work six weeks later he lifted an air conditioner and tinnitus and vertigo immediately returned to preoperative levels. As before, any exertion worsened the vertigo and brought on nausea.

He continued to experience CSF hypertension with absence of ophthalmic signs. In July 1996 his pure-tone audiogram demonstrated random changes (generally decreasing in threshold by 5 to 10 dB) between 125 and 2000 Hz when he assumed a Trendelenburg position. Tinnitus increased in the head-down position. These findings indicated probable CA patency. A CA blockade was performed through the posterior cranial fossa in July 1996. Following the surgery the patient's hearing has remained stable between 125 and 1000 HZ with a 15 to 20 dB decline above 1000 Hz. Discrimination was 64% preoperatively and has declined to 44% postoperatively. His tinnitus declined immediately to nearly negligible levels and has remained like this to the present (May 1997).

Interestingly, coincident with the surgery this patient experienced a temporary hearing decline of 20 dB in all frequencies in his right (opposite ear), as well as some ataxia. The right-ear hearing returned to baseline within three weeks and his ataxia diminished (tendency to veer to the right with eyes closed whereas before surgery he veered left). As before the initial PLF repair, IM Lasix again resulted in complete resolution of vertigo. This led me (DJW) to suspect a hydrodynamic problem (PLF or hydrops) in the right ear. A ventriculo-peritoneal shunt was performed in September 1996 to relieve his CSF hypertension and to decrease ataxia. Postoperatively ataxia continued to be a minor problem but nausea on exertion stopped. In February 1997 a diagnosis of hypothyroidism was made and his remaining ataxia cleared completely with Synthroid administration.

Sismanis (1988) discussed the symptoms of tinnitus, vertigo and fluctuating hearing loss in patients with ideopathic CSF hypertension. Marchbanks (1996a, b) studied a group of patients with benign intracranial hypertension using the tympanic membrane displacement technique to determine CA patency (Reid et al. 1990). Marchbanks found that patients with CSF over pressure who did not have patent CAs did not develop symptoms of tinnitus, vertigo, fluctuating hearing and pressure, whereas patients who did have patent CAs did develop these symptoms. Effecting closure in this patient stopped the tinnitus, and along with the ventriculo-peritoneal shunt, has controlled his stress-induced vertigo as well. Hearing has been only slightly negatively affected.

Case 5

A 34-year-old woman presented to the DHMC in May 1996 with a two-month history of increasing vertigo with lifting, bending or straining. She also had intermittent high-pitched tinnitus, and intermittent right-ear pain lasting about 10 minutes and recurring five to six times a day. Seven years prior, after heavy lifting, she had trouble with almost constant vertigo, which was exacerbated by exercise. This continued until she was seen by Fitzgerald (1995) who diagnosed and repaired a right OW PLF. Following this treatment her symptoms of vertigo diminished by over 50% and she was able to function well in her job as a nurse. The tinnitus diminished in intensity and was noticed only occasionally.

In 1996 she came to the DHMC. Interestingly, increased right-ear pain (Weider 1988, 1992) was her first symptom, followed by a marked increase in vertigo and return of tinnitus.

On physical examination the patient had a positive subjective fistula test bilaterally, and she noticed more vertigo when the left TM was moved. Her standing Quix test and high-stepping Quix test veered to the left, as did her ECTT and ECWT. Her audiogram appeared grossly normal with 100% SDS bilaterally, though she had a symmetrical mild low-tone sensorineural hearing loss and a mild asymmetrical high-tone loss on the right. The SRT was 5 dB on the left and 15 dB on the right. On 5/20/96 her neurologic testing was repeated and she was given 40 mg of IM Lasix. One hour later she felt nearly asymptomatic, her Quix tests and ECWT completely normalized, and tinnitus diminished. Bilateral middle-ear explorations were performed to repair suspected PLFs because of a continuing tendency to drift to the left and a mildly positive fistula test on the left but with symptoms of pressure, fullness, pain and tinnitus on the right.

In August 1996 bilateral middle-ear explorations with reinforcement of OW and RW areas were performed. The left ear was explored first. An obvious OW fistula was seen with fluid repeatedly welling up in the OW niche after suctioning. Both windows were reinforced with areolar tissue. Exploration of the right ear revealed many adhesions from the previous repair. Despite using the KTP laser to lyse the adhesions a fistula could not be clearly identified. Postoperatively the patient experienced near-complete resolution of vertigo, pressure and tinnitus. These results lasted for four weeks, until one morning, a half-hour after waking, she began to feel pressure in the right ear. Right-ear tinnitus and stress-induced vertigo returned. A fistula test was subjectively negative on the left side but positive on the right.

Transtympanic electrocochleography in the right ear demonstrated an SP of 36%; upon doing a Valsalva it immediately increased to 80% (see Figure 32.2). This was repeated on four successive occasions with identical results. The EcoG on the left side was normal. A diagnosis of a patent CA was made, along with relative cochlear hydrops, probably secondary to a still-active PLF on the right. An ELS and repeat OW reinforcement on 12 December 1996 resulted in complete resolution of vertigo and greatly diminished tinnitus.

Performing an endolymphatic shunt can improve or eliminate symptoms of a PLF (Seltzer and McCabe 1986; Lehrer et al. 1987; Weider 1988, 1992). The fact that doing a Valsalva during an EcoG could immediately and repeatedly change a summating potential from 36% to 80% would indicate a wide-open CSF-perilymph connection, most likely via the CA. The wide-open CA most likely made this patient prone to developing a PLF and vulnerable to recurrence after initial repeat repair. The fact that the tinnitus greatly diminished after the ELS would indicate the restoration of a more normal cochlear physiology and anatomy, perhaps by adding necessary compliance or elasticity to the cochlear fluid system.

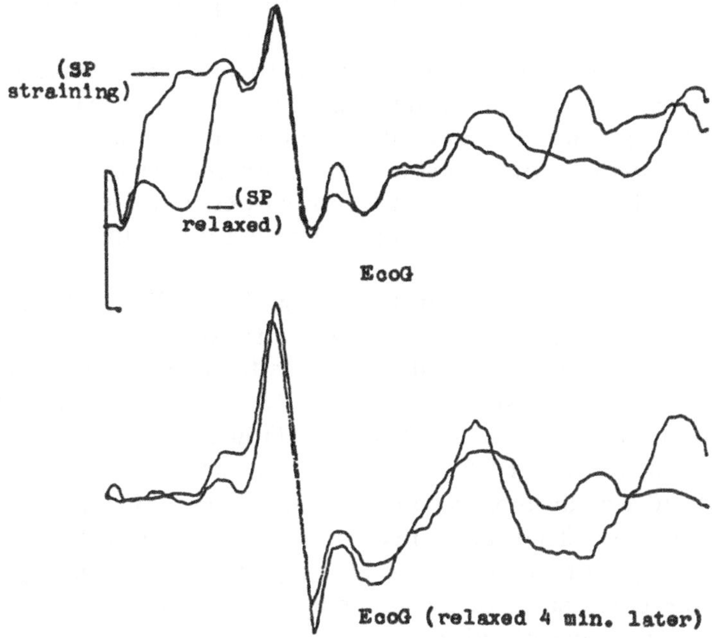

Figure 32.2: the EcoG in Case 5 demonstrating the increase in SP from 36% before straining to 80% after straining.

In late February of 1997 the patient fell, sustaining a badly compressed and fragmented lumbar 3-4 disc. A decompressive laminectomy and disc excision was performed in March 1997. This surgery was followed with an extreme amount of postoperative nausea and vomiting. Stress-induced vertigo and tinnitus returned minimally and intermittently. In July 1997 they became constant. In August of 1997 bilateral exploratory tympanotomies were repeated with reinforcement of oval and round windows in each ear. She has been vertigo-free for 6 months. Tinnitus is much diminished. CA blockade is contemplated should symptoms return.

Discussion

We would like to make three basic points. Firstly, there is a subgroup of patients who present with symptoms of PLF that do not respond with permanency to routine exploratory tympanotomy and reinforcement of the OW and RW with soft tissue grafting. This group may compose 10 to 15 percent of all patients with PLF. We feel that two main reasons for these failures are 1) pathologically elevated CSF pressure or pseudotumour cerebri as in our cases 2 and 4 (Weider and Saunders 1996), and 2) a widely patent cochlear aqueduct (Wlodyka 1978) allowing a significant CSF pulse pressure to be transmitted through the perilymph to a weakened OW or RW effectively reopening the repaired area.

When the above situations exist then the treating physician must do more than just repeatedly repair the same leak in the same manner. One can lower CSF pressure if it is elevated by performing a lumbar (Weider 1996) or ventriculo-peritoneal shunt (favoured because of fewer post-operative problems). Case 2 had a lumbar-peritoneal shunt which worked well for a time. Case 4 did well with a ventriculo-peritoneal shunt.

If the CA is too widely patent then it can be safely and effectively closed by way of the suboccipital route with ABR monitoring. Our experience with hearing preservation in cases 1–4 confirms the observations of Schuknecht and Kimura (1953) and Carlborg et al. (1983, 1996) in animals and of Farrior and Endicot (1971) in humans. This was the procedure of 'last resort' for cases 1–4. In each case hearing was preserved and actually improved in case 1. As witnessed in cases 2 and 3 a better method of effecting a permanent seal needs to be devised.

Lastly, the performance of an endolymphatic shunt (case 5) may increase the compliance of the inner ear by offsetting some of the pressure changes permitted to enter the inner ear through a widely patent CA. This would diminish some of the stress against the OW repair site. Additionally, a secondary cochlear hydrops (Ackley 1984) is probably relieved.

Acknowledgements

The authors would like to thank Richard L Sanders, MD, David Roberts, MD and Joseph Phillips, MD for their neurosurgical contributions to these cases and to Alicia Green for her help with the preparation of this manuscript.

References

Ackley RS (1984) Electrocochleography and histological result of perilymphatic fistula in guinea pigs. Ann Arbor, MI: University Microfilms.

Carlborg B, Farmer JC (1983) Transmission of cerebrospinal fluid pressure via the cochlear aqueduct and endolymphatic sac. Am J Otolaryngol 4: 273–82.

Carlborg B, Konr·dsson KS, Farmer JC (1996) Pressure relation between the labyrinthine and intracranial fluids: experimental study in cats. In Ernst A, Marchbanks R, Samii M (eds) Intracranial and Intralabyrinthine Fluids: Basic Aspects and Clinical Applications. New York: Springer-Verlag, pp. 63–72.

Farrior JB, Endicott JN (1971) Congenital mixed deafness: cerebrospinal fluid otorrhea – ablation of the aqueduct of the cochlea. Laryngoscope 81: 684–99.

Fitzgerald DC (1995) Persistent dizziness following head trauma and perilymph fistula. Arch Phys Med Rehab 76: 1017–20.

Lehrer JF, Pool DC, Sigal B (1980) Use of the glycerine test in the diagnosis of post-traumatic perilymph fistulas. Am J Otolaryngol 1: 207–10.

Lehrer JF, Quraishi AU, Poole DC (1987) The role of endolymphatic sac surgery in the management of secondary endolymphatic hydrops associated with perilymphatic fistulas: Preliminary observations. Am J Otol 8: 93–5.

Marchbanks R (1996a) Audiovestibular dysfunction in patients with confirmed intracranial hypertension. Proceedings of the British Society of Audiology Meeting.

Marchbanks RJ (1996b) Hydromechanical interactions of the intracranial and intralabyrinthine fluids. In Ernst A, Marchbanks R, Samii M (eds) Intracranial and Intralabyrinthine Fluids: Basic Aspects and Clinical Applications. New York: Springer-Verlag, pp. 51–61.

Reid A, Marchbanks RJ, Burge DM, Martin AM, Bateman D, Pickard JD, Brightwell AP (1990) The relationship between intracranial pressure and tympanic membrane displacement. Br J Audiol 24: 123–9.

Schuknecht HT, Kimura RS (1953) Functional and histologic findings after obliteration of the periotic duct and endolymphatic sac in sound conditioned cats. Laryngoscope 63: 1170.

Seltzer S, McCabe BF (1986) Perilymphatic fistula: the Iowa experience. Laryngoscope 94: 37–49.

Singleton GT, Post KN, Karian MS, Bock DG (1978) Perilymph fistulas. Diagnostic criteria and therapy. Ann Otol Rhinol Laryngol 87: 797–903.

Sismanis A (1988) Otologic manifestations of benign intracranial hypertension syndrome: diagnosis and management. Laryngoscope 184–96.

Weider DJ (1981) Use of the Williams microclip in various aspects of tympanoplastic surgery. Am J Otol 91: 2106–25.

Weider DJ, Johnson GD (1988) Perilymph fistula: A New Hampshire experience. Am J Otol 9:184–96.

Weider DJ, Saunders RL, Musiek FE (1991) Repair of a cerebrospinal fluid perilymph fistula primarily through the middle ear and secondarily by occluding the cochlear aqueduct. Otolaryngol Head Neck Surg 105: 35–9.

Weider DJ (1992) Treatment and management of perilymphatic fistula: a New Hampshire experience. Am J Otol 13: 158–66.

Weider DJ, Saunders RL (1996) Recurrent perilymphatic fistula as the initial and prime symptom of pseudotumor cerebri: diagnosis and management with lumbar-peritoneal shunt – report of three cases. In Ernst A, Marchbanks R, Samii M (eds) Intracranial and Intralabyrinthine Fluids: Basic Aspects and Clinical Applications. New York: Springer-Verlag, 293–8.

Weider DJ (in press) Tinnitus: report of ten cases with perilymphatic fistula and/or endolymphatic hydrops positively affected by surgery. Int Tinnitus J (in press).

Wlodyka J (1978) Studies on cochlear aqueduct patency. Ann Otolaryngol 87: 22–7.

Appendix 1: Lasix test

Patients with PLF, unlike patients with Ménière's disease, are mildly unstable most of the time, their vertigo increasing with increased physical activity. The presence of a hole in the OW or RW areas allows an abnormal hydrodynamic condition to exist within the cochlea. One of the effects of Lasix, a loop diuretic, is to decrease potassium ion production. If this occurs in the stria vascularis then endolymphatic pressure would be diminished, effecting a decrease in total endocochlear pressure. The observed result is improved patient balance as observed in platform posturography, Quix testing, ECWT and tandem Romberg. Rhenograffin, mannitol and urea have similar effects on these patients. Lasix is the easiest drug to give in an office setting and has few side effects. Oral Lasix is not effective. Lasix given IV has an almost immediate effect and works within 45 minutes given IM.

Correspondence and reprint requests to Dr. Dudley Weider, Section of Otolaryngology, Dartmouth-Hitchcock Medical Center, One Medical Center Drive, Lebanon, NH 03756-0001. Tel. 603/650-8122. Fax 603/650-4961.

Chapter 33
Effects of the atrial natriuretic peptide (ANP) on hearing and tinnitus in patients suffering from Ménière's disease

JÜRGEN LAMPRECHT, Alfried-Krupp-Krankenhaus, Essen, Germany

ABSTRACT. The atrial natriuretic peptide (ANP) has been shown to be effective in volume regulation in the kidney (hypertension), and the eye (glaucoma) via cyclic guanosine monophosphate (cGMP). The presence of ANP receptors in the inner ear led to the question, whether specific effects of the hormone can be found in endolymphatic hydrops too. In accordance with the glycerol test, we investigated hearing during ANP infusion and constant serum osmolality. Two out of 10 patients showed hearing improvement, one of them reported that tinnitus ceased. We need further investigations to determine whether ANP can be of value for therapy in Ménière's disease.

Key words: tinnitus, Ménière's disease, atrial natriuretic peptide, hearing loss, dizziness, glycerol test.

Introduction

Tinnitus is not a disease, it is a symptom. One of the underlying pathologies can be Ménière's disease. It comes along with fading hearing, tinnitus, dizziness and a sensation of pressure in the ear. The aetiology of Ménière's disease remains unclear. However, the hypothesis of an endolymphatic hydrops based upon a dysregulation of electrolyte concentrations in the inner-ear fluid has been widely accepted.

The human atrial natriuretic peptide (ANP) is a potent cardiovascular hormone (de Bold et al. 1981) that promotes renal sodium secretion in response to increased extracellular fluid volume and atrial pressure. Specific receptors for ANP have been identified in various tissues. These include kidney, adrenal glands, the ciliary process of the eye, blood vessels, the plexus chorioideus of the cerebral ventricles and the inner ear. Effects of this hormone have been demonstrated in hypertension (Espiner and Richards 1989) and in glaucoma (Diestelhorst and Kreigstein 1989). Using autoradiographic methods in 1988, Lamprecht first demonstrated the existence of ANP-receptors in the inner ear of mammals (Lamprecht and Meyer zum Gottesberge 1988). The issue of the following studies has been, whether specific effects of the hormone can be found in endolymphatic hydrops in humans.

Methods

In a clinical pilot study, 48 patients suffering from sensorineural hearing impairment have been investigated (Lamprecht and Plum 1990). In 14 of these the Klockhoff Glycerol Test (Klockhoff and Lindblom 1966) was positive. Thirty-four of the patients did not show this item. Before and every hour while the glycerol test was carried out, plasma concentrations of ANP and cGMP were recorded. cGMP is considered to be the intracellular messenger that is activated by a stimulation of guanylate cyclase. As expected before plasma concentrations of ANP increased with plasma osmolality. For this parameter there was no difference found between patients suffering from Ménière's disease and patients suffering from other cochlear disturbances. For cGMP there was a dissociation between the different groups (Figure 33.1). Baseline values of cGMP differed significantly in the respective populations. This difference remained stable for the entire observation period.

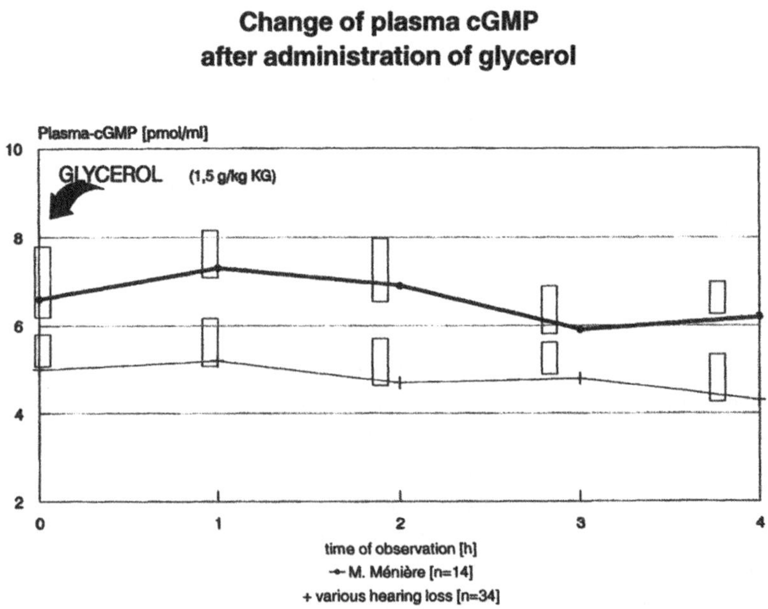

Figure 33.1: change of plasma cGMP after administration of glycerol, plasma-cGMP (pmol/ml) v time of observation (hours) for patients with Ménière's disease and various hearing losses.

Aim of the study

In the first place it was not clear whether the improvement in hearing was induced by a mere change of osmolality. There might have been an indirect effect on the inner-ear receptor by the mobilization on ANP. To investigate this effect, only Ménière patients were included in a follow-up trial. Instead of glycerol, human ANP was administered by a perfusor pump. The dosage was 10 pg/kgBWmin in six patients or 20 pg/kgBWmin in two patients resp. Serum osmolality was registered and kept constant through substitution of electrolytes and fluid volume (Figure 33.2).

Hearing was investigated by audiometry before and during administration of ANP. Simultaneously tinnitus was scored according to Klockhoff's and Lindblom's criteria.

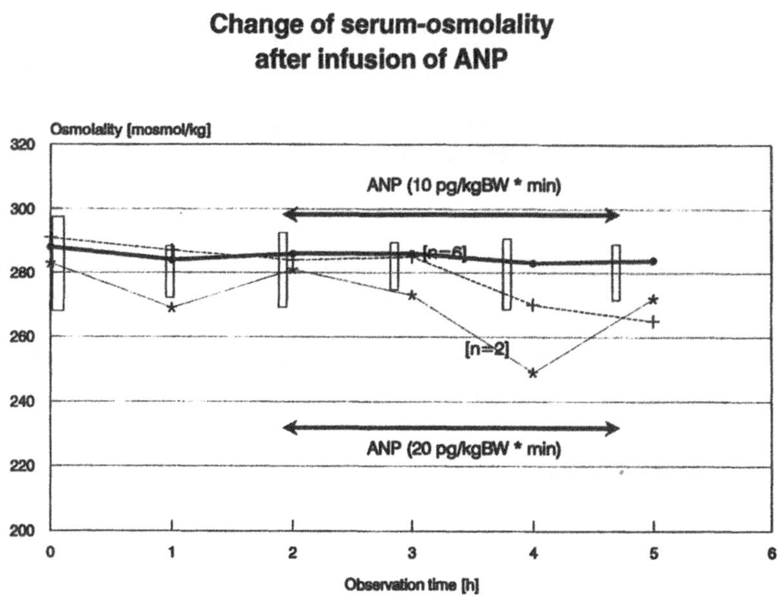

Figure 33.2: change of serum-osmolality after infusion of ANP, osmolality (mosmol/kg) v time (hours).

Results

In two out of eight patients with previous positive glycerol test hearing increased significantly during ANP application (10 pg/kgBWmin) without glycerol and during constant serum osmolality. In one more patient hearing improved and tinnitus ceased under ANP perfusion. The higher dosage of ANP application (20 pg/kgBWmin) had no better effect in patients without hearing improvement.

Discussion

The results mentioned above must be confirmed by investigations including large patient populations. Nevertheless, the following conclusions can be made:

- cGMP-plasma concentration could be another parameter to increase the reliability of the clinical diagnosis of Ménière's disease. Further studies need to be performed.
- by analogy to the glycerol-test there could be an ANP-test for the diagnosis of Ménière's disease;
- we need further investigation to determine whether ANP can be of value for the therapy of tinnitus caused by endolymphatic hydrops.

References

de Bold AJ, Borenstein HB, Veress AT, Sonnenberg H (1981) A rapid and potent natriuretic response to intravenous injection of atrial extracts in rats. Life Sci (28) 89–94.

Diestelhorst M, Kriegstein GK (1989) Zur augendrucksenkenden Wirkung von humanem atrialen Peptid (HAP) Fortschr Ophthalmol (86) 89–91.

Espiner EA, Richards AM (1989) Effects of low dose atrial natriuretic peptide in essential hypertension. The Lancet 707.

Klockhoff I, Lindblom U (1966) Endolymphatic hydrops revealed by glycerol test. Acta Otolaryngol (61) 459–62.

Lamprecht J, Meyer zum Gottesberge A-M (1988) The presence and localisation of receptors for atrial natriuretic peptide in the inner ear of the guinea pig. Arch Otorhinolaryngol (245) 300–1.

Lamprecht J, Plum J (1990) Plasmaspiegel des atrialen natriuretischen Peptids (ANP) nach oraler Glycerolbelastung Z Laryngol Rhinol Otol 31–4.

Correspondence to Professor Jürgen Lamprecht, Alfried-Krupp-Krankenhaus, Alfried-Krupp-Str 21, 45117 Essen, Germany. Tel (49) 434-2672. Fax (49) 434-2562.

Index

Printed and bound by CPI Group (UK) Ltd, Croydon, CR0 4YY

13/09/2026

14961030-0004